AF249117

QUEUEING

QUEUEING
Basic Theory And Applications

Walter C. Giffin

Department of Industrial and Systems Engineering
The Ohio State University

Grid Inc., Columbus, Ohio

To
Bev, Steve, and Becky

CONTENTS

APPENDICES

PREFACE

This book is concerned with the fundamentals of queueing theory which the author feels are most important for fledgling systems analysts and engineers. The objectives are 1) to provide the reader with an understanding of the power and limitations of a small class of Markov-based queueing models, 2) to provide the tools necessary to analyze transient as well as steady-state conditions, and 3) to teach model-building skills in the sense of adaptating well-known results to new and novel situations.

The book is an interpreter which seeks to bring proven techniques into reach of practitioners with modest mathematical skills. Rigorous mathematical proofs are often bypassed for more intuitively appealing if less precise arguments. Computer programs are provided to perform numerical calculation for large classes of stationary problems. Programs are also provided for obtaining numerical approximations to time dependent solutions for certain classes of nonstationary systems.

The math background necessary to understand the material includes a knowledge of basic calculus and probability theory. Basic transform techniques, particularly the use of Laplace and geometric transforms, would be helpful but not essential. An appendix containing the rudiments of the necessary techniques is included for readers with no background in that area.

Chapters 1 through 6 are devoted mostly to steady-state solutions for stationary systems. The discussion begins in Chapter 1 with a brief history of the origins of queueing theory. Chapters 2 and 3 provide basic information on the theory and manipulation of elementary Markov chains. Chapter 4 extends the Markov arguments to continuous time processes. The birth and death postulates and the resulting general model for a wide class of queueing systems are discussed in Chapter 5. Chapter 6 then applies those general results to obtain the steady-state solutions for several standard queueing system configurations. Computer programs to solve the general steady-state problem are contained in Appendix B.

Chapters 7 and 8 provide an introduction to transient analysis. They treat the time-varying properties of queue length distributions. Chapter 7 is devoted to exploring the mathematical properties of transient solutions for (M/M/1) and (M/M/1/k) systems. The closed-form expressions presented depend upon having time-invariant arrival and service rates. Chapter 8 suggests methods for relaxing the constant arrival and service rate assumptions. In that chapter numerical solutions are proposed for multi-channel systems with time-varying arrival and service rates. Computer programs designed to perform these calculations are provided in Appendices C and D.

Chapters 9 and 10 make extensive use of imbedded Markov chain arguments to model systems which are essentially non-Markovian. General arrival and service distributions are treated in Chapter 9. Chapter 10 treats similar systems in which arrivals or service may occur in batches.

Chapters 11 and 12 are devoted to some specialized subjects in queueing theory. Here one will find such topics as networks of queues, priority systems and optimization discussed.

The final two chapters are devoted to statistical questions which must be answered by a designer of waiting line systems. Chapter 13 treats the problems of data analysis and initial model selection. Chapter 14 provides information for estimating parameters, stating confidence intervals and designing hypothesis tests once a prototype model has been selected.

The book concludes with a set of appendices. Appendix A provides information on the probability implications of basic transform techniques. Appendices B, C, and D contain computer programs which can be used to solve a number of the more complicated and hence more useful and interesting problems.

Most of the material contained in the text has been used by the author in combined undergraduate and graduate level courses in Industrial and Systems Engineering at The Ohio State University.

The hoped-for contribution is to provide a useful guide for both students and practitioners on ways to analyze and design those systems with properties of queues. The method of presentation and the techniques suggested tend to be pragmatic. The philosophy is: if it works, use it.

ACKNOWLEDGMENTS

Appreciation is hereby expressed to all the named and unnamed students, teachers, and authors who molded my outlook over the years. They deserve any credit this work may reflect upon their contributions to my education. The defects belong to the author.

Thanks also to my colleagues in the Department of Industrial and Systems Engineering at The Ohio State University. In addition to providing technical consultation they also assisted me in the performance of many duties so I would have the time to complete this textbook.

My special thanks goes to Dr. William T. Morris, Dr. John B. Neuhardt, and James J. Mager. Dr. Morris provided expert editorial services and important words of encouragement when the task seemed too large. Dr. Neuhardt was especially helpful in monitoring my attempts to bring statistical techniques to bear on queueing analysis. Jim Mager provided technical support by preparing computer codes and more importantly by patiently reading every page on the first draft of this manuscript. Many errors were corrected and presentation was clarified thanks to his effort.

Among my former students who made identifiable contributions to this book through their thesis efforts are William W. Bundy, Michael G. Hartman, and Anderson H. Walters. They and their fellow students have made teaching an exciting and rewarding task.

Members of the secretarial staff of the Department of Industrial and Systems Engineering deserve special note for their patience in deciphering an illegible scrawl and their astuteness in correcting spelling and equation numbers. The bulk of the typing was done by Lois Graber, assisted by Susan Hesselbrock and Carol McDonald.

Finally let me thank my wife, Beverly, and children, Steven and Rebecca. They suffered the deprivation of many weekend outings and a summer vacation to see this effort through. Their faith in the ultimate success of the venture is the most important of all.

1

INTRODUCTION

The subject matter of this textbook concerns the analysis and design of waiting line systems. The terms "waiting lines" or "queues" are used to characterize a broad class of processes sharing a common mathematical structure but which may include many diverse physical realizations. For example, the obvious queue of customers awaiting service at the grocery store checkout counter may be described by the same set of equations as the movement of aircraft from holding patterns to final approach course in major airport terminal areas. Both are stochastic processes with three characteristics which can be used to describe the system:

1. **An input process.** This may be the flow of customers from private homes to the grocery or the flow of jet liners from various enroute sectors into the airport terminal area. Both processes may involve a degree of uncertainty concerning the exact arrival time of units seeking service. To describe the input process one needs to specify such things as the source of the arrivals, the size of the customer population, the grouping of customers and the inter-arrival times. Any or all of these characteristics may be best described by random variables.
2. **A service mechanism.** This may be the checkout counter where prices are totaled and the groceries bagged or an airport runway which must be cleared of landing traffic before the next aircraft can touch down. Major elements to be specified here include the number of servers, the number of customers which can be in service simultaneously and the length of service intervals. Any or all of these characteristics may be best described by random variables.
3. **A queue discipline.** The discipline defines the rules of behavior within the queue. The grocery line may operate on a first-in first-out principle while the queue of airliners may operate on a group-by-speed-class type priority rule. Customer behavior characterized by balking, reneging, or jockeying may also require specification.

The presence of uncertainty is the thing which makes these systems challenging to analyze and design. If there were no random events everything could be scheduled without error and there would be no queue. However, randomness in one or more of these three characteristics is the rule rather than the exception. Furthermore, the distributions used to describe these events may be nonstationary which adds further complication to their study. Our goal is to assist the design of such systems.

INTENT OF THE TEXTBOOK

There are a number of textbooks available which discuss in great detail the mathematical structure of the differential-difference equations evolving from attempts to model waiting line systems. Furthermore technical journals in fields such as Operations Research, Management Science, and Industrial Engineering continue to publish new research results in queueing theory at a dizzying pace. However, it has been the experience of the author that many of these efforts fall short in one of two ways. First, in many of the advanced, and hence most realistic models, the math is so formidable that applications-oriented practitioners are discouraged from attempting to use the results. Indeed, they may even be incapable of understanding the results. Second, in presentations designed for undergraduate students and nonresearch oriented practitioners, the other extreme of handbook type formulas for stationary systems is sometimes offered. These are of limited usefulness in the analysis of realistic systems because of the assumed distribution type and stationary characteristics.

There seems to be a large gap between the theoretician and the practitioner. Many models exist which could be extremely useful to the analyst if he understood them well enough to apply them. Furthermore, useful results can often be obtained by judicious approximations used in place of the more formally correct but computationally intractable models which populate the research literature. The problem is how to identify those situations in which the approximations are reasonable.

It is the goal of this textbook to narrow the gap between the theoretician and the practitioner. We will trace the development of the most useful standard models while continually emphasizing how these formulations can be modified and adapted to new systems. We will present methods for obtaining transient solutions to those systems with time-varying demand and service patterns. These solutions will be numerical approximations as contrasted with closed-form functional representations of system distributions. However, such solutions provide the designer with what he really needs: numbers to evaluate the relative merits of alternative system designs.

The reader should not falsely assume that our attempts to develop a pragmatic approach to the study of queues implies that he can avoid immersion into some higher mathematics. We must meet the theoretician half-way. We will make generous use of methods from transform theory, linear algebra, and numerical analysis. The desired result is to impart to the student the capability to interpret existing models, modify those models to fit new situations, and to enhance his intuition about the analysis and design of any system with the characteristics of a waiting line. Our presentation is not designed for the research mathematician. Most of the models we will discuss have existed for many years. The goal of this textbook is to make those models a realistic and useful part of the fledgling systems analyst's tool bag.

HISTORICAL PERSPECTIVE

Before anyone becomes overly enchanted with the study of a "new technology" it is important to learn something about the origins of that technology. In the case of queueing theory, the problems have been actively researched and papers published since 1907. This fact is often ignored by people who tend to associate queueing theory with the relatively new field of operations research, which had its origins in the 1940–1950 decade. Operations research and management science practitioners have certainly made use of the early models and in some cases extended their range of applicability, but in no way did they invent the topic. An excellent guide to the major developments prior to 1969 is contained in the paper by Bhat (8).

The first published paper on queues seems to be that of Johannsen, "Waiting Times and Number of Calls," which appeared in 1907 (54). However, the most important work, in terms of its long-range impact on the study of queues, appeared in a series of articles by A. K. Erlang beginning in 1909 and covering the next twenty years (14, 36). Erlang is credited with developing the foundation for the Poisson and exponential distributions which continue to play the pivotal role in many queueing models. In addition he developed other fundamental concepts such as the idea of statistical equilibrium and a method for writing state balance equations which remain in use today. Erlang was a Danish telephone engineer who developed his ideas while trying to solve the operational problems of a telephone system.

Other significant work by early authors includes that of E. C. Molina and T. C. Fry. In 1927 Molina wrote for the *Bell System Technical Journal* on the "Application of the Theory of Probability to Telephone Trunking Problems" (67). In 1928 Fry wrote a book titled *Probability and Its Engineering Uses* which contained much important information on congestion theory (39).

The important thing to note about such early authors is that these men were engineers seeking solutions to practical real-world problems. The resulting models were judged on the basis of their usefulness in solving those problems rather than on the theoretical elegance of the proofs used to establish their logical consistency.

In the 1930 and 1950 era authors such as Crommelin (30), Khintchine (58), Kolomogorov (63), Palm (72), and Pollaczek (75) were prominent in published queueing theory results. These men were primarily theoreticians. They found that the basic structure of queueing systems offered a fertile field for fundamental mathematical research on stochastic proc-

4

esses. They sought to develop more general models which could cover a wide variety of complex situations. The contributions in this period were extremely important in terms of providing a firm theoretical understanding of this class of stochastic processes.

The trend toward the analytical investigation of the basic stochastic processes associated with queueing systems has continued until the present time, as a quick glance at recent issues of journals such as *Operations Research* will verify. In addition to further study of more traditional steady-state solutions, authors such as Bailey (4), Bhat (9), Cox (24), Kendall (57), Kielson and Kooharian (59), and Takacs (87) have carried theoretical developments forward in the areas of time-dependent solutions and Markov analysis. Countless other authors have also contributed to the embellishment and increasing elegance of the models collectively referred to as "queueing theory." The text by Gross and Harris (42) provides an excellent source for the state of the art from a theoretical point of view.

The plea of the nonresearch-oriented practitioner throughout this period of expansive theoretical development has been for techniques which he can understand and demonstrated applications for the results of the theory. There seems to be no other area of study in the field of operations research in which the theoretical developments have so far outdistanced the applications. One can speculate that there are many reasons why this may be so. However, the most commonly mentioned one is that the resulting equations from many theoretical investigations are simply too complex for the average practitioner to understand or too cumbersome to be useful manipulative devices.

From the point of view of the applied scientist, perhaps the most important developments in recent years involve approximations and numerical techniques for solution of nonsteady-state systems. Nonsteady-state systems are those which may never stabilize in a probability sense. Arrival rates and service rates may be constantly shifting over time so that the state distribution must be expressed as a function of elapsed time since the system began operation. Such systems can be contrasted with so-called steady-state systems in which arrival and service patterns are such that the state probability distribution will soon stabilize. This permits one to predict long-term behavior in terms of a single distribution independent of elapsed system operating time. Even those systems which ultimately reach steady state may have transient behavior from certain starting conditions which may be of vital importance to the system designer. It is the time varying character of the queue, whether the system is stable or not, which should be analyzed. Authors such as Kingman (60), Koopman (64), and Newell (71) have reported significant advances in nonsteady state-analysis. Unfortunately, the numbers of such papers appearing in the literature is small relative to the total number of queueing articles. Much more work is needed in the way of demonstrating the robustness of simple models and in providing easily used approximate solutions to complex problems. A model is robust when it can provide useful results even though the system being analyzed may violate assumptions used to construct the model. This may not be the frontier from the viewpoint of a theoretical mathematician, but it is the necessary ingredient for moving the study of realistic queueing phenomena from academia back to the physical problems which should be the focal point.

Most of the discussion above and indeed most of the literature, relates to what Bhat calls behavioral problems of the system (8). The focus is on the use of mathematical models to seek understanding of a particular process. The other problems which Bhat discusses are classed as statistical problems and operational problems. Statistical problems refer to the analysis of empirical data and tests of hypotheses regarding queue processes. Operational problems refer to the design, testing, and control of real life systems. He notes that most advances have been on the behavioral problems, fewer on the statistical problems, and still fewer on the operational problems. This occurs in spite of the avowed interest of applied operations researchers in operational problems of systems. The relative portions of available information in each of these areas further emphasizes the gulf between the theoretician and the practitioner.

APPLICATIONS

Lest the reader become discouraged by the recounting of the seeming imbalance of theory and practice in queueing publications, let us hasten to add that there have been notable applications reported over the years.

The general area of communications systems has been a source for many applications. As previously mentioned, much of the early development in queueing theory was related to the analysis of telephone traffic (14). The telephone service industry continues to be a major user and source of new model developments today (7). In addition, other forms of communications traffic such as telegraph, post office, and radio communication systems have been analyzed by queueing models (15).

There is also a growing literature on computer applications of queueing systems (61). Most of these applications are related to the design of software for time-sharing systems and computer-communications networks. It is the network structure and priority scheduling rules for these systems which provide the research challenge.

The transportation field provides another fertile area of applications for queueing models. Automobile traffic moving past toll booths in the New York area was an award-winning application reported by Edie (34). In addition there have been many attempts to use queueing theory to analyze the flow of traffic through intersections and along congested areas of freeways. Arrivals and departures of aircraft at major air terminals have also provided opportunities for the application of queueing models (64).

Industrial processes including maintenance, tool crib operation, machine interference, inventory control, and assembly line balancing are common sources of application for queueing concepts familiar to the industrial engineer. In addition to the scattered articles appearing in the literature concerning industrial process applications, it is the hypothesis of this author that very large numbers of others exist which have never been published. This is especially true of industrial applications due to 1) the proprietary nature of some studies, 2) the (false) perception that such applications are relatively routine and of no interest to a professional audience, and 3) the lack of strong incentives for industrial practitioners to publish the results of their efforts.

Another area of application sometimes overlooked by engineers and management scientists is that of the biological sciences. Models of epidemic processes, population growth, and the psychological flow of nervous impulses bear strong resemblence to those we call queueing models (6). The same modeling techniques we will use throughout this text have been successfully applied to these problems.

Finally we should note the most obvious area of application, that of queueing for service. Waiting lines exist at banks, grocery stores, theaters, restaurants, hospitals, bus stops, and at countless other points where human beings position themselves in a queue awaiting service. Questions concerning the effects of limited waiting space, increased service capability, variations in flow rates, and many others have been addressed by the application of queueing models.

The list of applications above is not intended to be exhaustive. It is presented only as a means of impressing upon the reader the very broad nature of phenomena which have been shown to have the common characteristics we identify with a queue. It is not the lack of potential applications which disturbs the serious practitioner. It is the level of complexity and hence reality which one is able to capture and understand within existing model structures which is of real concern. We will attempt to develop guidelines for model adaptability and robustness which may ease the transition from theory to practice. The often unattainable ideal is to develop useful, simple models for realistic complex systems. The systems are complex; the best models are not.

MODELING APPROACH

The concept of specifying the structure for a queueing system is simple. We stated earlier that all one requires is a proper description of the input process, the service mechanism, and the queue discipline. Developing the "proper description" and the interaction among those major elements is then the role of model building. The goal is to develop expressions which will give some measure of the number of customers present in the waiting line at time t and some measure of delay experienced by customers passing through the system.

Although one may gain an appreciation for the structure of queues by examining deterministic systems, we will not study such systems on the grounds that they are seldom encountered and if they are, the presence of queues is not a problem. It is the uncertainty present in most real systems which makes the model building a challenging task. The most realistic characterization one may be able to accomplish is to describe the input, service, and queue-discipline properties in terms of probability distributions. The basic problem is to use these distributions to develop probability statements concerning line length and delay times.

Obviously, there is virtually a limitless number of variations one could encounter in describing such systems. However, there is a solid core of basic prototype models and techniques which have proven useful for a surprising range of applications. We will investigate some of the core items while trying to maintain a posture of objectivity relative to the limits of their applicability.

Since the study of queues as we have defined it is essentially the study of a special class of stochastic processes, we will first turn to that field for the development of our fundamental theory. Moving one step beyond the description of simple independent trials, the most widely researched and easily manipulated stochastic models are associated with Markov processes. A Markov process is one which exhibits one stage dependence; the probability distribution for future system states can be developed from a knowledge of the existing state distribution without regard to any other past history. Queues are not inherently Markovian. However, those models which have proven most easy to construct and manipulate seek to preserve that property where possible.

The queueing systems which will receive most of our attention can be structured as Markov processes. For that reason the entire next chapter will be devoted to the presentation of elementary Markov chains in preparation for future model development. Those readers with an understanding of Markov chains may want to skip that material and proceed directly to the continuous-time process model discussion.

The study of continuous-time processes will begin with the development of the general birth and death equations. These equations follow directly from the Markov chain discussion. In the context of queueing systems a "birth" generally refers to the arrival of a customer to be served, and a "death" refers to the departure of a customer after service or for other reasons. The resulting set of differential-difference equations are extremely versatile, having served as the basis for innumerable models simply by varying the state-dependent birth and death rates. We will illustrate how several of the simple system prototype models under steady-state conditions follow directly from those equations. At the same time the reader will be encouraged to design other variations of this theme through numerous examples and problem exercises.

The power of the Markov assumption is so great that we will attempt to make a system look as if it is Markovian even when it is not. This is the tactic pursued under the title of imbedded Markov chains. The idea here is to define regeneration points such that the system can be described as a Markov process over those points even though it may be decidedly non-Markovian if one attempts to model it on a continuous time scale.

So-called steady-state solutions will be developed for a wide range of systems including such variations as bulk arrivals, bulk service, priority rankings, and networks of queues. Steady state implies that the system has run long enough to stabilize in the sense that the state probability distribution no longer changes with time. The analysis will draw heavily

8

upon a variety of transform techniques which are discussed in some detail in Appendix A.

Systems with time-varying demand and time-varying service capability are of great practical importance and of great difficulty to analyze using the more formal techniques of algebra and operational calculus. Our approach to those problems will be to use the power of the digital computer to find numerical solutions to large but finite sets of differential equations. In terms of realistic design information for systems with time dependencies or service interruptions, this approach is one of the most promising ones available in today's state-of-the-art. We will spend considerable time exploring numerical techniques.

An alternative method for handling the more complex systems is that of computer simulation. We will discuss that technique briefly and compare results of such efforts with numerical solutions to a comparable set of differential equations.

Finally, we will complete the queueing discussion by suggesting how one might go about selecting the proper model and once validated how one might use that model in a prescriptive sense as opposed to its more traditional descriptive role.

NOTATION

There is often an unfortunate amount of confusion on the part of students brought about by inconsistent notation among authors writing on queueing theory. Several of the technical societies whose journals contain such articles have recognized the problem and held a joint conference at Northeastern University in May 1971 to discuss standardization of terminology and notation. We will attempt as far as possible to follow the recommendations of that conference as reported by the Operations Research Society of America.

The exact nature of the assumptions underlying the models presented is specified in the following notation:

(arrival process/service process/number of parallel servers)

When necessary an appendage may be added of the form

—/limit on the number in the system/number in the source/ queue discipline)

When the appendage is omitted it is assumed that there is no limit on system capacity, customers are drawn from an infinite population, and the queue discipline is first come, first served.

The arrival and service processes are described by their interarrival and service process probability distributions. The number of servers or channels in parallel is an integer. Standard queue disciplines are identified by appropriate abbreviations. The following notation is recommended:

M Exponential (which we will show is *M*arkovian)

E_k *E*rlangian (gamma) with k identical phases

D	Constant (*Deterministic*)
G	*General*
IFR	Distributions with *i*ncreasing *f*ailure *r*ates
DFR	Distributions with *d*ecreasing *f*ailure *r*ates
MR	*M*arkov *r*enewal process
c	Number of servers or *c*hannels in parallel
FCFS	*F*irst *c*ome, *f*irst *s*erved
LCFS	*L*ast *c*ome, *f*irst *s*erved
RSS	*R*andom *s*election for *s*ervice
PR	*P*riority.

To illustrate the notation, among the systems we will discuss the following:

(M/M/c)	*M*arkovian input and service with c servers
(M/G/1)	*M*arkovian input, *g*eneral service distribution, and one server
(M/M/1/n/m)	Single-channel *M*arkov system with capacity n and m potential customers.

In developing the models the following mathematical notation is recommended:

$1/\lambda$ Mean interarrival time

$1/\mu$ Mean service time

λ Mean arrival rate

μ Mean service rate

ρ Traffic intensity $= \dfrac{\text{Mean service time}}{\text{Mean interarrival time}}$

$$= \dfrac{\text{Mean arrival rate}}{\text{Mean service rate}}$$

For single channel systems $\rho = \lambda/\mu$

For c-channel systems $\rho = \dfrac{\lambda}{c\mu}$

L	Expected number in the system, including the units being served, under steady state conditions
L_q	Expected number in the queue, excluding the units being served, under steady state conditions
W	Expected time spent in the system, including service time, by a unit at steady state
W_q	Expected time spent in the queue, excluding service time, by a unit at steady state
$N(t)$	Number of units in the system at time t
$N_q(t)$	Number of units in the queue at time t

With the notation now in hand we are ready to begin the development of our queue analysis skills.

SUMMARY

Our goal is to develop analytical skills in adapting prototype models to novel situations. To enhance these skills and maintain a problem orientation, both transient and steady-state analyses will be introduced from a common mathematical structure. This book is a primer on queues. It is intended for practitioners rather than theoretical mathematicians. The worth of a technique will be measured by its usefulness in solving real world problems rather than by its mathematical elegance.

The range of application for these techniques is extremely wide. Consider for example the problem faced by a staff industrial engineer who must analyze fork truck utilization in a truck body assembly plant as contrasted with an air traffic control specialist who must project the ripple affect caused by temporarily closing a runway at a major air terminal. Both analysts can use the results of queueing theory to be presented in later chapters.

The fork truck problem arises from the trade-off between the cost of direct labor operations, which may be delayed while awaiting service by a fork truck, and the cost of operating the fork trucks. Suppose for example that it costs twenty dollars per direct labor hour in our assembly plant and sixty dollars per day to operate a fork truck. Service time per call for a fork truck is a random variable with a mean of four minutes. The number of calls for service each day is also a random variable whose mean varies with the number of work stations assigned to the pool of fork trucks. The questions to be answered are 1) What is the optimum utilization of a single fork truck? and 2) How many stations should be assigned to a fork truck to minimize the cost of the product produced? We will use simple steady-state queueing models to analyze this problem in future chapters.

The air traffic control problem arises from the cost of delays to airline traffic and the provision of alternate landing sites when a terminal reaches saturation. Suppose for example that a terminal has holding space for sixty aircraft and an arrival pattern which varies by hour of the day with peak demand of seventy aircraft arrivals occuring between 5:00 and 6:00 PM. Under normal operating conditions the service rate is forty-five aircraft per hour. However, when a thunderstorm moves over the approach path the service rate may be effectively zero until the storm has passed. During the life of the storm, arriving aircraft must enter a racetrack-like holding pattern until service can be resumed. When the number of aircraft holding reaches sixty, all new arrivals must be diverted to other airports in the area. The questions to be answered are 1) For a given length of storm delay how many aircraft must be diverted? 2) For a given length of storm delay how will the queue length vary once the system is restored to full operational status? and 3) What is the optimum service rate given the cost of new facilities, the cost of delay to aircraft, and the probability of service interruption due to meteorological conditions at various hours of the day? Numerical techniques for the analysis of time varying queues can be used to solve this problem.

These problems are typical of those posed by real-world systems of concern to us. The problem context is broad, but the mathematical structure which can be used to describe such systems is surprisingly narrow. Our plan is to develop an understanding of the structure and sufficient model-building skills to be able to adapt that structure to operational problems which exhibit the characteristics of a waiting line. Ideally, when our study is complete we can analyze many such problems regardless of the context in which they may arise.

PROBLEMS

1.1 Discuss the following systems in terms of the basic characteristics of queueing systems.

a) Approach and departure control at a major air terminal.

b) Baggage handling at an airline terminal.

c) Ferryboat service across a large lake.

d) Emergency room service at a hospital.

e) Police radio dispatching.

f) Production control in a small job shop.

g) A telephone switchboard in a business office.

h) A series of flood control dams in a major river district.

i) Spread of a new strain of flu.

1.2 Construct a list of scientific research and applications journals in which one might reasonably expect to find references to queueing studies. Justify the inclusion of each journal by citing at least one article from that journal which concerns queues.

1.3 Consult the following four journals over the past year:

> *AIIE Transactions*
> *Industrial Engineering*
> *Management Science*
> *Operations Research*

a) Prepare a list of articles from these journals which treat queueing problems.

b) Classify each article as behavioral, statistical, or operational.

1.4 Prepare a critique of a queueing article found in current literature.

1.5 Give four examples of queueing situations which fall within your personal experience. Describe each in terms of the basic characteristics of queues. Suggest how you think the systems might be improved.

1.6 Survey the past ten years of the journal *Operations Research*.

a) List all queueing articles published during that time which have titles alluding to applications.

b) Of those articles listed tabulate the number which include data.

1.7 Prepare a list of properties which you feel a model for queueing systems must possess to qualify as a "practical tool."

1.8 Consider a deterministic queueing system in which the arrival rate λ exceeds the service rate μ and capacity is limited to n customers in the system. At time zero no customers are present.

a) Describe this system in terms of the standard queue notation.

b) Develop an equation for the number of customers in the system at time t, $N(t)$.

c) Develop an equation for the waiting time in the queue which will be experienced by the nth customer who enters.

d) Find the number of customers who will be denied service in the interval $[0, t]$.

1.9 Construct a table showing the expression for the probability density (or mass) function, the mean and variance for each of the following:

a) Poisson	*d*) k-Erlang	*g*) Negative Binomial
b) Exponential	*e*) Chi-Square	*h*) Binomial
c) Gamma	*f*) Geometric	*i*) Uniform

1.10 Find an appropriate transform for each of the distributions listed in problem 1.9 (see Appendix A). Use the transforms to prove that the means and variances are as you listed them in your table.

2

ELEMENTARY MARKOV CHAINS

All the material in this textbook concerns the study of stochastic processes. The term stochastic implies that a probability function is generating an ordered sequence of events. In the queueing context it is common to relate that ordering to time. For example, a time series plot of the number of customers waiting at a bus stop at the time each bus arrives for pick-up represents a stochastic process. The number of customers present is a time indexed random variable. The number at time t, say $N(t)$ may or may not be related to the number at time $t + r$, say $N(t + r)$. Furthermore, if we were to collect data over many days' operations, each day would give a unique time series, the time series itself being a random function. If one were to develop probability distributions for each hour of the day from these data, he might find that the distribution for number of customers at 8:00 AM for a random day would differ markedly from the distribution for number present at 2:00 PM. Systems with such time varying random outcomes are typical of the stochastic processes of concern to us.

MARKOV PROPERTY

In an experiment involving independent trials the random portion of the outcome of any trial is in no way dependent upon the outcomes of past trials. In the bus stop example, the series of indexed random variables representing customers present at time t might be viewed as outcomes of independent trials if all customers who are not served by an up-coming vehicle leave the system. The group size at the time of arrival of the next bus would then in no way depend upon the group size at the time of the previous bus arrival. If we assume that buses arrive on the hour, probability statements concerning group size at 9:00 can be made without regard to any information concerning group size at 8:00. The joint probability for the sequence or "chain" of events of having groups of g_1, g_2, g_3, g_4 for the first four pick-ups of the day could be written as

$$P[g_1, g_2, g_3, g_4] = P(g_1)\,P(g_2)\,P(g_3)\,P(g_4) \qquad (2.1)$$

where $P(g_i)$ is the probability that a random variable assumes a value g_i.

At the opposite extreme from independent trials one might imagine an experiment in which the sequence of events is such that the entire past history of the experiment is needed to make probability statements about the next trial in the sequence. In the bus stop example this means that one would need to account for the potential number of customers present at every pick-up time beginning with the first run in the morning to be able to predict the probability that N customers will be present at 2:00 PM. To calculate the joint probability for a particular n-pick-up sequence would require an equation such as

$$P[g_1, g_2, \ldots, g_n] =$$
$$P(g_1)\, P(g_2|g_1)\, P(g_3|g_1, g_2) \ldots P(g_n|g_1, g_2 \ldots g_{n-1}) \quad (2.2)$$

In realistic systems the determination of the various conditional probability density functions in (2.2) may be extremely difficult. What we now seek is a simplification of this expression which still seems to fit a great number of practical situations.

The Markov property can be viewed as a first-order generalization in moving from the concept of independent trials toward total history dependence. Here we assume one stage dependence in which the outcome at any trial depends upon the outcome of the immediately preceding trial but is independent of all others. A system exhibiting this characteristic is said to have no memory. Information concerning how it reached a known state is not needed to modify the probabilities for the next stage. In the context of the bus stop the system exhibits the Markov property when we can estimate the probability distribution for the number of customers present at 2:00 PM if we know the number present at the previous (1:45 PM) pick-up. How we got to the present state is irrelevant. In a Markov process all we need to know is the present state and appropriate transition probabilities to be able to predict future outcomes.

To construct a Markov model of a system one needs to understand the concepts of "state" and "state transition." The term "state" implies all we need to know to completely describe the system at any particular instant. The calculations are greatly simplified if we can minimize the number of variables in the state description. In the bus stop example the state could mean simply the number of customers present at a particular instant. The single real number $N(t)$ completely describes the state of the system. In more complicated situations, such as a transit system with multiple classes of customers, the state description might be a vector in which element i represents the number of customers of class i present at that instant.

State transitions imply that changes of state take place. In many of our models we will associate these transitions with time as we attempt to capture the dynamic nature of systems. Time is not however an inherent part of the description of Markov processes since state transitions can be caused by events other than the passage of time. In the bus stop example if there are twenty passengers waiting at 1:45 PM and fifteen at 2:00 PM, the system has undergone a "state transition" from twenty to fifteen.

DISCRETE TIME SYSTEMS

Both continuous and discrete parameter processes can be modeled under the Markov assumption. However, for purposes of this chapter we will confine our attention to discrete parameter processes. Markov processes in which the state variable is integer valued and in which transitions occur at discrete points in time are referred to as *Markov chains*. Such chains are the topics for discussion in this chapter.

Suppose that the outcome of a trial in a series of experiments can be described by a single state variable. The random variable associated with the nth transition then takes on a particular value k with some frequency:

$$Pr\,(x_n = k) = v_k(n) \tag{2.3}$$

The term $v_k(n)$ is the marginal probability that $x_n = k$. The Markov property implies that a description of the transition expected on the nth trial can be reduced to

$$Pr[x_n = j\,|\,x_1 = k_1, x_2 = k_2 \ldots x_{n-1} = i] = Pr[x_n = j\,|\,x_{n-1} = i] \tag{2.4}$$

These transition probabilities may or may not be related to the number of trials which have taken place. If they depend upon both the state and the trial number, we will write

$$Pr[x_n = j\,|\,x_{n-1} = i] = p_{ij}(n) \tag{2.5}$$

The term $p_{ij}(n)$ is then the conditional probability that the system will be in state j after n transitions, given that it occupied state i after $n - 1$ transitions. When these transition probabilities are not functions of the stage (or transition) number n, the system is said to be stationary. Under stationary conditions the notation will be shortened to

$$p_{ij}(n) = p_{ij} \quad \text{for all } n$$

Suppose now that one wishes to describe a particular "chain" of events. Here we take the viewpoint of an observer who sees the process at time zero but does not observe any intermediate states. The goal is to forecast the occurrence of a particular set of events. A time series for one day's activities at our prototype bus stop is such a chain. Consider calculating the joint probability that for the first four pick-ups of the day, the driver will find 3, 4, 7, and 9 passengers waiting. Treating the first pick-up as trial zero, the probability of achieving this particular chain of events is

$$Pr(x_0 = 3, x_1 = 4, x_2 = 7, x_3 = 9) = v_3(0)\,p_{34}(1)p_{47}(2)p_{79}(3) \tag{2.6}$$

If the system is stationary this expression reduces to

$$Pr(x_0 = 3, x_1 = 4, x_2 = 7, x_3 = 9) = v_3(0)p_{34}p_{47}p_{79} \tag{2.7}$$

Note that all that is necessary to completely define such a Markov chain is an estimate of the initial probability distribution and the appropriate single stage transition probabilities.

We can construct conditional probability distributions for transitions over long intervals from those over short intervals by means of the Chapman-Kolmogorov equation. For stationary processes let $p_{ij}{}^n$ represent the conditional probability that the system will occupy state j after n transitions, given that it was initially in state i. Then

$$p_{ij}{}^n = \sum_k p_{ik}{}^r p_{kj}{}^{n-r} \qquad (2.8)$$

for any $r < n$. Intuitively one can picture the system as moving from state i to some intermediate state k in r transitions then from k on to the final state j in the remaining $n\text{-}r$ transitions. Summing over all possible intermediate values k then yields the n-stage transition probability $p_{ij}{}^n$.

The most important application of this equation occurs when $r = 1$. If we have a stationary system which starts in state i at time zero, the probability that it will occupy state j at any time n can be calculated in recursive fashion from the single-stage transition probabilities. Beginning with $n = 1$, we note that $p_{ij}{}^1$ is simply the single stage transition probability p_{ij}. Moving to $n = 2$ it follows from equation (2.8) that

$$p_{ij}{}^2 = \sum_k p_{ik}{}^1 p_{kj}{}^{2-1} = \sum_k p_{ik} p_{kj}$$

For $n = 3$ the results are

$$p_{ij}{}^3 = \sum_k p_{ik}{}^1 p_{kj}{}^2$$

This procedure generalizes to

$$p_{ij}{}^n = \sum_k p_{ik}{}^1 p_{kj}{}^{n-1} \qquad (2.9)$$

which can then be solved in recursive fashion.

These recursive relationships could be used in the bus stop example to calculate the probability that the fourth bus of the day would find 9 passengers waiting, given that the first one found 3, that is $p_{39}{}^3$. Such a calculation in effect compresses the time scale by condensing all intermediate transitions into a single conditional probability covering three time periods.

We could develop similar relationships for more general nonstationary Markov chains. Let

$$Pr(x_t = j \,|\, x_r = i) = p_{ij}(r, t), \ t > r$$

then the Chapman-Kolmogorov equations take the form

$$p_{ij}(r, t) = \sum_k p_{ik}(r, s)\, p_{kj}(s, t), \ r < s < t \qquad (2.10)$$

Recall that in the nonstationary case the transition probability changes as a function of time. The probability for a chain of events $(x_1 = i,$

$x_2 = j, \ldots, x_n = z)$ is given by

$$Pr(i, j, k, \ldots, z) = v_i(0)\, p_{ij}(1)\, p_{jk}(2) \ldots p_{yz}(n)$$

It therefore follows that one could write as a particular case of equation (2.10)

$$p_{ij}(0, 2) = \sum_k p_{ik}(0, 1)\, p_{kj}(1, 2)$$

$$= \sum_k p_{ik}(1)\, p_{kj}(2) \tag{2.11}$$

MATRIX NOTATION

It is convenient from a computational point of view to use vectors and matrices to represent the various probability distributions associated with Markov chains. The vectors and matrices are merely bookkeeping devices which facilitate calculations. They add nothing to the underlying concepts introduced in the last section.

Let us define the state probability distribution for an m-state chain after n transitions as

$$\mathbf{V}(n) = (v_1(n), \ldots, v_m(n)) \tag{2.12}$$

The elements of this vector are simply the probabilities that the system will occupy any given state after n transitions. For example,

$$\mathbf{V}(2) = (0.2, 0.3, 0.5)$$

indicates that we have a three state system which can be found in state 1 after two transitions with probability 0.2. Since the vector $\mathbf{V}(n)$ represents a probability distribution, it is obvious that all of its elements must be nonnegative and that their sum must be unity. That is

$$\sum_j v_j(n) = 1 \tag{2.13}$$

and $\qquad\qquad v_j(n) \geq 0 \quad$ for all j and n

Let us define the single stage transition matrix for an m-state system on the nth transition as

$$\mathbf{P}(n) = \begin{bmatrix} p_{11}(n) & p_{12}(n) & \cdots & p_{1m}(n) \\ p_{21}(n) & p_{22}(n) & \cdots & p_{2m}(n) \\ \vdots & & & \\ p_{m1}(n) & p_{m2}(n) & \cdots & p_{mm}(n) \end{bmatrix} \tag{2.14}$$

Element i, j of this matrix is the conditional probability that the system will occupy state j after the nth transition given that it occupied state i after the $n - 1^{\text{st}}$ transition. Since $p_{ij}(n)$ is a conditional probability, it follows that all elements are nonnegative and the sum over all j for any fixed value of i must be unity. That is

$$\sum_j p_{ij}(n) = 1 \quad \text{for all } i$$

and
$$p_{ij}(n) \geq 0 \quad \text{for all } i, j \text{ and } n$$

This is equivalent to stating every row in the matrix $\mathbf{P}(n)$ must sum to unity. Each row is itself a conditional probability distribution. Let us designate row i as the vector

$$\mathbf{P}_i(n) = [p_{i1}(n), p_{i2}(n) \ldots p_{im}(n)] \tag{2.15}$$

This vector gives the conditional probability distribution for the state of the system on trial n given that it occupied state i on trial $n - 1$. The transition matrix can then be written in alternative fashion as

$$\mathbf{P}(n) = \begin{bmatrix} \mathbf{P}_1(n) \\ \mathbf{P}_2(n) \\ \vdots \\ \mathbf{P}_m(n) \end{bmatrix} \tag{2.16}$$

The number of rows or vectors in the matrix $\mathbf{P}(n)$ is equal to the number of states the system may occupy.

The matrix $\mathbf{P}(n)$ then accounts for all possible one stage transitions on trial n for an m-state system. We can summarize its characteristics as follows:

1. All elements are nonnegative since they represent probability values.
2. The number of rows and columns are equal since we must account for all possible transitions from any feasible initial state to any feasible final state.
3. The sum of all elements in any row is unity since each row represents a conditional probability distribution.

Any matrix which meets these specifications is said to be a *stochastic matrix*.

When the system is stationary the transition probabilities do not change over time. It then follows that

$$\mathbf{P}(1) = \mathbf{P}(2) = \ldots = \mathbf{P}(n)$$

For the stationary case we will drop the functional argument and write the single-stage transition matrix as

$$\mathbf{P} = \begin{bmatrix} p_{11} & p_{12} & \cdots & p_{1m} \\ p_{21} & p_{22} & \cdots & p_{2m} \\ \vdots & & & \\ p_{m1} & p_{m2} & \cdots & p_{mm} \end{bmatrix}$$

$$= \begin{bmatrix} \mathbf{P}_1 \\ \mathbf{P}_2 \\ \vdots \\ \mathbf{P}_m \end{bmatrix} \tag{2.17}$$

A stationary Markov chain is completely defined by an initial probability vector $\mathbf{V}(0)$ and the single stage transition matrix $\mathbf{P}$.

Example.   Consider the operation of an inventory control system for desk calculators which employs the following ordering policy. Every Friday at closing time the storekeeper checks his supply of calculators. If he has more than two on hand he does nothing. If he has two or less he telephones the local distributor to order enough stock to bring his supply up to five. The store is open five days a week and the distributor can provide Monday morning delivery for any order placed on Friday night. No returns are allowed. A study of past demand for calculators at this store indicates that the number of customers per week can be described by a Poisson probability distribution with a mean of 1.5. The problem is to develop a model for this system which will enable the storekeeper to estimate variations in his inventory of calculators from week to week.

This inventory policy can be modeled as a stationary Markov chain. The state of the system is presented by the number of calculators on hand at closing time Friday night. This means that there are six possible states denoted by the integers zero through five. State transitions result from the acts of receiving calculators from the distributor and selling them to consumers. The time between transitions is one week, the elapsed time between reviews of inventory status. This system is analogous to a queueing system in which calculators form the queue and customers provide the service mechanism.

Suppose that the inventory is now being reviewed with the resulting count of three calculators on hand. The current state of the system is then represented by the vector

$$\mathbf{V}(0) = (0, 0, 0, 1, 0, 0)$$

The elements of the one-week transition matrix can be calculated by combining demand and shipping event probabilities. For example, a transition from state 3 to state 1 can only occur under the stated policy if two calculators are sold since no shipments are requested with an initial inventory greater than two. The probability of that event comes from the Poisson distribution.

The Poisson distribution is given by

$$p_n = \frac{m^n e^{-m}}{n!}$$

where n is the number of units demanded and m is the mean or expected value of the distribution. For a mean weekly demand of 1.5 calculators it then follows that the transition probability p_{31}, implying two demands, is given by

$$p_{31} = \frac{(1.5)^2 \, e^{-1.5}}{2!}$$

A transition from state 1 to state 2, on the other hand, results from a combination of events. The initial inventory level of one will trigger an order for four additional units to be delivered on Monday morning. The total available supply for next week is then five calculators. Depleting that supply to a level of two calculators by next Friday will require a demand for three units. It then follows that

$$p_{12} = \frac{(1.5)^3 \, e^{-1.5}}{3!}$$

Let us denote the probability that n units will be demanded in one week's time as $f(n)$. The elements of the one-week transition matrix for our simple inventory scheme can then be obtained from the Poisson distribution with a mean of 1.5 by evaluating that distribution as the indicated values of n.

$$\mathbf{P} = \begin{array}{c|cccccc}
 & \text{To} \\
\text{From} & 0 & 1 & 2 & 3 & 4 & 5 \\
\hline
0 & \sum_{i=5}^{\infty} f(i) & f(4) & f(3) & f(2) & f(1) & f(0) \\
1 & \sum_{i=5}^{\infty} f(i) & f(4) & f(3) & f(2) & f(1) & f(0) \\
2 & \sum_{i=5}^{\infty} f(i) & f(4) & f(3) & f(2) & f(1) & f(0) \\
3 & \sum_{i=3}^{\infty} f(i) & f(2) & f(1) & f(0) & 0 & 0 \\
4 & \sum_{i=4}^{\infty} f(i) & f(3) & f(2) & f(1) & f(0) & 0 \\
5 & \sum_{i=5}^{\infty} f(i) & f(4) & f(3) & f(2) & f(1) & f(0)
\end{array} \tag{2.18}$$

The reader is urged to justify each of the elements of the **P** matrix above.

The specification of $\mathbf{V}(0)$ and $\mathbf{P}$ now provide us with all the information necessary to predict weekly fluctuations in inventory in a probability sense. We will explore the mechanics of that calculation in the next section.

COMPUTATION TECHNIQUES

By stating the Markov chain problem in the form of vectors and matrices, we gain a convenient bookkeeping device. The techniques of matrix algebra can now be brought to bear in the analysis of the system.

Consider the problem of finding the state distribution after one transition of a stationary Markov chain. Beginning with state 1 we could simply enumerate all of the mutually exclusive and collectively exhaustive ways one might enter that state, given knowledge of the initial state distribution.

$$Pr(x_1 = 1) = \sum_{i=1}^{m} Pr(x_1 = 1, x_0 = i)$$

$$\begin{aligned} &= Pr(x_0 = 1)\, Pr(x_1 = 1 | x_0 = 1) \\ &+ Pr(x_0 = 2)\, Pr(x_1 = 1 | x_0 = 2) \\ &+ \ldots + Pr(x_0 = m)\, Pr(x_1 = 1 | x_0 = m) \end{aligned} \qquad (2.19)$$

Rewriting these joint, marginal and conditional probability statements in terms of our standard notation yields

$$v_1(1) = v_1(0)\, p_{11} + v_2(0)\, p_{21} + \ldots + v_m(0)\, p_{m1}$$

$$= \sum_{i=1}^{m} v_i(0)\, p_{i1} \qquad (2.20)$$

Equation 2.20 is nothing more than the row into column product of the vector $\mathbf{V}(0)$ and the first column of the $\mathbf{P}$ matrix. In similar fashion, the marginal probability of finding the system in any arbitrary state j after one transition, given by

$$v_j(1) = \sum_{i=1}^{m} v_i(0)\, p_{ij} \qquad (2.21)$$

is simply the product of the vector $\mathbf{V}(0)$ and the jth column of the $\mathbf{P}$ matrix. It then follows that the total state distribution after one transition could be obtained from the product of the vector $\mathbf{V}(0)$ and the entire matrix $\mathbf{P}$. The jth element of the resulting vector is the probability that the system will occupy state j after a single transition.

$$\mathbf{V}(1) = \mathbf{V}(0)\, \mathbf{P} \qquad (2.22)$$

Recall that the defining characteristic of a Markov chain is its one stage dependence. It therefore follows that we should be able to calculate the state distribution after two transitions if we know the distribution after the first transition. The calculation is identical to that of equation (2.22) except that we now treat $\mathbf{V}(1)$ as the initial state distribution. For the second transition

$$\mathbf{V}(2) = \mathbf{V}(1)\, \mathbf{P} \qquad (2.23)$$

But $\mathbf{V}(1)$ is itself a function of $\mathbf{V}(0)$ and $\mathbf{P}$ by virtue of equation (2.22). It therefore follows that an equivalent expression is

$$\mathbf{V}(2) = \mathbf{V}(0)\, \mathbf{P}^2 \qquad (2.24)$$

The extension to n stages should now be apparent. The general recursion formula for Markov chains can be written in matrix notation as

22

$$\mathbf{V}(n) = \mathbf{V}(n-1)\,\mathbf{P}(n) \tag{2.25}$$

which reduces to $\mathbf{V}(n) = \mathbf{V}(n-1)\,\mathbf{P}$ for stationary systems. Alternatively the state distribution forecast for n stages in the future can be calculated from the product of the current state distribution and the product of n single stage transition matrices

$$\mathbf{V}(n) = \mathbf{V}(0)\prod_{i=1}^{n}\mathbf{P}(i) \tag{2.26}$$

which becomes $\mathbf{V}(n) = \mathbf{V}(0)\,\mathbf{P}^n$ for stationary systems.

We are now in a position to restate the Chapman-Kolmogorov equations (2.8) and (2.10) in matrix form. For stationary systems it follows that

$$\mathbf{P}^n = \mathbf{P}^r\mathbf{P}^{n-r} \tag{2.27}$$

For the more general nonstationary case covering an arbitrary period from r to t we can write

$$\mathbf{P}(r,t) = \mathbf{P}(r,s)\,\mathbf{P}(s,t) \quad r < s < t \tag{2.28}$$

where the elements of $\mathbf{P}(r,t)$ constitute the set of multistage transition probabilities

$$\{p_{ij}(r,t)\} = \mathbf{P}(r)\,\mathbf{P}(r+1)\ldots\mathbf{P}(t-1)$$

For example if $r = 1$, $s = 3$, and $t = 5$ we could calculate

$$\mathbf{P}(1,3) = \mathbf{P}(1)\,\mathbf{P}(2)$$
$$\mathbf{P}(3,5) = \mathbf{P}(3)\,\mathbf{P}(4)$$
$$\mathbf{P}(1,5) = \mathbf{P}(1)\,\mathbf{P}(2)\,\mathbf{P}(3)\,\mathbf{P}(4) = \mathbf{P}(1,3)\,\mathbf{P}(3,5)$$

Equations (2.25) through (2.28) form the basis for all of our future calculations involving Markov chains.

As an interesting aside concerning stationary processes, we might think of the matrix $\mathbf{P}$ as an operator which takes a vector $\mathbf{V}(n)$ at time n and delivers a vector $\mathbf{V}(n+1)$ at time $n+1$. Diagrammed as an input-output system we have

$$\mathbf{V}(n) \rightarrow \boxed{\mathbf{P}} \rightarrow \mathbf{V}(n+1)$$

When treated from the viewpoint of classical systems analysis, the Markov chain behaves as a simple feedback loop in which the element in the loop is a delay element multiplied by the matrix $\mathbf{P}$. In geometric transform notation the corresponding flow-graph is

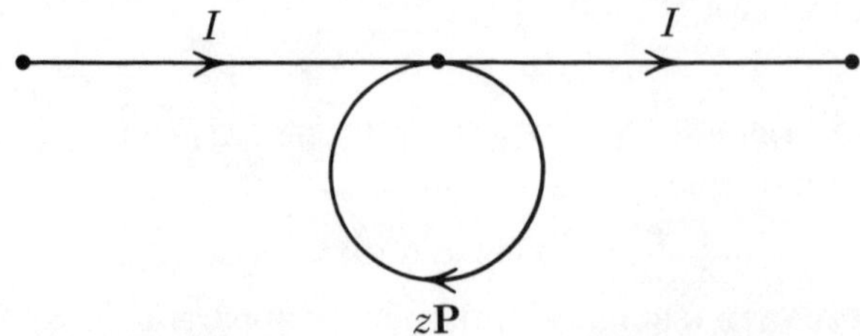

The transfer function implied is then

$$H(z) = (I - z\mathbf{P})^{-1} \tag{2.29}$$

where the negative exponent represents the inverse matrix operation. The transfer function $H(z)$ represents the transform of the impulse response of the system or equivalently the ratio of the transform of the output to the transform of the input to the system, given that it is initially at rest. The variable z is the usual transform variable for geometric transforms. The transfer function can be used to predict the output of the system when subjected to new driving signals. We will have more to say about this equation in future sections.

Example. Reconsider the inventory control problem which we characterized by the transition matrix of equation (2.18). We can provide numerical values for the elements of that matrix by consulting a table for the Poisson distribution with a mean of 1.5 as indicated by the statement concerning customer demand. The values

$$f(n) = \frac{(1.5)^n e^{-1.5}}{n!}$$

lead to the following numerical entries in $\mathbf{P}$.

	0	1	2	3	4	5
0	.0186	.0471	.1255	.2510	.3347	.2231
1	.0186	.0471	.1255	.2510	.3347	.2231
2	.0186	.0471	.1255	.2510	.3347	.2231
3	.1912	.2510	.3347	.2231	0	0
4	.0657	.1255	.2510	.3347	.2231	0
5	.0186	.0471	.1255	.2510	.3347	.2231

$$\mathbf{P} = \tag{2.30}$$

Suppose that the current Friday night inventory level is three calculators on hand. This condition is represented by the vector

$$\mathbf{V}(0) = (0, 0, 0, 1, 0, 0)$$

The problem now is to forecast inventory levels for future weeks based upon current information.

Let us now apply the basic Markov recursion equation (2.25). The projected inventory distribution one week in the future is

$$\mathbf{V}(1) = \mathbf{V}(0)\mathbf{P} = (0, 0, 0, 1, 0, 0) \begin{pmatrix} .0186 & .0471 & .1255 & .2510 & .3347 & .2231 \\ .0186 & .0471 & .1255 & .2510 & .3347 & .2231 \\ .0186 & .0471 & .1255 & .2510 & .3347 & .2231 \\ .1912 & .2510 & .3347 & .2231 & 0 & 0 \\ .0657 & .1255 & .2510 & .3347 & .2231 & 0 \\ .0186 & .0471 & .1255 & .2510 & .3347 & .2231 \end{pmatrix}$$

$$= (.1912, .2510, .3347, .2231, 0, 0)$$

This indicates that the probability of finding an inventory level of three calculators again next Friday is only .2231.

Forecasting two weeks into the future leads to

$$\begin{aligned} \mathbf{V}(2) &= \mathbf{V}(1)\mathbf{P} \\ &= (.1912, .2510, .3347, .2231, 0, 0)\,\mathbf{P} \\ &= (.0571, .0926, .1722, .2448, .2600, .1733) \end{aligned}$$

At this point the probability is .26 that we will have four calculators on hand.

Continuing on to a three week forecast we have

$$\begin{aligned} \mathbf{V}(3) &= \mathbf{V}(2)\,\mathbf{P} \\ &= (.0571, .0926, .1722, .2448, .2600, .1733)\,\mathbf{P} \\ &= (.0731, .1174, .2093, .2659, .2238, .1105) \end{aligned}$$

Note that there has been very little change between the forecasts for week 2 and week 3.

The forecast for the fourth week is

$$\begin{aligned} \mathbf{V}(4) &= \mathbf{V}(3)\,\mathbf{P} \\ &= (.0731, .1174, .2093, .2659, .2238, .1105)\,\mathbf{P} \\ &= (.0750, .1189, .2092, .2623, .2207, .1139) \end{aligned}$$

which to two significant decimals is identical to the forecast for the third week.

Continuing for a five week forecast we have

$$\begin{aligned} \mathbf{V}(5) &= \mathbf{V}(4)\,\mathbf{P} \\ &= (.0750, .1189, .2092, .2623, .2207, .1139)\,\mathbf{P} \\ &= (.0743, .1179, .2081, .2621, .2223, .1153) \end{aligned}$$

It is obvious that the difference between $\mathbf{V}(n)$ and $\mathbf{V}(n+1)$ is becoming smaller with each iteration. If we were to continue this calculation through even more iterations we would eventually reach a point at which $\mathbf{V}(n+1) = \mathbf{V}(n)$. To the nearest one hundredth, the long range state distribution for calculator inventory on Friday nights is given by

$$\mathbf{V} = (.07, .12, .21, .26, .22, .12) \tag{2.31}$$

This indicates that approximately 7 percent of the time the inventory will be zero, 12 percent of the time it will be one unit, etc.

CHARACTERISTICS

The example of the last section can be explicitly categorized as a *finite, irreducible* Markov chain with stationary transition probabilities. The term finite refers to the number of states the system may occupy. The inventory example had six possible states. A Markov chain is said to be irreducible if the chain can eventually make transitions from any state to any other state. The move from any given state to another need not occur in a single transition. In the matrix of equation (2.30), for example, it is impossible to move from state 3 to state 5 in a single transition.

However, the system could move from state 3 into state 2 and hence on to state 5 in a second transition. The irreducible nature of the chain becomes apparent if one calculates $\mathbf{P}^n$ for several values of n. If the chain is irreducible every cell will eventually contain a nonzero entry. Chains which do not have this property are said to be *nonirreducible*. The system has stationary transition probabilities whenever the p_{ij} values are the same for all stages n.

Other terms used to characterize Markov chains include, *absorbing*, *transient* and *periodic* states, *closure* and *recurrence*. A state i is said to be *absorbing* if $p_{ii} = 1$. This indicates that once the system enters that state it will be trapped there and can never transition to another state. A *transient* state is one in which the limiting state probability goes to zero. State j is periodic if when the system starts in state j, subsequent occupation of that state can only occur at times $t, 2t, 3t \ldots$ where t is an integer representing the period of that state. A set of states S is *closed* if no state outside of S can be reached from any state in S. The smallest set containing S is called the *closure* of S. Note that an absorbing state is a closed set with a single element. A *recurrent* chain is analogous to the concept of a closed set. Transient states may lead the sysem into a set of states that are connected by possible transitions within the set but it can never jump from that set. Every Markov process has at least one recurrent chain. When it has only one recurrent chain the process is said to be *ergodic*. The inventory control process of the last section is an ergodic process.

The following transition matrix illustrates these properties:

$$
\mathbf{P} = \begin{vmatrix}
0.5 & 0.5 & 0 & 0 \\
0.6 & 0.4 & 0 & 0 \\
0.2 & 0.3 & 0.4 & 0.1 \\
0 & 0 & 0 & 1.0
\end{vmatrix}
\tag{2.32}
$$

The Markov process represented by this matrix has two recurrent chains. The single state 4, which is an absorbing state can be classed as a recurrent chain. The set of states [1, 2] constitutes a closed set and forms a recurrent chain. State 3 is a transient state. If the system starts in state 3 it will sooner or later transition into state 4 or the closed set [1, 2] where it will be trapped. If the system is projected far enough into the future, the probability of finding it in state 3 will be zero.

There is an extensive literature concerning the formal properties of Markov chains as they relate to long-range system behavior. The reader is urged to consult texts such as Feller (37) or Cox and Miller (28) for a formal treatise of this topic. Our approach will be more pragmatic in that we will operate in true engineering fashion by manipulating the basic Markov equations to obtain numerical results without worrying too much about the formalism which girds the theoretical development justifying such calculations.

PROBLEMS

2.1 Construct a list of five processes you encounter in your daily activities which you feel might possess the Markov property. Explain why you hypothesize that each may be Markovian and what caveats you would observe in attempting to test your hypotheses.

2.2 Consider each of the following processes:

a) Customer withdrawals from a savings institution

b) Stock price movements for a single company

c) Daily changes in the Dow Jones Industrial Average

d) Brand switching by consumers of soft drinks

e) Progress toward graduation for a college student

In each case list reasons both for and against modeling it as a Markov process. Assuming that the Markov assumption will be made, describe the state variable, state transition and time between transitions for each.

2.3 The following transition matrix has been developed for a stationary Markov process.

$$\mathbf{P} = \begin{pmatrix} 0 & 0 & 1 \\ 0.3 & 0.3 & 0.4 \\ 1 & 0 & 0 \end{pmatrix}$$

a) Is the chain:

    i) Finite?

    ii) Irreducible?

    iii) Aperiodic?

Explain.

b) Suppose the system is in state two at time zero.

    i) Find the probability that it will be in state two after five transitions.

    ii) Estimate the limiting probability that it will be in state 2.

    iii) Suppose it started in state one. What would be your answers for parts (*b-i*) and (*b-ii*)?

2.4 Consider the following single-stage transition matrix.

$$\mathbf{P} = \begin{pmatrix} 1 & 0 & 0 \\ 0.2 & 0.3 & 0.5 \\ 0 & 0 & 1 \end{pmatrix}$$

a) Is the chain

    i) Finite?

    ii) Irreducible?

    iii) Aperiodic?

Explain.

b) Suppose the system starts in state two.

    i) Find the probability that it will be in state two after four transitions.

    ii) Estimate the limiting probability that it will be in state two.

    iii) Estimate the limiting probability that it will be in state one.

2.5 A marketing study has been conducted concerning the brand loyalty of customers for three competing makes of automobiles. Two-door sedans are the focus of the analysis. The following observations during a three-month period in a single metropolitan area have been made:

Trade-in Make	No. traded for A-2 door	No. traded for B-2 door	No. traded for C-2 door	No. traded for non-2 door
A	240	96	64	100
B	100	150	50	60
C	40	90	70	30
non-2 door	30	70	50	150

a) Marketing staff wants to develop a Markov chain forecasting model in an attempt to predict market shares over future time periods. Comment on the viability of such an approach and defend your position.

b) Assuming that a Markov approach is reasonable, find the probability that a car-A owner will still have brand A after four trades.

c) Suppose that every owner trades once a year and that there are currently 400 A-owners, 200 B-owners, 100 C-owners, and 300 non-2-door owners in the area. How many of each would you forecast for year 5?

2.6 A large metropolitan automobile dealer is interested in the chain of automobile trades triggered by a new car trade-in. From his new car sales orders involving trades, he has established the following age distribution for trade-in cars.

Age	Percent
1–2 years	30
3–4 years	30
5–6 years	20
7–8 years	20

The dealer never handles a car which is over 8 years old. Each trade-in goes to the used car lot where it may also generate a trade. However, it is the dealer's policy to accept only older cars on trades. His records show that 20 percent of the 1–2 year old cars on his lot are sold without trade, 20 percent involve a trade for a 3–4 year old car and the balance of the trades are equally divided among the remaining older age classes. The 3–4 year old cars are sold for cash 20 percent of the time and involve trades spread equally across the older age classes the rest of the time. The 5–6 year old cars receive trades 60 percent of the time. All 7–8 year old cars are wholesaled and hence involve no trades.

a) Structure this problem as a Markov chain.

b) Obtain the probability distribution for the number of transactions generated by a new car trade-in as a function of the age of the trade-in. A chain of transactions is broken when a dealer accepts cash for a car.

c) Determine the average number of transactions generated by a new car trade.

2.7 Prepare a short user instruction manual for the program in Appendix D outlining the model and indicating its potential usefulness in the analysis of Markov chains.

2.8 Consider the following transition matrix

$$\mathbf{P} = \begin{vmatrix} 0.4 & 0 & 0.4 & 0.2 & 0 \\ 0 & 1 & 0 & 0 & 0 \\ 0.8 & 0 & 0.1 & 0.1 & 0 \\ 0 & 0 & 0.5 & 0.5 & 0 \\ 0.2 & 0.2 & 0.3 & 0.2 & 0.1 \end{vmatrix}$$

a) Identify each state as absorbing, transient, or periodic.

b) How many recurrent chains are contained in this process? What are they?

c) Is the process ergodic?

d) What would you expect the limiting state distribution to be?

e) Calculate the probability of achieving the chain of events $\{E_5, E_5, E_4, E_3, E_1\}$ if the system is initially in state 5.

2.9 A moving average statistic is often used to forecast customer demand. Suppose that demand on day i is represented by D_i. A ten-day moving average calculated at the end of day n would then be

$$\bar{D}_n = \sum_{i=n-9}^{n} D_i/10$$

An analyst has suggested that the sequence $\{\bar{D}_n\}$ be modeled as a Markov process. Prove whether or not this is a reasonable approach.

2.10 The shipping department at a gas-line compressor assembly plant has storage space for four compressors. Compressor assembly is a complicated process, the net result of which is that the number of units assembled per day is essentially a random variable which can be described by a Poisson distribution with a mean of one per day. At the end of each day rail shipment of a single compressor is made if any are available. Units arriving at the shipping department when four are already there must be shipped by truck. Rail shipment costs $1200 per unit. Truck shipment cost $1600 per unit.

a) Show that the number of units on hand immediately following rail shipment time can be described as a Markov chain.

b) Write the transition matrix for this process.

c) Estimate current daily shipping costs.

d) How much could the company afford to pay to expand the shipping area by one compressor storage space? (Assume that all costs must be amortized over a two-year planning horizon.)

2.11 Consider the problem faced by a tourist agency selling tickets for a single flight to the South Seas. The aircraft has a capacity of 240 passengers. The flight leaves in five days. There are currently 230 reservations on file for the flight. Daily demand for the next five days is estimated to be Poisson distributed at a rate of twenty per day. Customers holding reservations may cancel at any time. The probability that a single customer will cancel *j* days prior to departure is estimated to be

j	$p(j)$
5	.05
4	.05
3	.04
2	.03
1	.02
0	.02

Cancellations occur at the beginning of each business day. The net profit for each passenger served is $150. The penalty for overbooking as established by the CAB is $200 for every passenger who has a reservation at the time of departure but cannot be accommodated. The company is attempting to evaluate two booking policies. Policy A says do not accept reservations beyond the capacity of the aircraft. Policy B says take reservations up to $(240 + 10(j))$ if there are *j* days remaining prior to departure. Compare the expected net profit under each of these booking policies.

2.12 Consider the problem faced by an antiaircraft missile launcher firing at a single reconnaissance aircraft which has intruded protected airspace. The launch control officer has *n* missiles available. If the previous shot was a hit, there is a relatively high probability that the next shot will also be a hit. If the previous shot was a miss, he must adjust his aim and consequently reduce the probability of a hit on the next shot. Field data show that the first shot will hit 20 percent of the time. Subsequently a hit will follow another hit 70 percent of the time and a hit will follow a miss 30 percent of the time.

a) Model the system as a simple Markov chain.

b) Find the probability of recording a hit on launch *j*.

c) Suppose two hits are required to down the aircraft. Find the probability that the aircraft will escape if there are five missiles to launch.

2.13 An escort service is open for business on Monday, Wednesday and Friday each week. The service maintains a list of escorts available for hire. The state of the system is given by X_n, the number of escorts available at the beginning of the business day *n*. Due to municipal licensing requirements the maximum number they can have available is four. The calls for escorts are screened to protect both escort and client. On Monday an incoming call for an escort will be acceptable to the agency with probability 0.3. On

Wednesday this rises to 0.6 and on Friday to 0.9. Replacements for those escorts who receive an assignment are obtained by telephoning members of a reserve list. All replacements are available at the start of the business day. There are ten escorts on the reserve list. The probability that any one of them will be available for service when called is 0.2. Customer demands on day i are distributed by $f_i(d)$ where

d	$f_M(d)$	$f_W(d)$	$f_F(d)$
0	0.1	0	0
1	0.1	0.3	0
2	0.2	0.4	0.1
3	0.2	0.2	0.2
4	0.1	0.1	0.3
≥ 5	0.3	0	0.4

a) Model the business as a Markov process.

b) Show how to find the probability of having three escorts available at the end of the day on Friday if they started with one on Monday.

c) Is this a stationary process? If not can you suggest a way of redefining portions of the problem so that a stationary process does exist.

2.14 A local transit service is considering offering bus service with a six-seat jitney for students at the university library. The jitney will pass the library every t minutes at which time they will load all the passengers present up to the capacity of the jitney. It then proceeds to the dormitory where all passengers are discharged. The library stop has a capacity of twelve passengers. Student arrivals are assumed to be Poisson distributed at a rate of fifteen per hour.

a) Develop a model to describe the behavior of the system as viewed by an arriving jitney.

b) Suppose that the cycle time is $t = 36$ minutes. Explain how you would estimate the portion of time that the jitney would have a full load and the number of students per hour denied service.

2.15 In the political arena it is often hard to identify the political beliefs of members of congress. At the start of a legislative session the house was equally divided between liberals and conservatives. At the end of each day's debate one man from each side defects to the opposition. Defections occur at random and past political beliefs have no influence on the probability of successful swaying of a member. Each party thus maintains a nominal voting strength of N. However, the liberal leader has suddenly become quite interested in the pre-session loyalty of his current voting strength. He defines his state E_j to mean that he has j "original liberals" in his voting block.

a) Model the liberal voting block as a Markov chain.

b) What can you tell the liberal leader about the distributon of possible states if this session continues indefinitely?

c) Suppose that $N = 6$. Determine the liberal voting composition after the fifth day of debate.

2.16 A judge receives requests for hearings which he must process within a seven-day period. At any day within the life of the request he can take no action, grant the request or reject the request. If he does not act by the end of the seven-day period, the request is automatically rejected. From past experience it appears that requests which are less than three days old are granted with probability 0.2 on any given day. They are never rejected at that age. Requests which are three, four, or five days old are equally likely to be granted or rejected and may be allowed to age one more day with probability 0.2. A six or seven-day-old request will be granted with probability 0.1.

a) Model this system as Markov process. (Decisions are made at times $0, 1, \ldots, 7$)

b) Calculate the probability that a new request will be granted at the end of day three.

c) Calculate the probability that a three-day-old request will ultimately be granted.

d) Calculate the average number of days a request will remain in the system before a decision is made.

2.17 An educational consultant is attempting to structure a model for evaluation of high school grading policies for sophomores, juniors, and seniors. He has access to data which he thinks can be used to estimate the portion of each class which will fail each year. He calls this variable f_i. Only students who fail are prone to drop out of school. The portion of students failing in class i who withdraw because of that failure is estimated to be w_i. Others who fail repeat the grade failed. Students who successfully complete their senior year leave via graduation.

a) State the assumptions necessary to be able to structure this system as a Markov process. Identify the states of the system.

b) Develop a transition matrix which can be used to predict relative enrollments over the next ten years.

c) Suppose that each class now has 200 students

$$f_i = \begin{cases} 0.3 & i = \text{soph.} \\ 0.2 & i = \text{jr.} \\ 0.1 & i = \text{sr.} \end{cases}$$

and

$$w_i = \begin{cases} 0.5 & i = \text{soph.} \\ 0.4 & i = \text{jr.} \\ 0.3 & i = \text{sr.} \end{cases}$$

i) Find the numbers and grade distribution for these classes three years from now.

ii) Find the probability that an entering sophomore will withdraw before his senior year.

iii) Find the probability that a student who enters as a sophomore will ultimately graduate.

3

MARKOV CHAIN COMPUTATIONS

The formal Markov chain models of the last chapter provide the basis for many important calculations we will be asked to perform in the analysis of queueing systems. What we now wish to explore are some manipulative skills which may be useful in achieving numerical answers for systems which can be described by Markov chains.

STEADY-STATE SOLUTIONS

In many applications involving stationary transition mechanisms we will be interested in the long-run behavior of the process. We would like to make statements concerning the state-probability distribution at some distant time in the future after the system has stabilized in a probability sense. Alternatively we would like to be able to calculate the relative frequency with which the system might occupy a given state over a lengthy time period. Both questions are answerable by the so-called steady-state solution for Markov chains.

Our focus now is on the limiting behavior of the system. The steady-state probabilities we seek can be expressed as

$$v_j = \lim_{n \to \infty} v_j(n) \tag{3.1}$$

In the queue of subassemblies at a work station, for example, we may be interested in the long-run effect more than the transients which may occur when the production line is first started. This long term effect is described by the probabilities v_j.

To reach steady state implies that the system has run sufficiently long that the effects of initial conditions are no longer present. This implies the existence of an equilibrium distribution which is stationary. That is

$$v_j(n) = v_j \tag{3.2}$$

for all n. If these limiting probabilities do not depend upon the initial state of the system, it seems reasonable to also interpret them as the limit of the n-stage transition probabilities, $p_{ij}{}^n$. That is

$$v_j = \lim_{n \to \infty} p_{ij}{}^n \tag{3.3}$$

indicates that the probability of moving from any initial state i to a final state j in the limit is the same for all i. Denoting the steady-state distribution as $\mathbf{V}$ we are then led to

$$\lim_{n \to \infty} \mathbf{P}^n = \begin{pmatrix} \mathbf{V} \\ \mathbf{V} \\ \vdots \\ \mathbf{V} \end{pmatrix} \tag{3.4}$$

where all rows of the matrix are identical.

Computing the elements of the vector $\mathbf{V}$ is facilitated by returning to the basic Chapman-Kolomogorov equation (2.25). Since

$$\mathbf{V}(n) = \mathbf{V}(n-1)\mathbf{P}$$

it follows that in the limit as $n \to \infty$ we have

$$\mathbf{V} = \mathbf{VP} \tag{3.5}$$

In some sense steady state has been reached when the vector $\mathbf{V}(n)$ operated on by the matrix $\mathbf{P}$ again yields the vector $\mathbf{V}(n)$. Performing the indicated multiplication for a finite chain with m states yields a set of m equations in the m unknown probabilities v_j.

$$v_1 = \sum_{k=1}^{m} v_k p_{k1}$$

$$v_2 = \sum_{k=1}^{m} v_k p_{k2}$$

$$\vdots$$

$$v_m = \sum_{k=1}^{m} v_k p_{km} \tag{3.6}$$

These equations are homogeneous. To obtain a unique solution meeting the requirements of a probability distribution we must add to this set the condition

$$\sum_{j=1}^{m} v_j = 1 \tag{3.7}$$

Let us summarize this discussion by indicating the steps necessary to find the limiting distribution for Markov chains.

1. Form a set of m homogeneous equations by using equation (3.5).
2. Replace any one of the resulting equations with the probability condition equation (3.7).

3. Solve the resulting set of m nonhomogeneous equations for the unknown probabilities v_j.

For example, consider the transition matrix

$$\mathbf{P} = \begin{array}{ccc} 0.3 & 0.2 & 0.5 \\ 0.4 & 0.3 & 0.3 \\ 0.2 & 0.6 & 0.2 \end{array}$$

Three homogeneous equations are obtained from

$$\mathbf{V} = \mathbf{VP}$$

or

$$(v_1, v_2, v_3) = (v_1, v_2, v_3)\,\mathbf{P}$$

Equating like elements on both sides of the equation,

$$v_1 = 0.3v_1 + 0.4v_2 + 0.2v_3$$
$$v_2 = 0.2v_1 + 0.3v_2 + 0.6v_3$$
$$v_3 = 0.5v_1 + 0.3v_2 + 0.2v_3$$

Replacing the third equation with the probability condition $\sum_{i=1}^{m} v_i = 1$

leads to

$$-0.7v_1 + 0.4v_2 + 0.2v_3 = 0$$
$$0.2v_1 - 0.7v_2 + 0.6v_3 = 0$$
$$v_1 + v_2 + v_3 = 1$$

which when solved simultaneously yield

$$v_1 = 0.304 \qquad v_2 = 0.368 \qquad v_3 = 0.328$$

In the calculations above we have been operating under the assumption that the limiting probabilities do exist. It can be shown that for the special case of a finite, irreducible, aperiodic Markov chain, the limiting vector $\mathbf{V} = (v_1, v_2, \ldots v_m)$ does exist and is the unique stationary probability vector of the process. The calculation $\mathbf{V} = \mathbf{VP}$ is however very robust. Anytime a solution to the resulting equations can be obtained it is meaningful. If the process is periodic, in which case the idea of a limit in the sense of equation (3.1) is meaningless, the solution obtained represents the relative frequency with which each state is occupied over a long period of transitions. If the process is not one with a single, closed set of states as required for the existence of a limiting stationary distribution which is independent of initial conditions, the attempted solution of relationship (3.5) will lead to an indeterminate set of equations rather than to an incorrect solution. The net result is that we can use $\mathbf{V} = \mathbf{VP}$ both to prove the existence of a limiting distribution which is independent of initial conditions and to obtain its values.

TRANSIENT ANALYSIS

In the analysis and design of many service systems it may be vital to know how the system will respond to sudden changes in input or service capabilities. For example, in studying the behavior of a conveyor loading station one might like to know how long it would take the storage bank distribution to stabilize again if work stoppage at a prior work station temporarily restricts the flow of goods to the loading station. In a similar vein, if one seeks to study the flow of air traffic in a major terminal area, it is the time-varying nature of arrivals and service which are important in analyzing system performance. The flow of arrivals at 8:00 AM is much higher than that at 11:00 PM. Furthermore service interruptions may occur in the form of thunderstorms or other meteorological phenomena which may temporarily restrict the ability of the terminal to process air traffic.

Steady-state analysis is inadequate for both the conveyor and the air traffic control problem. Each requires a consideration of the time-varying nature of the queue and its associated delays. If these systems can be modeled as finite Markov chains, it is a simple matter to extract the desired time-varying performance by harnessing the computational power of the modern digital computer.

It is important to note that the limitation to a finite number of states is not unrealistic. To the contrary, one might argue that the many formal queueing models with their assumed unlimited capacity queues are the unrealistic ones. In the air traffic control problem, for instance, the system is limited by the number of aircraft which can be accommodated in holding patterns while providing reasonable assurance of separation. It is much more realistic to model that system assuming a capacity of, say twenty-five aircraft than it is to assume that one has infinite holding space for arrivals.

The reason that many existing models assume an infinite number of states is that this assumption often permits one to obtain a nice compact equation expressed in terms of well-known probability distributions. In the context of finite Markov chains, however, we are more interested in numerical solutions. We want a number for the probability that no aircraft will be in the holding pattern at 5:00 PM. We really do not care if that number happens to be the definite integral of some exotic density function. It is its numerical value which is most important.

For those processes with a small number of states we can obtain the desired transient solutions by straightforward matrix multiplication. All that is required is the specification of an initial probability distribution and an appropriate transition matrix for each time interval being considered. Equation (2.26) can then be used in straightforward fashion to compute the state distribution at every time interval in question. Recall that from the basic recursion formula

$$\mathbf{V}(1) = \mathbf{V}(0)\,\mathbf{P}(1)$$
$$\mathbf{V}(2) = \mathbf{V}(1)\,\mathbf{P}(2)$$
$$\vdots$$
$$\mathbf{V}(n) = \mathbf{V}(n-1)\,\mathbf{P}(n)$$

An Air Traffic Control Example

Koopman (64) and Hartman (43) have treated the time-dependent air traffic control problem. The problem is one of analyzing how queues build and dissipate during the day at major air terminals.

A typical aircraft arrival pattern might look like that shown in Figure 3-1. Demand for service, in terms of aircraft needing access to the runway for landing, is very low during the period from midnight to 7:00 AM. At that time the demand rate rises steeply and reaches a peak sometime in mid-morning after which it may decline slightly, then be nearly stationary for several hours. As the normal business day comes to a close, people again seek transportation in large numbers. Consequently the airlines will schedule many late afternoon flights which may reach their destination in the neighborhood of 6:00 PM. From that time on till midnight traffic gradually diminishes.

The maximum potential service rate may also fluctuate in the air traffic control system. The rate may be constant at a low level during nighttime operations due to reduced tower crew size or to the tendency to increase separation in hours of darkness. It may increase during daylight hours as a full crew of controllers comes on board. In addition, service rate may change suddenly because of equipment failures or because a thunderstorm moves across the approach path. Figure 3-1 also diagrams a possible service rate pattern.

An analysis of the air traffic control system which is limited to stationary processes and unlimited queues may overlook many important aspects of system performance. For example, what will happen to the aircraft holding times if a thunderstorm closes the field for thirty minutes during rush hour? What would be the effect of load leveling by rescheduling some peak hour flights into less congested time periods? A simple Markov chain model can be used to answer such questions.

To begin the analysis let us make the following assumptions:

1. During any fifteen minute time interval the arrival rate and service rate can be approximated by a constant.
2. The arrival process is Poisson with a time-varying parameter.
3. The service time during any given fifteen minute period is a constant; i.e., service time has zero variance.
4. The system has a maximum capacity of m aircraft at one time. Any aircraft arriving when the system is saturated are diverted to other airports.

The first assumption suggests that the arrival and service patterns of Figure 3-1 can be approximated by a series of step functions as illustrated by Figure 3-2. Fifteen minutes was selected simply for illustration. The step could be shorter or longer depending upon the rate of change in the curve being approximated. We are replacing a single 24-hour period containing varying arrival and service rates with 96 periods of fifteen minutes each. During any fifteen-minute period the rates are constant, but the rates may vary from period to period.

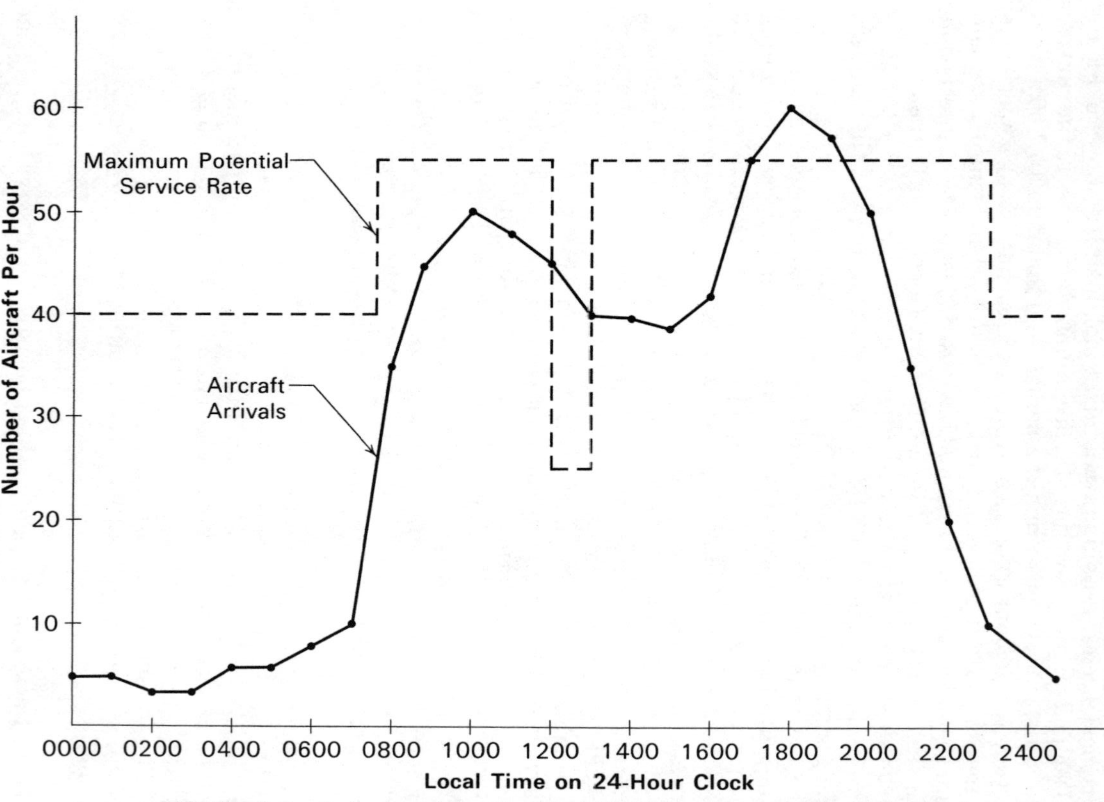

FIGURE 3-1 Arrival and Service Rate Functions for a Typical Airport

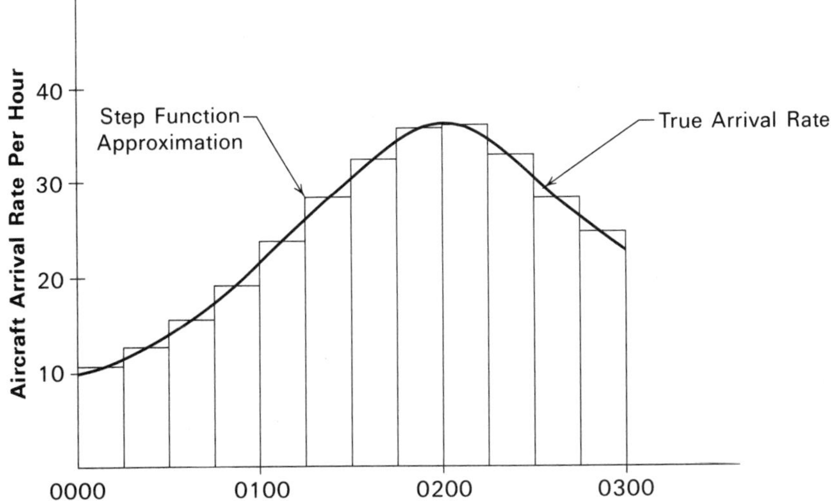

FIGURE 3-2 Approximating a Continuous Arrival Rate Function

The second assumption has been shown to be realistic in a variety of empirical studies of air traffic movement. The Poisson distribution with a mean appropriate for the hour of the day being considered can then be used to fill the elements of our transition matrices.

The third assumption is an obvious approximation; service time is rarely a constant in such a system. However, both Koopman and Hartman have compared results for the constant service time and exponentially distributed service time models with surprisingly close values for the usual measures of performance. We will have more to say on this topic in future chapters.

The fourth assumption fits the realities of air traffic control and permits us to use the iterative formula for finite Markov chains. A capacity in the neighborhood of 60 is all that is available even at our largest air terminals.

A further simplifying computational assumption will be made which provides that service may be initiated on waiting units only at the beginning of a service epoch. The length of a service epoch is the length of time it requires for service to take place. For example, if service time is three minutes, service can take place only at times $0, 3, 6, 9, \ldots$. The approximation is inaccurate whenever an arriving unit appears midway in a service epoch while the system is empty. Even though no one is waiting, the assumed behavior requires that the arriving unit be delayed until the start of the next service epoch. The approximation is weak only for those systems with a high probability of being empty.

The solution progresses in straightforward fashion. The state probability vector at time t, $\mathbf{V}(t)$, where t is a multiple of fifteen minutes, is multiplied by an appropriate single stage transition matrix covering one service epoch. The result is a new vector $\mathbf{V}(t + 1/\mu)$ for a "$1/\mu$"-minute service epoch. This calculation continues in iterative fashion until the end of a fifteen-minute period. At that time new values for both arrival rate and service rate are introduced, a new transition matrix is prepared, $\mathbf{V}(t + 15)$ is used as the new initial probability vector and the computation proceeds for the next fifteen-minute interval. Whenever there is not an integral number of service epochs in a fifteen-minute interval, the length of the epoch overlapping two intervals is determined by the service rate during the interval in which the epoch started.

Symbolically, for an arrival rate λ and service time $1/\mu$

$$\mathbf{V}(t + 1/\mu) = \mathbf{V}(t)\,\mathbf{P}(t|\lambda, 1/\mu) \tag{3.8}$$

The equation yields the probability distribution for the number of aircraft awaiting service at time $(t + 1/\mu)$. This distribution can then be used to calculate expected number in the system, variance for number in the system and expected holding time.

The transition matrix is constructed in exactly the same manner as we previously illustrated for the inventory control problem. The probability of x aircraft arrivals during a $1/\mu$-minute service epoch, given that the arrival rate is λ aircraft per minute is from the Poisson distribution,

$$pr\,(x|\lambda, \mu) = \frac{(\lambda/\mu)^x\, e^{-\lambda/\mu}}{x!}, \; x = 0, 1, 2\ldots \tag{3.9}$$

The numerical values for the elements of the transition matrix for a single service epoch are then determined from various values of $pr(x|\lambda, \mu)$ as noted in equation (3.10). Recall that transitions occur from the point immediately preceding a service epoch to the point immediately preceding the next service epoch. Moving from state i to state j where $0 < i < j < m$ then requires that $j - i + 1$ arrivals occur during a single service epoch.

$$\mathbf{P}(t|\lambda, \mu) = \begin{array}{c|ccccc}
 & 0 & 1 & 2 & & m \\
\hline
0 & pr(0) & pr(1) & pr(2) & \cdots & \sum_{i=m}^{\infty} pr(i) \\
1 & pr(0) & pr(1) & pr(2) & \cdots & \sum_{i=m}^{\infty} pr(i) \\
2 & 0 & pr(0) & pr(1) & \cdots & \sum_{i=m-1}^{\infty} pr(i) \\
\vdots & & & & & \\
m & 0 & 0 & 0 & \cdots & \sum_{i=1}^{\infty} pr(i)
\end{array} \tag{3.10}$$

where $pr(x) = \dfrac{(\lambda/\mu)^x\, e^{-\lambda/\mu}}{x!}$

Although the iterative procedure proposed here for time-varying systems may seem very awkward and time consuming if viewed as a hand calculation, it is a natural for the digital computer. Hartman (43) used this technique programmed in FORTRAN to analyze a variety of air traffic control configurations. Typical computation time for a twenty-four hour time period with time-varying arrival and service rates, fifteen-minute step approximations and a capacity of twenty-five aircraft was twenty-one seconds on the IBM 360/75. His program included a subroutine for internally generating the matrices of equation (3.10) as required, printout of the entire probability distribution every fifteen minutes, and calculation of a full array of performance measures. A similar program for the readers' use is reproduced in Appendix D.

ROLE OF TRANSFORMS

For the portion of time during which a Markov chain is stationary one can use transform techniques to advantage. Transforms are used to solve the basic difference equation which defines the Markov iteration. Howard has an extensive discussion concerning this approach (46).

The basic difference equation for a stationary system can be written as

$$\mathbf{V}(n + 1) = \mathbf{V}(n)\,\mathbf{P} \tag{3.11}$$

Although this equation involves vectors and matrices, it poses no conceptual difficulties for a transform solution. The transform of a matrix is another matrix formed by transforming every element in the original matrix. The matrix $\mathbf{P}$ of real numbers can be treated as a constant when forming the subsidiary equation. Using geometric transforms and the table of transform pairs in the Appendix, the subsidiary equation for (3.11) is obtained by first taking transforms of both sides. That is

$$G[\mathbf{V}(n + 1)] = G[\mathbf{V}(n)\,\mathbf{P}]$$

The transform on the left can be replaced by a function of the transform of $\mathbf{V}(n)$ from the tables as

$$G[\mathbf{V}(n + 1)] = z^{-1}[G[\mathbf{V}(n)] - \mathbf{V}(0)]$$

The transform on the right is simply the transform $G[\mathbf{V}n]$ multiplied by the constant matrix $\mathbf{P}$. It then follows that the subsidiary equation is given by

$$z^{-1}[G[\mathbf{V}(n)] - \mathbf{V}(0)] = G[\mathbf{V}(n)]\,\mathbf{P} \tag{3.12}$$

where $G[\mathbf{V}(n)]$ is the single unknown. The elements of the vector $G[\mathbf{V}(n)]$ are transforms which when inverted give the desired probability distribution after any number of transitions n.

Solving equation (3.12) for the unknown vector, taking care to observe the rules of matrix algebra, yields

$$G[\mathbf{V}(n)] = \mathbf{V}(0)[\mathbf{I} - z\mathbf{P}]^{-1} \tag{3.13}$$

Here $\mathbf{I}$ represents the $m \times m$ identity matrix and the negative exponent indicates a matrix inversion operation.

Recall that an alternate form of the Markov iteration beginning with the distribution at time zero is

$$\mathbf{V}(n) = \mathbf{V}(0)\,\mathbf{P}^n \tag{3.14}$$

Taking transforms of both sides yields

$$G[\mathbf{V}(n)] = \mathbf{V}(0)G(\mathbf{P}^n) \tag{3.15}$$

Comparing the results of equations (3.13) and (3.14) suggests that

$$[\mathbf{I} - z\,\mathbf{P}]^{-1} = G(\mathbf{P}^n) \tag{3.16}$$

It therefore follows that we can obtain a closed-form expression for the nth power of the single stage transition matrix by inverting the transform $[\mathbf{I} - z\,\mathbf{P}]^{-1}$. Since $\mathbf{P}^n$ contains all of the desired transient information we now have a quick way to find the state distribution at any time n as a function of $\mathbf{V}(0)$ without going through the iteration of previous sections. It is interesting to note that this is the same expression for the transform of $\mathbf{P}^n$ as equation (2.29) which was developed from feedback control arguments.

Howard (46) makes several observations about the closed-form expression for $\mathbf{P}^n$. In general

$$\mathbf{P}^n = \mathbf{S} + \mathbf{T}(n) \tag{3.17}$$

where $\mathbf{S}$ is a stochastic matrix representing the limiting behavior of the system and $\mathbf{T}(n)$ is a set of matrices representing time varying perturbations around the limiting state distribution.

The matrix $\mathbf{S}$ arises from inverting a term in the transform of the form $(1/(1 - z))$. This is equivalent to asserting that $\mathbf{P}$ has an eigenvalue of one.[1] If the process consists of a single closed set, all of the rows of $\mathbf{S}$ will be identical. Their values will then be the same as those obtained from the steady state calculation of equation (3.5). If the process has more than a single closed set of states, each row of $\mathbf{S}$ may be unique. The proper interpretation of row i of $\mathbf{S}$ in this case is that of a limiting conditional distribution. Row i gives the limiting state distribution given that the system initially occupies state i. Attempting to find a solution for multi-chained systems by the usual steady-state arguments will lead to an indeterminant set of equations. The transform solution suffers no such difficulty.

The matrix $\mathbf{T}(n)$ may be factorable into a sum of several differential matrices. (A differential matrix is one whose rows sum to zero.) The terms of $\mathbf{T}(n)$ will always have coefficients of the form c^n, nc^n, n^2c^n, etc. where $|c| \leq 1$.

Example. Consider the matrix

$$\mathbf{P} = \begin{pmatrix} 0.2 & 0.8 \\ 0.6 & 0.4 \end{pmatrix}$$

---

[1]The eigenvalues of the matrix $\mathbf{P}$ are those values of λ which satisfy the characteristic equation of $\mathbf{P}$ defined by the determinant $|\mathbf{P} - \lambda\mathbf{I}| = 0$. For an $(m \times m)$ matrix $\mathbf{P}$ there will be m roots or eigenvalues which satisfy the characteristic equation. If $\mathbf{P}$ is a stochastic matrix one of those values must be unity.

By the transform relationship

$$[\mathbf{I} - z\,\mathbf{P}] = \begin{pmatrix} 1 - 0.2z & - 0.8z \\ - 0.6z & 1 - 0.4z \end{pmatrix}$$

and

$$[\mathbf{I} - z\mathbf{P}]^{-1} = \frac{1}{(1 - z)(1 + 0.4z)} \begin{pmatrix} 1 - 0.4z & 0.8z \\ 0.6z & 1 - 0.2z \end{pmatrix}$$

By partial fraction expansion

$$[\mathbf{I} - z\,\mathbf{P}]^{-1} = \frac{1}{1 - z} \begin{pmatrix} \dfrac{3}{7} & \dfrac{4}{7} \\ \dfrac{3}{7} & \dfrac{4}{7} \end{pmatrix} + \frac{1}{1 + 0.4z} \begin{pmatrix} \dfrac{4}{7} & \dfrac{-4}{7} \\ \dfrac{-3}{7} & \dfrac{3}{7} \end{pmatrix}$$

From the table of transform pairs it follows that

$$\mathbf{P}^n = \begin{pmatrix} \dfrac{3}{7} & \dfrac{4}{7} \\ \dfrac{3}{7} & \dfrac{4}{7} \end{pmatrix} + (-0.4)^n \begin{pmatrix} \dfrac{4}{7} & \dfrac{-4}{7} \\ \dfrac{-3}{7} & \dfrac{3}{7} \end{pmatrix}$$

AN ALGEBRAIC APPROACH

The time dependent solution for stationary finite Markov chains involves finding appropriate powers of the single-stage transition matrix $\mathbf{P}$. We have demonstrated how this can be accomplished by brute force matrix multiplication and by using geometric transforms to find a general solution to the linear difference equation. The transform solution is awkward for large-scale problems because of the difficulty of finding the proper inverse transforms. The brute force multiplication is adequate for modest sized matrices and large-scale computers but can be made more efficient by using certain properties of nonnegative square matrices. The purpose of this section is to introduce some of those properties.

The point of the ensuing analysis is that the nature of a finite chain is determined by the properties of the eigenvalues of $\mathbf{P}$. The eigenvalues of the matrix $\mathbf{P}$ are determined from the characteristic equation

$$|\mathbf{P} - \lambda\mathbf{I}| = 0 \tag{3.18}$$

The vertical bars here denote the determinant of the matrix $\mathbf{P} - \lambda\mathbf{I}$. The determinant yields a polynomial in λ, the roots of which are the eigenvalues of the matrix $\mathbf{P}$.

For example suppose that

$$\mathbf{P} = \begin{bmatrix} 0.2 & 0.8 \\ 0.4 & 0.6 \end{bmatrix}$$

The characteristic equation is

$$|\mathbf{P} - \lambda\mathbf{I}| = \begin{vmatrix} 0.2 - \lambda & 0.8 \\ 0.4 & 0.6 - \lambda \end{vmatrix} = 0$$

which leads to the polynomial

$$\lambda^2 - 0.8\lambda - .20 = 0$$

The roots of this equation give the eigenvalues

$$\lambda_1 = 1$$
$$\lambda_2 = -0.2$$

The numerical values above illustrate two important points about the eigenvalues for a stochastic matrix.

1. Every stochastic matrix has at least one eigenvalue whose value is unity.
2. Every eigenvalue of a stochastic matrix will have an absolute value less than or equal to unity.

Other properties related to the eigenvalues of $\mathbf{P}$ are

3. If a finite Markov chain is irreducible and aperiodic $\mathbf{P}$ has a single eigenvalue 1 which is larger than the absolute value of all other eigenvalues.
4. If a finite chain is irreducible and periodic with period t the matrix $\mathbf{P}$ has exactly t eigenvalues with absolute value 1 and all other eigenvalues have smaller absolute values.
5. When $\mathbf{P}$ is reducible, the number of closed sets of recurrent states is equal to the multiplicity of the eigenvalue 1.

For proof of these and many other theorems concerning the algebra of Markov chains, the reader is urged to consult Cox and Miller (28:123–132).

As an aside, it is worthwhile to note that the eigenvalues of $\mathbf{P}$ can be related to the coefficients of z in the partial fraction expansion required to achieve a solution with geometric transforms. Recall from equation (3.16) that the transform of $\mathbf{P}^n$ is written as $[\mathbf{I} - z\mathbf{P}]^{-1}$. One technique for finding the inverse of a matrix is to use the determinant and adjoint as

$$[\mathbf{I} - z\mathbf{P}]^{-1} = \frac{\text{adj}\,[\mathbf{I} - z\mathbf{P}]}{|\mathbf{I} - z\mathbf{P}|} \tag{3.19}$$

Multiplying a row (or column) of a determinant by a scalar k causes the determinant to be multiplied by k. Suppose we now multiply every row of the m row determinant $|\mathbf{I} - z\mathbf{P}|$ by $(-z^{-1})$. The result is

$$(-z^{-1})^m|\mathbf{I} - z\mathbf{P}| = |\mathbf{P} - (1/z)\mathbf{I}| \tag{3.20}$$

Let us now compare the determinant on the right with the determinant in the characteristic equation.

$$|\mathbf{P} - (1/z)\mathbf{I}| : |\mathbf{P} - \lambda\mathbf{I}|$$

From this comparison it is obvious that the inverse of the eigenvalues λ_i, which are roots of the characteristic equation, are equivalent to z_i, which are the roots of $|\mathbf{I} - z\mathbf{P}| = 0$, (ignoring the extraneous multiple root zero introduced by our scalar multiplication). The roots z_i determine the coefficients present in a partial fraction expansion of the transform $(\mathbf{I} - z\mathbf{P})^{-1}$. For example with the matrix

$$\mathbf{P} = \begin{pmatrix} 0.5 & 0.5 \\ 0.4 & 0.6 \end{pmatrix}$$

$$(\mathbf{I} - z\mathbf{P})^{-1} = \frac{\begin{pmatrix} 1 - 0.6z & 0.5z \\ 0.4z & 1 - 0.5z \end{pmatrix}}{\begin{vmatrix} 1 - 0.5z & -0.5z \\ -0.4z & 1 - 0.6z \end{vmatrix}}$$

$$= \frac{1}{(1 - z)(1 - 0.1z)} \begin{pmatrix} 1 - 0.6z & 0.5z \\ 0.4z & 1 - 0.5z \end{pmatrix}$$

where the polynomial in the denominator of each term has been factored into a product of terms of the form $(1 - az)$. The transform in this case is said to have first order poles at $z = 1$ and $z = 10$. The poles are the values of z which make the denominator zero. This leads to a time function of the form

$$\mathbf{P}^n = (1)\, C_1 + (0.1)^n C_2$$

where the constants C_1 and C_2 are determined by partial fraction expansion. The coefficients a_i are nothing more than the eigenvalues λ_i of the $\mathbf{P}$ matrix.

The matrix $\mathbf{P}$ is similar to a matrix $\mathbf{\Lambda}$ if there exists a matrix $\mathbf{Q}$ such that

$$\mathbf{Q}\mathbf{P}\mathbf{Q}^{-1} = \mathbf{\Lambda} \tag{3.21}$$

It is well known that if the eigenvalues of $\mathbf{P}$ are distinct, $\mathbf{Q}$ can be found which transforms $\mathbf{P}$ into a diagonal matrix $\mathbf{\Lambda}$. The diagonal elements of $\mathbf{\Lambda}$ are the eigenvalues λ_i of $\mathbf{P}$. When the eigenvalues are not distinct $\mathbf{P}$ may still be transformed to a nearly diagonal form called the Jordan canonical form. For details of that calculation see Cox and Miller (28:121).

Limiting ourselves for the moment to finite, irreducible, aperiodic Markov chains the transition matrix $\mathbf{P}$ can be diagonalized where the elements of the diagonal matrix $\mathbf{\Lambda}$ are the eigenvalues of $\mathbf{P}$. Existing computer programs are available for finding the eigenvalues for large scale problems. Rewriting equation (3.21)

$$\mathbf{P} = \mathbf{Q}^{-1}\mathbf{\Lambda}\mathbf{Q} \tag{3.22}$$

which by repeated multiplication yields

$$\mathbf{P}^2 = (\mathbf{Q}^{-1}\mathbf{\Lambda}\mathbf{Q})(\mathbf{Q}^{-1}\mathbf{\Lambda}\mathbf{Q})$$

$$= \mathbf{Q}^{-1}\mathbf{\Lambda}^2\mathbf{Q}$$

and in general

$$\mathbf{P}^n = \mathbf{Q}^{-1}\,\boldsymbol{\Lambda}^n\mathbf{Q} \tag{3.23}$$

Since $\boldsymbol{\Lambda}$ is a diagonal matrix it is a trivial matter to evaluate $\boldsymbol{\Lambda}^n$. The elements of $\boldsymbol{\Lambda}^n$ are simply the nth power of the corresponding elements of $\boldsymbol{\Lambda}$. The row vectors of matrix $\mathbf{Q}$, $\mathbf{q}_i$ must satisfy the equation

$$\mathbf{q}_i\mathbf{P} = \lambda_i\mathbf{q}_i \tag{3.24}$$

Any vector $\mathbf{q}_i$ which satisfies this equation is called a *left eigenvector* of the matrix $\mathbf{P}$ belonging to the eigenvalue λ_i. The column vectors of $\mathbf{Q}^{-1}$, $\mathbf{q}^j$ must satisfy

$$\mathbf{P}\mathbf{q}^j = \mathbf{q}^j\lambda_j \tag{3.25}$$

Any vector $\mathbf{q}^j$ which satisfies this relationship is called a *right eigenvector* of $\mathbf{P}$. In determining the eigenvectors, correct proportionality constants must be introduced to assure that $\mathbf{Q}^{-1}\mathbf{Q} = \mathbf{I}$.

To illustrate the technique consider the matrix

$$\mathbf{P} = \begin{pmatrix} 0.2 & 0.8 \\ 0.4 & 0.6 \end{pmatrix}$$

We previously determined the eigenvalues of this matrix to be $\lambda_1 = 1$, $\lambda_2 = -0.2$. The row vectors for the matrix $\mathbf{Q}$ are determined from the set of equations

$$\mathbf{q}_1\mathbf{P} = 1\mathbf{q}_1$$

and

$$\mathbf{q}_2\mathbf{P} = -0.2\mathbf{q}_2$$

from which

$$0.2q_{11} + 0.4q_{12} = q_{11}$$
$$0.8q_{11} + 0.6q_{12} = q_{12}$$

and

$$0.2q_{21} + 0.4q_{22} = -0.2q_{21}$$
$$0.8q_{21} + 0.6q_{22} = -0.2q_{22}$$

Neither of these two sets of equations is independent which prevents us from determining the elements q_{ij} uniquely. However, they will determine those elements to within a multiplicative constant. Let us arbitrarily assign values $q_{11} = 1$ and $q_{21} = 1$. The equations can then be solved to yield $q_{12} = 2$ and $q_{22} = -1$. The entire matrix is then

$$\mathbf{Q} = \begin{pmatrix} 1 & 2 \\ 1 & -1 \end{pmatrix}$$

from which

$$\mathbf{Q}^{-1} = \begin{pmatrix} \dfrac{1}{3} & \dfrac{2}{3} \\[2ex] \dfrac{1}{3} & \dfrac{-1}{3} \end{pmatrix}$$

From equation (3.23) it follows that

$$\mathbf{P}^n = \mathbf{Q}^{-1}\,\mathbf{\Lambda}^n\mathbf{Q}$$

$$= \begin{pmatrix} \dfrac{1}{3} & \dfrac{2}{3} \\[2mm] \dfrac{1}{3} & \dfrac{-1}{3} \end{pmatrix} \begin{pmatrix} 1 & 0 \\ 0 & (-0.2)^n \end{pmatrix} \begin{pmatrix} 1 & 2 \\ 1 & -1 \end{pmatrix}$$

$$= \begin{pmatrix} \dfrac{1}{3} + \dfrac{2}{3}(-0.2)^n & \dfrac{2}{3} - \dfrac{2}{3}(-0.2)^n \\[3mm] \dfrac{1}{3} - \dfrac{1}{3}(-0.2)^n & \dfrac{2}{3} + \dfrac{1}{3}(-0.2)^n \end{pmatrix}$$

PROBLEMS

3.1 Consider the single stage transition matrix.

$$\mathbf{P} = \begin{pmatrix} 0.2 & 0.3 & 0.5 & 0 \\ 0 & 0.4 & 0.4 & 0.2 \\ 0.1 & 0.7 & 0 & 0.2 \\ 0 & 0 & 0.5 & 0.5 \end{pmatrix}$$

  a) Prior to solving any equations can you tell whether or not there exists a steady-state distribution for this process? Explain.

  b) Suppose that the system starts in state 4.

   i) Find the probability that it will occupy state 1 after five transitions.

   ii) Find the long-run frequency with which it will occupy state 3.

3.2 Consider the average aircraft arrival and service rates represented by Figure 3-1. Assume that service times are deterministic and that arrivals are Poisson distributed at the indicated rates. Capacity is limited to twenty-five aircraft in the system.

  a) Write the general model which describes the number of aircraft present at the end of each fifteen-minute interval.

  b) Use the program in Appendix D to estimate the daily operating profile for this system.

3.3 Consider an auto wash with a constant service time of three minutes per car and a capacity of twenty cars. Cars arrive for washing in Poisson fashion at the rate of fifteen per hour.

  a) Develop a model, the solution to which would give the steady-state distribution for number of cars waiting at the start of the nth wash cycle.

  b) Suppose that the brushes jam after the system has been running long enough to stabilize (in a probability sense). It will require forty-five minutes to repair the brushes. Calculate the expected line length both immediately prior to and immediately after the breakdown. How long will it be before the system will again stabilize? (Use the program in Appendix D to assist you in these calculations.)

3.4 Consider a conveyor system which has containers that accomodate up to four subassemblies. The conveyor is timed so that an empty container will be available at the loading station once every five minutes. Subassemblies become available for loading at a constant Poisson rate of thirty per hour. Capacity at the station is ten subassemblies. Any arriving when the system is saturated must be moved by fork truck.

a) Develop a model to describe the probability distribution for number of subassemblies ready for loading at the time of arrival of an empty container.

b) Suppose that the system begins operation at 8:00 AM with no subassemblies present. At 10:00 AM the production rate is expected to increase to a rate of ninety per hour for a thirty-minute period. After that it is expected to return to the normal rate.

   i) Find the distribution of subassemblies present both before and after the increased load period.

   ii) Estimate the time necessary for the system to stabilize after receiving such a pulse. (Use the program in Appendix D to help answer this part.)

3.5 Consider a simple Markov chain with the single stage transition matrix.

$$P = \begin{pmatrix} 0.2 & 0.8 \\ 0.5 & 0.5 \end{pmatrix}$$

Use transform techniques to find a closed-form expression for the conditional probability that the system will be in state j after n transitions given that it started in state i.

3.6 Solve the problem posed in 3.5 using the algebraic approach in which

$$P^n = Q^{-1}\Lambda^n Q$$

3.7 The final assembly line for an appliance manufacturer, who produces washers, dryers, and dishwashers, has the following set of rules:

   Two dishwashers can never follow each other down the line because their assembly time is too long.

   A dryer must be followed by a washer to balance the line.

   A washer can be followed by either a dishwasher or a dryer but not by a washer.

The production engineer has proposed that the system be modeled as a Markov chain with states defined as:

   1 = washer
   2 = dryer
   3 = dishwasher

a) A study of adjacent units on the line during the last month reveals the following data:

Entry Unit \ Next Unit	Washer	Dryer	Dishwasher	
Washer	0	100	300	
Dryer	300	0	0	Observations
Dishwasher	100	200	0	

Use these data to estimate the probability that the nth unit on the line will be item j if the first is item i.

b) Next month the manufacturer has a contract to deliver 1000 appliances at the end of the month. Of these, approximately 300 must be washers, 100 must be dryers, and 600 must be dishwashers. Can they meet the contract commitment?

c) Suppose that the contract calls for 400 washers, 200 dryers, and 400 dishwashers. Devise a production rule which will permit that contract to be honored.

3.8 Aqua Cloud Seeding Company has the responsibility of maintaining an irrigation water supply. Aqua performs seeding flights every day. Aqua's analyst has studied their effectiveness for many months. He has the following empirical data on daily rainfall (in inches):

Inches	Days Observed
0	80
1	30
2	60
3	20
4	10

The irrigation reservoir has a capacity which can be converted to an equivalent of three inches of rainfall. Daily demands are such that, provided the reservoir is not empty, an equivalent of one inch of rainfall is released to consumers at the end of each day. Any overflow is regarded as lost.

a) Model this system as a Markov chain. State all assumptions.

b) Find the expected level of water in the reservoir over a long period of time.

3.9 Referring again to the scenario of problem 3.8, suppose that the matrix turned out to be

$$
\mathbf{P} = \begin{pmatrix} .4 & .4 & .1 & .1 \\ .3 & .4 & .2 & .1 \\ .0 & .4 & .2 & .4 \\ .0 & .0 & .6 & .4 \end{pmatrix}
$$

where now i implies i inches of water in the reservoir beginning with zero.

a) The reservoir now has three inches of water. One customer has vowed to sue Aqua the next time he is forced to pump water from the well because of a dry reservoir.

   i) What is the probability that Aqua will not be sued?

   ii) Find the expected time to filing of the first suit.

   iii) If Aqua plans to cease operating in ten days, what is the probability that they will be leaving the job without being sued?

b) Each time the reservoir hits the one inch mark, Aqua's lawyer rushes to the site in anticipation of a suit. A trip to the site costs Aqua $150.

i) Find the expected cost of the lawyer's trips over the remainder of Aqua's operations.

ii) Find the variance of his trip costs.

c) Calculate the expected amount of rainfall lost during the next three days (i.e., starting with reservoir full) using the data given in the original problem statement.

3.10 A land-use planning agency has identified four major land uses for the county area

    residential
    industrial
    commercial
    rural

In studying past land use data they have developed the following information on changes over a five-year period:

Initial Number of Parcels	Residential	Final Usage Industrial	Commercial	Rural
400 residential	300	20	80	0
200 commercial	20	80	100	0
100 industrial	0	80	10	10
50 rural	15	5	5	25

a) What assumptions are necessary to justify developing a Markov model to describe the development of the county?

b) Write a five-year period transition matrix for the county.

c) Find the limiting distribution for land use in the county.

d) If 60 percent of the county is now rural, what percent will remain rural after twenty years?

3.11 The operator of a fleet of vehicles follows a very rigid inspection policy. Each year in October all vehicles are inspected. After inspection, those which fail are replaced by new vehicles. However, no vehicle is kept beyond three years after purchase. Historical data for a sample of 500 vehicles reveal that the first inspection was conducted at an average cost of $50 per vehicle. The 80 percent retained had repairs performed which cost an average of $200 per vehicle. The remainder were traded at an average trade-in price of 60 percent of list. The second inspection of the original vehicles, costing $36,000 for inspection and $120,000 for repairs, found an additional 40 percent ready for trade-in at 40 percent of list. At the end of three years the remaining vehicles were traded at 30 percent of list without inspection or repair. The new vehicle list price is $3000.

a) Structure this problem as a Markov chain with rewards in which the state variable is age at inspection time, i.e., develop a one-stage transition matrix and a reward (cost?) matrix.

b) Find the average age of the fleet after the company has been in business a very long time.

c) If the fleet consists of 200 vehicles, on the average how many will be traded each year?

d) Calculate the total cost vector for the fleet operator who intends to liquidate his entire fleet of 200 after three years. Use this information to find the total expected cost for his present fleet make-up which consists of 100 one-year old, 50 two-year old and 50 three-year old vehicles.

3.12 Consider a three-state Markov process which always moves from state 1 to 2, state 2 to 3, and state 3 to 1.

a) Develop the single stage transition matrix for this process.

b) Find the general expression for $\mathbf{P}^n$ by using transform techniques.

c) Repeat part (*b*) using algebraic methods.

3.13 Given:

$$\mathbf{P} = \begin{pmatrix} a & b \\ c & d \end{pmatrix}$$

Find:

a) Proper values for *a*, *b*, *c* and *d* given that
Long Range Frequency Distribution $= \mathbf{V} = (0.4, 0.6)$.

b) Suppose that *b* must be zero. Discuss the character of the limiting state distribution as a function of values assumed by the remaining variables.

c) If $a = 0.5$, $b = 0.5$, $c = 0.8$, and $d = 0.2$, find $\mathbf{P}^n$ using eigenvalues.

3.14 An aircraft maintenance officer classifies the status of all aircraft in his squadron at the completion of each mission as

1—No maintenance required

2—Minor discrepancy (maintenance optional)

3—Major discrepancy (ground the aircraft)

During the past year he has collected data on 1000 aircraft missions. The number arriving in each status were

1—438

2—178

3—384

Further study indicates that 16 percent of the aircraft currently in status 1 will require minor maintenance after the next mission and 30 percent will require major maintenance. Of the aircraft currently in status 3, 34 percent will require no maintenance and 50 percent will require major maintenance after the next mission.

a) Construct a single stage transition matrix for this system.

b) Four aircraft currently are in status 1, 1, 2, and 3, respectively.

   i) What is the expected number which will need major repair after the next mission?

   ii) What is the probability that the fourth aircraft will again need major repair after its second mission?

3.15 A glider pilot has entered a competition in which both maximum altitude and endurance are to be judged. Due to airspace restrictions no one is permitted to go over 5000 feet. Having flown in this area many times before he has developed data on the number of times and amount his altitude changed. His estimates are based on five minute time intervals and altitude changes to the nearest 1000 feet.

Initial altitude	Altitude 5 minutes later	No. of observations
1000	0	150
	1000	40
	2000	10
2000	0	30
	1000	60
	2000	60
	3000	90
3000	1000	20
	2000	80
	3000	80
	4000	20
4000	2000	10
	3000	20
	4000	60
	5000	10
5000	3000	35
	4000	10
	5000	5

Suppose that he is given a tow to 2000 feet.

a) Develop a single period transition matrix.

b) What is his "steady-state" altitude?

c) What is the probability that he will ever reach 5000 feet?

d) What is the expected duration of his flight?

3.16 A dairy farmer raises milking cows. All of his dairy cows are bred in December to calve the following September. Calves are either marketed or retained to replenish the herd. At the end of her first producing year a cow is judged for her milk production. If she produces over 400 gallons, she will be used as a dairy cow for a total of four years. Sixty percent of past cows met this criterion. If she does not produce 400 gallons, she is tagged but kept in production one more year, since she has already been bred again. At the end of the second year these marginal producers are marketed for beef if they are fat enough or retained two more years as dry cows to fatten them for market. Fifty percent are marketable at the end of their second production year. When the milk cows reach the end of their four-year productive life, 70 percent of them must be retained for a fifth year as dry cows for fattening before marketing. The remainder can be immediately marketed. Whenever a cow is marketed it is replaced by a heifer who was retained from the last batch of calves, bred, and is ready to produce milk.

a) Formulate the operation of this herd as a Markov process. (You should have eight states.) State all assumptions.

b) Over a long period of time, what portion of the herd requires milking accomodations?

c) What portion of new-born calves should be kept to replenish the herd? Is this number feasible?

d) If a farmer begins operation with a herd of 200 fresh heifers, how many cows can he expect to market at the end of his fifth year of operation?

4

CONTINUOUS TIME PROCESSES

The Markov chain models of the previous chapters can be viewed as discrete time processes. The state of the system was known only at instants n, $n + 1$, $n + 2$, etc. In the inventory control example our transition matrix was defined over a one-week period. We could forecast the inventory level on hand for each Friday night but could say nothing about it on Wednesday, for example. What we now seek is a way to describe systems on a continuous-time scale. The resulting time-dependent probability distributions will describe the system status at every instant, not only for integer values of t.

ARRIVAL PATTERNS

Let us begin our study of continuous-time processes by examining typical arrival patterns.

The simplest physical pattern one might imagine is one in which customers arrive one at a time at equally spaced time intervals. Such a pattern might exist on assembly lines fed by conveyor systems or in appointment systems with very punctual customers. If the scheduled time between arrivals is a minutes, the rate at which customers arrive is $\lambda = 1/a$. Even though such a deterministic system is the simplest physically, it is not the easiest one to handle mathematically.

Poisson Process

At the opposite extreme from the constant interarrival time is one which is completely random. The concept of complete randomness implies that there is a constant probability of an arrival event occuring between t and $t + \Delta t$. Furthermore, whatever happens in $(t, t + \Delta t)$ is statistically independent of the arrival or nonarrival of customers in any other nonoverlapping time interval. This concept can be formalized by the following set of postulates:

1. The probability of a single arrival during a small interval of time Δt is $\lambda \Delta t$ where

$$\lim_{\Delta t \to 0} \frac{\lambda \Delta t}{\Delta t} = \lambda$$

2. The probability of more than a single arrival during a small interval of time is $O(\Delta t)$ where

$$\lim_{\Delta t \to 0} \frac{O(\Delta t)}{\Delta t} = 0$$

To summarize, we are postulating a system in which customers arrive one at a time, the probability of an arrival is proportional to elapsed time, and time intervals can be selected which are small enough to negate the possibility of more than a single arrival. Such behavior is typical of those systems which draw from a very large group of customers all behaving independently from each other.

Let us describe the state of the system by the number of customers who have appeared by time t. We might think of this as a queueing system in which customers arrive at random and are held but never served. The associated probability distribution is

$$p_n(t) = \text{probability of } n \text{ customers present at time } t$$

The problem now is to develop a recursion equation which will enable us to calculate $p_n(t)$ for any desired time t.

If we consider the system to be making transitions between time t and $t + \Delta t$ we can structure it as a simple Markov chain. The probability of moving into state n at time $t + \Delta t$ depends only upon the state at time t and the constant-arrival probability $\lambda \Delta t$. Future state transitions are independent of the path followed in reaching the present state. What we need now is a simple one-stage transition matrix covering the interval $(t, t + \Delta t)$. Since we are not limiting the capacity of the queue, this matrix will have an infinite number of rows and columns but the concept is identical to that of our finite Markov chains of Chapter 2.

The matrix is

		0	1	2	3	$\cdots$
	0	$1 - \lambda \Delta t - O(\Delta t)$	$\lambda \Delta t$	$O(\Delta t)$	0	
$\mathbf{T} =$	1	0	$1 - \lambda \Delta t - O(\Delta t)$	$\lambda \Delta t$	$O(\Delta t)$	$\cdots$
	2	0	0	$1 - \lambda \Delta t - O(\Delta t)$	$\lambda \Delta t$	$\cdots$
	$\vdots$	$\vdots$	$\vdots$	$\vdots$	$\vdots$	

$$(4.1)$$

The entries are easily developed by arguing that a move from state i to $i + 1$ occurs if a single arrival takes place in Δt. The probability of that event is $\lambda \Delta t$. The system moves from state i to state j where $j > i + 1$ only if multiple arrivals occur in Δt. By our postulates the probability of such an event is $O(\Delta t)$. The system will remain in state i only if no arrivals occur. This can be calculated as one minus the probability of one or more arrivals in Δt. Since customers cannot leave the system once admitted, the probability of moving from i to j where $j < i$ is zero.

Applying the basic Markov recursion equation we can calculate the distribution at time $t + \Delta t$ as a function of the distribution at time t and the transition matrix.

$$\mathbf{P}_n(t + \Delta t) = \mathbf{P}_n(t)\mathbf{T} \tag{4.2}$$

Carrying out this multiplication yields the set of equations

$$p_0(t + \Delta t) = p_0(t)[1 - \lambda\,\Delta t - O(\Delta t)]$$
$$p_1(t + \Delta t) = p_0(t)\lambda\,\Delta t + p_1(t)[1 - \lambda\,\Delta t - O(\Delta t)]$$
$$\vdots$$
$$p_n(t + \Delta t) = p_{n-2}(t)O(\Delta t) + p_{n-1}(t)\lambda\,\Delta t + p_n(t)[1 - \lambda\,\Delta t - O(\Delta t)]$$

rearranging terms and dividing by Δt we are led to $\tag{4.3}$

$$\frac{p_0(t + \Delta t) - p_0(t)}{\Delta t} = -\lambda p_0(t) - p_0(t)\frac{O(\Delta t)}{\Delta t}$$

$$\frac{p_1(t + \Delta t) - p_1(t)}{\Delta t} = \lambda p_0(t) - \lambda p_1(t) - p_1(t)\frac{O(\Delta t)}{\Delta t}$$

$$\vdots$$

$$\frac{p_n(t + \Delta t) - p_n(t)}{\Delta t} = \lambda p_{n-1}(t) - \lambda p_n(t)$$

$$+ [p_{n-2}(t) - p_n(t)]\frac{O(\Delta t)}{\Delta t}, n \geq 1 \tag{4.4}$$

Since we are interested in continuous-time processes, let us now take the limit of both sides of equations (4.4) as $\Delta t \to 0$. The definition of the first derivative of a function $f(t)$ is given by

$$\frac{df(t)}{dt} = \lim_{\Delta t \to 0}\frac{f(t + \Delta t) - f(t)}{\Delta t} \tag{4.5}$$

Furthermore, every term involving $O(\Delta t)/\Delta t$ will approach zero as $\Delta t \to 0$ by virtue of our postulates. It therefore follows that equations (4.4) reduce to

$$\frac{dp_0(t)}{dt} = -\lambda p_0(t),$$

$$\frac{dp_n(t)}{dt} = \lambda p_{n-1}(t) - \lambda p_n(t), n \geq 1 \tag{4.6}$$

The set of equations (4.6) is an infinite set of differential-difference equations. They are differential equations with respect to the continuous-time variable t. They are difference equations with respect to the discrete state variable n. Equations of this form are typical of those we will encounter in nearly every attempt to find time-dependent solutions to queueing problems.

These equations can be solved in recursive fashion by first solving for $p_0(t)$, then using $p_0(t)$ to solve for $p_1(t)$ and so forth until a general solution $p_n(t)$ can be found by induction. For example, by separating variables we have

$$\frac{1}{p_0(t)} dp_0(t) = -\lambda dt \tag{4.7}$$

Integrating both sides yields

$$\ln p_0(t) = -\lambda t + c$$

from which

$$p_0(t) = e^{-\lambda t + c}$$

The constant c is determined from the boundary conditions. Since we are assuming that the system starts with no customers present at time zero it follows that $p_0(0) = 1$, therefore

$$p_0(0) = e^c = 1$$

from which $c = 0$. The particular solution is then

$$p_0(t) = e^{-\lambda t} \tag{4.8}$$

Moving on to $n = 1$, we have from (4.6)

$$\frac{dp_1(t)}{dt} = \lambda p_0(t) - \lambda p_1(t)$$

Replacing $p_0(t)$ with the known result $e^{-\lambda t}$ we have

$$\frac{dp_1(t)}{dt} = \lambda e^{-\lambda t} - \lambda p_1(t) \tag{4.9}$$

This is a first order linear differential equation of the form

$$y' + r(t)y = q(t)$$

which has a well-known general solution

$$y = e^{-\int r dt} \int e^{\int r dt} q dt + c e^{-\int r dt} \tag{4.10}$$

Mapping that solution onto equation (4.9) yields

$$p_1(t) = e^{-\int \lambda dt} \int e^{\int \lambda dt} \lambda e^{-\lambda t} dt + c e^{-\int \lambda dt}$$
$$= e^{-\lambda t} \int \lambda \, dt + c e^{-\lambda t}$$
$$= \lambda t e^{-\lambda t} + c e^{-\lambda t} \tag{4.11}$$

The constant c is determined from the boundary condition $p_1(0) = 0$, if we assume that the system started with no customers present. Therefore

$$p_1(0) = 0 + c e^0 = 0$$

from which $c = 0$ and

$$p_1(t) = \lambda t e^{-\lambda t} \tag{4.12}$$

The reader is encouraged to continue this iterative process until the general form $p_n(t)$ becomes obvious.

Transform Solution

It is instructive to solve the system of equations (4.6) by the use of transform techniques. It is the author's view that not only is this approach more expedient for the problem at hand, but also it provides a map for solving more general infinite sets of linear differential-difference equations which one encounters throughout the queueing literature.

To begin the transform solution let us define the geometric transform of the function $p_n(t)$ as

$$G(z, t) = \sum_{n=0}^{\infty} p_n(t)z^n \tag{4.13}$$

The G-function is written as a function of both the transform variable z and the time variable t even though t can be treated as a constant at this stage of the calculation. We can view this as the transform of a probability distribution which exists at some particular time t. Note that the partial derivative of G with respect to t can be written as

$$\frac{\partial G(z, t)}{\partial t} = \frac{\partial}{\partial t} \sum_{n=0}^{\infty} p_n(t)z^n = \sum_{n=0}^{\infty} \frac{dp_n(t)}{dt} z^n \tag{4.14}$$

The first step in solving an infinite set of differential-difference equations is to reduce that set to a single equation in the geometric transform by appropriate multiplication and addition. Suppose that we multiply the equation defining $\dfrac{dp_n(t)}{dt}$ in (4.6) by z^n. That is

$$\frac{dp_0(t)}{dt}z^0 = -\lambda p_0(t)z^0$$

$$\frac{dp_1(t)}{dt}z^1 = \lambda p_0(t)z^1 - \lambda p_1(t)z^1$$

$$\vdots$$

$$\frac{dp_n(t)}{dt}z^n = \lambda p_{n-1}(t)z^n - \lambda p_n(t)z^n, \ n > 1$$

Adding these equations we have the single expression

$$\sum_{n=0}^{\infty} \frac{dp_n(t)}{dt} z^n = \lambda \sum_{n=0}^{\infty} p_n(t)z^{n+1} - \lambda \sum_{n=0}^{\infty} p_n(t)z^n$$

$$= \lambda z \sum_{n=0}^{\infty} p_n(t)z^n - \lambda \sum_{n=0}^{\infty} p_n(t)z^n \tag{4.15}$$

The expression on the left is the first partial derivative of the transform $G(z, t)$ from equation (4.14). Each term on the right includes the entire transform $G(z, t)$ from equation (4.13). Rewriting as a partial differential equation in $G(z, t)$ we have

$$\frac{\partial G(z, t)}{\partial t} = \lambda(z - 1)\, G(z, t) \qquad (4.16)$$

We can solve the resulting differential equation by using Laplace transforms. For this portion of the calculation the transform variable z is treated as a constant. Taking Laplace transforms of both sides of equation (4.16) gives the subsidiary equation

$$s\mathcal{L}(G(z, t)) - G(z, 0) = \lambda(z - 1)\mathcal{L}(G(z, t))$$

from which

$$\mathcal{L}(G(z, t)) = \frac{G(z, 0)}{s - \lambda(z - 1)} \qquad (4.17)$$

When first encountered this notation can be unsettling. We appear to have a Laplace transform of a geometric transform. However, for purposes of this part of the calculation the z-argument is a constant. The function $G(z, t)$ is then treated as we would treat any function of the single continuous variable t.

From our knowledge of initial conditions

$$p_0(0) = 1$$
$$p_n(0) = 0,\ n > 0$$

Therefore

$$G(z, 0) = \sum_{n=0}^{\infty} p_n(0)z^n = 1$$

and

$$\mathcal{L}(G(z, t)) = \frac{1}{s - \lambda(z - 1)} \qquad (4.18)$$

This transform is of the form

$$\mathcal{L}(f(t)) = \frac{1}{s - a}$$

which from the tables we know can be inverted to

$$f(t) = e^{at}$$

In the present context $a = \lambda(z - 1)$. It therefore follows that the inverse Laplace transform for equation (4.18) is

$$G(z, t) = e^{\lambda(z-1)t} \qquad (4.19)$$

Note that this expression is still a transform, albeit a geometric transform in this case. We must now invert the geometric transform to obtain the desired time function. Applying a power series expansion to the right side of equation (4.19) we have

$$G(z, t) = e^{-\lambda t} \sum_{n=0}^{\infty} \frac{(\lambda t z)^n}{n!} \tag{4.20}$$

Comparing this expression term by term to the definition of the geometric transform, we can ascertain that $p_n(t)$ is the coefficient of the nth power of z in the series. That is

$$p_n(t) = \frac{(\lambda t)^n}{n!} e^{-\lambda t} \tag{4.21}$$

This expression is the familiar Poisson distribution with a mean of λt. What began in our basic postulates as a porportionality constant λ turns out to be the mean rate at which events take place in a Poisson process. In constructing queueing models the assumption of a Poisson arrival process implies that we are treating the system as Markovian with a constant probability of event occurrence.

The idea that the Poisson process is completely random may violate the intuition of some persons. It somehow seems more comfortable to think of the presence of uniformly distributed random variables as indicative of complete randomness. The Poisson distribution is related to a uniformly distributed random variable in the following sense. If it is known that a arrivals generated by a Poisson process have occurred in the interval $[0, T]$, the times at which those arrivals occurred, $t_1 < t_2 < \cdots < t_a$, are distributed as the order statistics of a uniform random variables on $[0, T]$. A simple proof of this relationship is given by Gross and Harris (42:28).

Exponential Equivalent

From the viewpoints of data analysis and system understanding, it is important to be able to model the interarrival time distribution. We have already shown that one can describe a very general, and in the truest sense of the word, completely random arrival process by means of the Poisson distribution. That distribution describes arrivals in terms of number of customers in a fixed time period. What we now seek is the implied distribution for the time between arrival events for a Poisson process. Note that time between events is a different concept than the times at which arrivals occur. We have already stated that the latter is uniformly distributed.

Suppose that an arrival has just occurred. The time from this point to the time of the next arrival will be less than t if and only if one or more arrivals occur in the interval $(0, t)$. Since we are assuming a Poisson arrival process as depicted in equation (4.21), we can write an explicit expression for that probability as

$$F(t) = 1 - Pr \text{ (zero arrivals)}$$
$$= 1 - e^{-\lambda t} \tag{4.22}$$

Note that $F(t)$ is the cumulative interarrival distribution. The interarrival density function is developed by taking the first derivative of the cumulative distribution as

$$f(t) = \frac{dF(t)}{dt} = \lambda e^{-\lambda t} \tag{4.23}$$

This expression is the well-known negative exponential distribution with parameter λ. What we have shown is that our "completely random" arrival process, which eminates from a simple time-proportionality argument, can be characterized by a negative exponential distribution for time between events. To say that the number of events per time interval follows a Poisson distribution is equivalent to saying that the time between events is exponentially distributed.

We developed the Poisson distribution by means of a Markov chain argument. It is important to note that the exponential distribution also exhibits the Markov property. It is the only continuous distribution which exhibits a total lack of memory in the sense that the time to the next event is independent of the time elapsed since the previous event took place. A lack of memory is the property required of a Markov process. The move to a new state depends only upon the current state and is independent of the events which preceded the entry to the current state. Mathematically

$$Pr(t > t_1 + t_2 \,|\, t > t_1) = Pr(t > t_2) \tag{4.24}$$

describes a random variable t with the Markov property. Using the exponential density for $f(t)$ and the usual relationships between joint marginal and conditional probabilities, we have

$$Pr(t > t_1 + t_2 \,|\, t > t_1) = \frac{Pr(t > t_1 + t_2 \text{ and } t > t_1)}{Pr(t > t_1)}$$

$$= \frac{exp(-\lambda(t_1 + t_2))}{exp(-\lambda t_1)} = e^{-\lambda t_2} \tag{4.25}$$

But this is the same distribution we would have obtained from a Poisson process which started at $t = 0$. Therefore the fact that the process may have been running for some time t_1 has no influence on predicting the time to the next event.

What we have shown is that queueing systems in which the times between events follow exponential distributions are Markov systems. The events of interest in these systems are the arrival of customers and the completion of service for customers already in the system. In queueing parlance to assert that one has a single channel system with Poisson arrivals and exponential service is to assert that both the arrival and service mechanisms are Markovian, hence the notation (M/M/1).

Example. Consider the operation of a carnival ride which cycles every five minutes. An analysis of past data reveals that the time between successive customer arrivals can be described by an exponential distribution with a mean of thirty seconds. The ride has twelve seats available per cycle.

Suppose that a ride cycle has just begun and there are no customers waiting. What is the probability that the next cycle will be loaded to capacity? Thirty seconds between customers translates into a mean arrival rate of two customers per minute. Since the time between customer arrivals is exponentially distributed the number of customers arriving in the next five minutes will be Poisson distributed as

$$p_n(5) = \frac{[(2)(5)]^n e^{-10}}{n!}$$

A capacity load is obtained whenever there are twelve or more arrivals during the five minute cycle time. Therefore the probability of reaching capacity on the next cycle is

$$Pr\ (capacity) = \sum_{n=12}^{\infty} \frac{(10)^n e^{-10}}{n!}$$

$$= 1 - \sum_{n=0}^{11} \frac{(10)^n}{n!} e^{-10}$$

$$= 1 - .697 = .303$$

Now suppose that there are eleven customers present, the last one of which arrived twenty seconds ago, and you hope to ride on the next cycle which begins in two minutes. However, you must first put your coat in a public locker nearby which will require one minute. What is the probability that you can be included in the next cycle? Since the arrival process is Poisson, the time of the most recent arrival is irrelevant. The probability that you can be included in the next batch is simply the probability that the time until the next customer arrival will exceed one minute. From the complementary cumulative exponential distribution

$$Pr(t > 1) = e^{-2} = .135$$

GENERALIZATIONS OF THE POISSON PROCESS

There are a number of generalizations of the Poisson process and its related exponential distribution which we will find useful in constructing queueing models. In every case these generalizations in some way preserve the Markov property with its inherent computational advantages.

Pure Birth Process

In analyzing arrival processes the basic postulates which led us to the Poisson distribution carried the assumption that the probability of an arrival between t and $t + \Delta t$ was independent of the number of arrivals in $(0, t)$. Let us now drop that assumption and suppose that the probability of a new arrival is a function of the current state of the system. The revision of the Poisson postulates now leads to the following:

1. If the system is in state n, the probability of a single arrival during a small interval of time Δt is $\lambda_n \, \Delta t$ where

$$\lim_{\Delta t \to 0} \frac{\lambda_n \, \Delta t}{\Delta t} = \lambda_n$$

2. The probability of more than a single arrival during a small interval of time is $O(\Delta t)$ where

$$\lim_{\Delta t \to 0} \frac{O(\Delta t)}{\Delta t} = 0$$

This is a very powerful generalization which permits one to easily model many diverse arrival or birth mechanisms. All that is necessary is to properly define a functional relationship for λ_n, fill the transition matrix for the interval $[t, t + \Delta t]$ with functions of λ_n rather than λ, perform the basic Markov chain iteration, and use limiting arguments to achieve the desired set of differential-difference equations. The only thing different in this operation from that used to develop the Poisson distribution is the specification of λ_n. The pure birth equations are then

$$\frac{dp_0(t)}{dt} = -\lambda_0 \, p_0(t)$$

$$\frac{dp_n(t)}{dt} = -\lambda_n p_n(t) + \lambda_{n-1} p_{n-1}(t), \; n \geq 1 \qquad (4.26)$$

Example. To consider a very simple application of these equations, suppose that we have a conveyor loading system which has a capacity of four units. When the loader is full, any units which arrive for service are diverted to another station. The state of the system is defined as the number of units in the loader at time t, given that the loader was empty at time $t = 0$. Here we might suspect that the birth or arrival rate is constant as long as unused capacity exists, then drops to zero once the loader is full. Symbolically

$$\lambda_n = \begin{cases} \lambda, & 0 \leq n < 4 \\ 0, & n \geq 4 \end{cases}$$

This leads to a particular case of the pure birth equations of

$$\frac{dp_0(t)}{dt} = -\lambda p_0(t)$$

$$\frac{dp_n(t)}{dt} = -\lambda p_n(t) + \lambda p_{n-1}(t), \; 1 \leq n < 4$$

$$\frac{dp_4(t)}{dt} = \lambda p_3(t)$$

$$\frac{dp_n(t)}{dt} = 0, \qquad\qquad\qquad n > 4 \qquad (4.27)$$

The solution to this finite system of differential-difference equations under the condition that $p_0(0) = 1$, yields

$$p_n(t) = \frac{(\lambda t)^n e^{-\lambda t}}{n!}, \quad 0 \leq n < 4$$

$$p_4(t) = 1 - \sum_{i=0}^{3} \frac{(\lambda t)^i e^{-\lambda t}}{i!}$$

$$p_n(t) = 0, \qquad\qquad n > 4 \qquad\qquad (4.28)$$

A moment's reflection should convince the reader that the result for this truncated system is intuitively satisfying. Equations (4.28) could be developed directly from the Poisson distribution by arguing that customers demand service in Poisson fashion, whether or not they are able to enter the system. The probabilities for occupying states $0, 1, 2, 3$ are simply the corresponding Poisson frequencies. State 4 exists whenever customer demands equal or exceed four. The probability of this event is one minus the cumulative probability that demand is less than or equal to three.

Yule Process

Another well-known application for the pure birth equations comes from the mathematical theory of evolution. The Yule process is one in which each member of a population can split or give birth to a new member but cannot die. Now we suppose that each individual has a probability $\lambda \, \Delta t$ of producing a new member in time Δt. Assuming no interaction among members and a population of n at time t, the total probability of increase in $(t, t + \Delta t)$ is then $n\lambda \, \Delta t$. In the context of the pure birth equations this is equivalent to saying $\lambda_n = n\lambda$ for all n. The general differential equation is then

$$\frac{dp_n(t)}{dt} = -n\lambda p_n(t) + (n - 1)\,\lambda p_{n-1}(t), \; n \geq 1 \qquad (4.29)$$

If we assume an initial population of size i, the distribution $p_n(t)$, $n \geq i$ is given by

$$p_n(t) = \binom{n - 1}{n - i} e^{-i\lambda t} (1 - e^{-\lambda t})^{n-i} \qquad (4.30)$$

Proof of that relationship is left as an exercise for the reader.

Other applications of the pure birth processes in the physical and biological sciences range from studies of epidemics to description of radioactive emission in atomic physics. Each of these diverse applications contains the set of equations (4.26). The assorted models are distinguishable only by the context dictated values for λ_n. We will make extensive use of this technique in future sections.

Time Dependent Processes

In developing the equations for the pure birth process we noted that this approach could be viewed as a generalization of the Poisson process in which state independence of arrivals could be assumed. A second generalization of the Poisson process is one in which the arrival or birth rate is permitted to be a function of time, say $\lambda(t)$.

For the time dependent arrival rate the pure birth equations are modified to

$$\frac{dp_0(t)}{dt} = -\lambda(t)p_0(t)$$

$$\frac{dp_n(t)}{dt} = -\lambda(t)p_n(t) + \lambda(t)p_{n-1}(t), \ n \geq 1 \qquad (4.31)$$

Using transform arguments in the same manner demonstrated for the normal Poisson process, we are led to a differential equation much like that of equation (4.16).

$$\frac{\partial G(z, t)}{\partial t} = \lambda(t)(z - 1)G(z, t)$$

or

$$\frac{\partial G(z, t)}{\partial t} - (z - 1)\lambda(t)G(z, t) = 0 \qquad (4.32)$$

Since this differential equation has variable coefficients we must revert to the standard solution form shown in equation (4.10). It then follows that

$$G(z, t) = exp\left((z - 1) \int_0^t \lambda(u) \, du\right) \qquad (4.33)$$

The function $G(z, t)$ is of exactly the same form as the geometric transform for the Poisson distribution developed in equation (4.19). The only change is that we now have a mean value of

$$\mu = \int_0^t \lambda(u) \, du \qquad (4.34)$$

The end result for this dependent Poisson process is that

$$p_n(t) = \frac{\left(\int_0^t \lambda(u) \, du\right)^n}{n!} \, exp\left(- \int_0^t \lambda(u) \, du\right) \qquad (4.35)$$

Note that this process still implies that the number of events in $(t, t + \Delta t)$ is completely independent of happenings in $(0, t)$.

Example. Suppose that one is examining the traffic build-up during a rush-hour period at a toll booth. Further, suppose that the arrival rate varies in parabolic fashion beginning at zero, peaking at $t = 2$ hours and returning to zero at $t = 4$ hours. During the time interval $(0, 4)$ the time varying arrival rate is described as

$$\lambda(t) = 4t - t^2 \qquad (4.36)$$

The corresponding probability distribution for number of arrivals through time t is given by

$$p_n(t) = \frac{(2t^2 - t^3/3)^n}{n!} exp(t^3/3 - 2t^2) \qquad (4.37)$$

Multiple Arrivals

A third generalization of the Poisson process is one in which multiple arrivals may occur. The simplest of these systems is one in which arrivals occur in bulk with each batch containing the same number of customers.

Suppose that the batch size is b. If the system is in state n at the instant of an arrival event, defined as the arrival of a batch of b customers, it will transition to state $n + b$. The time between arrival events is still assumed to be exponentially distributed with mean $1/\lambda$. When the batch size is $b = 1$, this process reduces to the standard Poisson process. One could construct a transition matrix similar to equation (4.1). For example, with a batch size of two the transition matrix over $(t, t + \Delta t)$ would be

$$T = \begin{array}{c|ccccc} & 0 & 1 & 2 & 3 & 4 \\ \hline 0 & 1 - \lambda\Delta t - O(\Delta t) & 0 & \lambda\Delta t & O(\Delta t) & 0 & \cdots \\ 1 & 0 & 1 - \lambda\Delta t - O(\Delta t) & 0 & \lambda\Delta t & O(\Delta t) & \cdots \\ 2 & 0 & 0 & 1 - \lambda\Delta t - O(\Delta t) & 0 & \lambda\Delta t & \cdots \\ 3 & 0 & 0 & 0 & 1 - \lambda\Delta t - O(\Delta t) & 0 & \cdots \\ 4 & 0 & 0 & 0 & 0 & 1 - \lambda\Delta t - O(\Delta t) & \cdots \\ \vdots \end{array}$$

$$(4.38)$$

Performing the basic Markov chain calculation we have

$$p_0(t + \Delta t) = p_0(t)[1 - \lambda \Delta t - O(\Delta t)]$$
$$p_1(t + \Delta t) = p_1(t)[1 - \lambda \Delta t - O(\Delta t)]$$
$$p_2(t + \Delta t) = p_0(t)\lambda \Delta t + p_2(t)[1 - \lambda \Delta t - O(\Delta t)]$$
$$p_n(t + \Delta t) = p_{n-2}(t)\lambda \Delta t + p_n(t)[1 - \lambda \Delta t - O(\Delta t)], \, n \geq 2 \quad (4.39)$$

Dividing by Δt and taking limits as Δt goes to zero yields the differential-difference equations

$$\frac{dp_0(t)}{dt} = -\lambda p_0(t)$$

$$\frac{dp_1(t)}{dt} = -\lambda p_1(t)$$

$$\frac{dp_n(t)}{dt} = \lambda p_{n-2}(t) - \lambda p_n(t), \, n \geq 2 \qquad (4.40)$$

One could generalize this set of equations for an arbitrary batch size b as

$$\frac{dp_n(t)}{dt} = -\lambda p_n(t), \qquad 0 \le n < b$$

$$\frac{dp_n(t)}{dt} = \lambda p_{n-b}(t) - \lambda p_n(t), \; n \ge b \qquad (4.41)$$

If the system begins with no customers present at time zero and each arrival event brings b customers, it is obvious that the only possible states at time t are $(0, b, 2b, 3b, \ldots)$. Furthermore the number of arrival events in $(0, t)$ is Poisson distributed with mean λt. It therefore follows that the solution to equation (4.41) will yield

$$p_n(t) = \frac{(\lambda t)^{n/b} e^{-\lambda t}}{(n/b)!}, \; n \bmod (b)$$

$$0, \text{ elsewhere} \qquad (4.42)$$

COMPOUND POISSON PROCESSES

A very broad class of arrival processes can be modeled by taking one more step toward generalization of the Poisson process. Let us now consider multiple arrival events in which the batch size itself may be a random variable. Rather than a fixed number of customers b arriving at each arrival epoch, as considered in the last section, we may have 1, 2, 3 or more arrive with some known frequency distribution.

One such process, termed a multiple Poisson process, arises from combining several independent fixed batch size streams. Suppose that the event rate for the stream containing arrival batchs of size i is $\lambda^{(i)}$. Each individual stream is a Poisson process. If there are k such streams, the total process will have a Poisson event distribution with rate

$$\lambda = \sum_{i=1}^{k} \lambda^{(i)} \qquad (4.43)$$

Any particular batch size b then occurs in the combined stream with probability

$$f_b = \frac{\lambda^{(b)}}{\lambda} \qquad (4.44)$$

In this system the probability of a transition from state i to state j, $j > i$, during time Δt is obtained from the probability that a batch of size $(j - i)$ will arrive. The transition matrix for $(t, t + \Delta t)$ is then

		0	1	2	3	$\cdots$
	0	$1 - \lambda\Delta t - O(\Delta t)$	$\lambda^{(1)}\Delta t$	$\lambda^{(2)}\Delta t$	$\lambda^{(3)}\Delta t$	$\cdots$
$\mathbf{T} =$	1	0	$1 - \lambda\Delta t - O(\Delta t)$	$\lambda^{(1)}\Delta t$	$\lambda^{(2)}\Delta t$	$\cdots$
	2	0	0	$1 - \lambda\Delta t - O(\Delta t)$	$\lambda^{(1)}\Delta t$	$\cdots$
	3	0	0	0	$1 - \lambda\Delta t - O(\Delta t)$	$\cdots$

$$(4.45)$$

As a special case, suppose that the batch size is always either 1 or 2. The resulting differential-difference equations are then

$$\frac{dp_0(t)}{dt} = -\lambda p_0(t)$$

$$\frac{dp_1(t)}{dt} = \lambda^{(1)} p_0(t) - \lambda p_1(t)$$

$$\frac{dp_2(t)}{dt} = \lambda^{(2)} p_0(t) + \lambda^{(1)} p_1(t) - \lambda p_2(t)$$

$$\frac{dp_n(t)}{dt} = \lambda^{(2)} p_{n-2}(t) + \lambda^{(1)} p_{n-1}(t) - \lambda p_n(t), \; n \geq 2 \qquad (4.46)$$

It is also instructive to view the multiple Poisson process as a special case of a more general compound Poisson process. The strategy now is to treat the total number of arrivals in $(0, t)$ as a random sum of random variables. Suppose that at each arrival epoch b customers arrive where

$$Pr(b = i) = g(i) \qquad (4.47)$$

Here the distribution $g(i)$ can be any discrete probability mass function. The total number of customers arriving in $(0, t)$ could be described by the random variable n where

$$n = b_1 + b_2 + \ldots b_k \qquad (4.48)$$

The random variable b_j represents the number of customers arriving at arrival epoch j. The random variable k, which is assumed to be Poisson distributed, represents the number of arrival epochs occurring in $(0, t)$. Assuming that the b's are all identically distributed, independent random variables we can use geometric transforms to obtain the desired compound distribution.

Let us adopt the following notation for the necessary geometric transforms.

$$G_b(z) = G(g(b)) = \sum_{b=0}^{\infty} g(b)z^b \qquad (4.49)$$

$$G_k(z) = G(f_k(t)) = \sum_{k=0}^{\infty} f_k(t)z^k \qquad (4.50)$$

$$G_n(z) = G(p_n(t)) = \sum_{n=0}^{\infty} p_n(t)z^n \qquad (4.51)$$

Again the probability functions are written to include the argument t even though for purposes of the geometric transform t is constant.

If one treats the number of terms k in equation (4.48) as a constant, the random variable n and its associated distribution are conditioned on k. Furthermore, n is a fixed sum of independent, identically distributed random variables. The distribution for a sum of random variables is obtained by convolving the distribution of the component random variables. In this case, since all of the variables are identically distributed, we have the k-fold convolution

$$p_n|_k(t) = g(b_1) * g(b_2) * \cdots * g(b_k) = g(b)^{*k} \tag{4.52}$$

From the convolution property of the geometric transform it follows that the transform for this conditional distribution can be written as

$$G_n(z|k) = G(p_n|_k(t)) = [G_b(z)]^k \tag{4.53}$$

From basic probability theory we know that the marginal distribution $p_n(t)$ can be obtained from

$$p_n(t) = \sum_k p_n|_k(t) f_k(t) \tag{4.54}$$

Using this expression to define the transform for $p_n(t)$ we have

$$G_n(z) = \sum_{n=0}^{\infty} p_n(t) z^n = \sum_{n=0}^{\infty} \sum_{k=0}^{\infty} p_n|_k(t) f_k(t) z^n \tag{4.55}$$

By interchanging the order of summation

$$G_n(z) = \sum_k \left[\sum_n p_n|_k(t) z^n \right] f_k(t) \tag{4.56}$$

But the sum in brackets is nothing more than the transform for the conditional distribution from equation (4.53). Rewriting, we have

$$G_n(z) = \sum_k G_n(z|k) f_k(t) = \sum_k [G_b(z)]^k f_k(t) \tag{4.57}$$

At this point it is convenient to relate the expression for $G_n(z)$ to the formal definition of the geometric transform. Equation (4.57) is identical to the geometric transform $G_k(z)$ with the argument z replaced by another entire transform $G_b(z)$. That is

$$G_n(z) = G_k(z) \Big|_{z = G_b(z)} \tag{4.58}$$

where the vertical bar signifies that the argument z is to be replaced by the function $G_b(z)$. Using functional notation this could be written as

$$G_n(z) = G_k(G_b(z)) \tag{4.59}$$

This expression is not necessarily limited to Poisson processes but rather applies to a very broad class of random sums. For a detailed discussion of such nesting techniques see Giffin (40:87–105).

For the problem at hand the distribution for the number of arrival epochs k is known to be Poisson. It therefore follows that

$$G_k(z) = \sum_{k=0}^{\infty} \frac{(\lambda t)^k}{k!} e^{-\lambda t} z^k$$

$$= e^{-\lambda t(1-z)} \tag{4.60}$$

Substituting in equation (4.59) we have

$$G_n(z) = G_k(G_b(z)) = exp(-\lambda t(1 - G_b(z))) \qquad (4.61)$$

To summarize, one can obtain the geometric transform for the distribution of total arrivals in a Poisson process, in which the number of customers per arrival event is a random variable, by replacing the z-argument in the transform for the Poisson distribution by the transform for the distribution of number of customers per arrival event. The function $p_n(t)$ is recovered by inverting the transform, perhaps by numerical techniques (40:138–41). Any desired number of factorial moments can be obtained by taking derivatives with respect to z evaluated at $z = 1$. In general

$$E(n^{(r)}) = \left. \frac{d^r G_n(z)}{dz^r} \right|_{z=1} \qquad (4.62)$$

where the factorial moments are related to power moments by

$$E(n^{(r)}) = E(n(n - 1)(n - 2) \cdots (n - r + 1)) \qquad (4.63)$$

For the special cases of mean and variance it is easy to show that if

$$n = b_1 + b_2 + \cdots + b_k$$

then

$$\mu_n = \mu_k \mu_b \qquad (4.64)$$
$$\sigma_n{}^2 = \mu_k \sigma_b{}^2 + \mu_b{}^2 \sigma_k{}^2 \qquad (4.65)$$

In our Poisson arrival process it therefore follows the mean and variance for number of customers arriving in $(0, t)$ are related to the mean and variance of arrival batch size by

$$\mu_n = (\lambda t)\mu_b \qquad (4.66)$$
$$\sigma_n{}^2 = (\lambda t)(\sigma_b{}^2 + \mu_b{}^2) \qquad (4.67)$$

Example. Consider an order-processing system in which the time between receipt of customer orders is exponentially distributed with a mean of twenty minutes. A customer may order one or more units. The probability distribution for number of units per order is estimated to be a shifted geometric where

$$g(b) = 0.8(0.2)^{b-1}, \; b = 1, 2, \ldots$$

The problem is to estimate the number of units ordered in the interval $(0, t)$.

The geometric transform for number of units per order is

$$G_b(z) = \sum_{b=1}^{\infty} 0.8(0.2)^{b-1}z^b$$
$$= \frac{0.8z}{1 - 0.2z}$$

Since the time between orders is exponentially distributed with a mean of twenty minutes, arrivals are Poisson distributed at a rate of three orders per hour. Measuring time in hours we can then use equation (4.61) to write the geometric transform for the distribution of number of units on order as

$$G_n(z) = exp[-\lambda t(1 - G_b(z)]$$
$$= exp\left[-3t\left(1 - \frac{0.8z}{1 - 0.2z}\right)\right]$$
$$= exp[-3t(1 - z)/(1 - 0.2z)]$$

The mean and variance for number of units per order are from the moments of the geometric distribution

$$\mu_b = 1.25, \ \sigma_b{}^2 = 0.3125$$

The corresponding moments for the total number of units ordered in $(0, t)$ are then calculated from equations (4.66) and (4.67).

$$\mu_n = (\lambda t)\mu_b = 3.75t$$
$$\sigma_n{}^2 = \lambda t(\sigma_b{}^2 + \mu_b{}^2) = 5.625t$$

The transform for $p_n(t)$ can with some effort be inverted. The resulting distribution is

$$p_n(t) = (0.2)^n e^{-3t} \sum_{i=1}^{n} \frac{1}{i!} \binom{n-1}{i-1} (12t)^i, \ n = 1, 2, \ldots$$

This distribution is sometimes referred to as the stuttering Poisson distribution.

RENEWAL PROCESSES

The Poisson streams we have been discussing are special cases of a more general class of problems called renewal processes. The generalization is one which considers a counting process for which the interarrival times are independent and identically distributed with an arbitrary distribution, as contrasted with the exponential interarrival distribution of the Poisson process. Any such sequence of independent, identically distributed random variables each of which has a positive range is called a *renewal process*. In a sense the process starts over or is renewed after each arrival event. Problems in renewal theory for both discrete and continuous time have been thoroughly discussed in the literature. The interested reader is referred to Feller (37: Ch. 13), Ross (81: Ch. 3) or Cox (26) for a complete discussion of renewal theory.

SUMMARY

We have explored a number of models for stochastic processes in the context of arrival streams of customers. Most of those models are directly related to the Poisson process. Even though the focus so far has been on arrivals, it is reasonable to expect that similar process descriptions are possible for the service side of queueing systems. Combining arrival and service descriptions for continuous time processes is the task of the following chapter.

The basic manipulative skills demonstrated for solution of models in this chapter will be carried forward to future discussion. Markov chain manipulation, formation of differential-difference equations, solution of those equations by a variety of techniques and generalization of basic models by relaxing assumptions are the steps we will follow as our queueing models become ever more realistic and complex.

PROBLEMS

4.1 *a*)Discuss the relationship between the Poisson, exponential, and uniform distributions. Clearly indicate the corresponding parameters.

b)How might one exploit this relationship when analyzing data for a queueing system?

4.2 Prove that when a arrivals are generated by a Poisson process in the interval $[0, T]$, the times at which those arrivals occurred are distributed as the order statistic of a uniform random variables.

4.3 In the textbook discussion it was shown that the time between events for a Poisson process is exponentially distributed. The proof assumed that an arrival had just occurred. Prove that the time to the next arrival is exponentially distributed regardless of our assumed observation point.

4.4 Equation (4.25) demonstrates the forgetfulness property of the exponential distribution. Prove that the geometric distribution (discrete) also possesses this Markov property.

4.5 Describe five processes encountered in your daily activities in which you feel that the Poisson distribution may be a reasonable model. Justify your statement.

4.6 A hospital has four emergency generators which they can use in the event of a community power failure. All four are required to maintain a full level of service. However, reduced services are possible with less than four generators. When all generators are out, the hospital must be evacuated. The generator manufacturer claims that the mean time between failures for single units is twelve hours. The hospital consultant is willing to assume that the failure time distribution is exponential.

a) Suppose that all four generators are put on line when a power failure occurs. Write the set of differential-difference equations the solution to which will describe the number still operating t hours after the power failure occurs.

b) Find the probability that the hospital must be evacuated if a power failure lasts 24 hours.

c) Suppose that the generators are put on line one at a time (e.g., when a power failure occurs a single emergency generator is activated, when that one fails a second is activated, etc.). What is the probability of evacuation under a 24-hour power failure with this policy?

4.7 An auditor has the task of auditing the account books for five motor vehicle registrars. The average time to complete a single audit is sixteen man hours. The distribution for audit time for a single registrar is assumed to be exponential.

a) Develop a set of differential-difference equations which when solved will yield the probability that n audits remain after time t.

b) Find the probability that he will have completed at least two of his audits at the end of a 40-hour week.

4.8 Prove that the solution to the Yule process is given by equation (4.30).

4.9 Consider a delay device in a production line designed to batch units for subsequent operations. Units are received in Poisson fashion at a rate of twenty per hour. The delay device is set to release batches of five units.

a) Determine the probability density function for time between batches.

b) Because of a poor sensing device this unit has recently been releasing batches of 4, 5, and 6 with frequencies of 0.2, 0.4, and 0.4, respectively. Find the mean and variance for time between batches under these conditions.

4.10 Consider the differential equation (4.29).

a) If the system has an initial population of one, show that the probability generating function can be written as

$$G(z, t) = \frac{ze^{-\lambda t}}{1 - z + ze^{-\lambda t}}$$

b) Find $G(z, t)$ if the system begins with an initial population of $n = i > 1$.

c) Find $E(n(t))$.

4.11 A memorial grove of ten elm trees has had one of its trees infected by Dutch elm disease. Once a tree is infected it cannot be cured and it serves as a carrier which can infect other trees. The probability of a healthy tree becoming infected in time Δt is directly proportional to the current number of infected trees in the grove. With one tree infected it is estimated that the time to infection of a second tree will be exponentially distributed with a mean of 0.1 years.

a) Develop a set of differential-difference equations which describe the probability distribution for number of healthy trees in the grove at time t.

b) Estimate the expected time it will take for all of the trees to become infected.

4.12 After studying industrial accident data for steel foundaries an insurance company analyst has formulated the hypothesis that accident rates increase in direct proportion to accident history. Suppose that the mean time to the first accident in a newly insured foundary is $1/\lambda$ months. The mean time between the first and second accident is $1/(\lambda + a)$ and between the ith and $(i + 1)^{\text{st}}$ is $1/(\lambda + ia)$.

a) Develop a set of differential-difference equations to describe this process.

b) Find the probability distribution for number of accidents in $(0, t)$ for a newly insured foundary under the assumption that the analyst's hypothesis is correct.

4.13 Consider the time-varying aircraft arrival pattern of Figure 3-1. Use the techniques suggested in this chapter to develop estimates for the distribution of the number of aircraft which will have arrived between midnight and 8:00 AM. Suggest a general solution strategy for solving such problems from empirical information.

4.14 A casualty insurance company is attempting to predict fire insurance claims related to lightning strikes in a single county region. The number of claims filed per year appears to be Poisson distributed with a mean of 42. The dollar amount of individual claims can be approximated by $(300 + m)$ where m is exponentially distributed with a mean of \$200. Find the mean and variance for the annual dollar amount of claims from this county.

4.15 The Chamber of Commerce provides information packets to any civic group who requests them. Past data indicate that the number of inquiries per day is approximately Poisson distributed with a mean of ten. The number of packets requested by each inquirer is distributed by

$$f(x) = 0.7(0.3)^{x-1}, x = 1, 2, \ldots$$

It takes seven days for the Chamber to replenish their supplies

a) Find the geometric transform for the number of packets requested during one replenishment cycle.

b) Find the probability that more than four packets will be requested per day.

4.16 The operator of a hamburger stand can identify three distinct groups of clientele. He serves motorcyclists who arrive one per bike at a rate of fifteen per hour, strolling couples who arrive on foot at the rate of twelve couples per hour and families of four who arrive by auto at the rate of thirty per hour. He would like to find the probability distribution for the number of customers served in any interval $(0, t)$.

a) Model this as a multiple Poisson process by developing a proper set of differential-difference equations.

b) Model the system as a random sum of random variables and develop an appropriate geometric transform.

c) Find the mean and variance for customers served per hour.

BIRTH-DEATH PROCESS IN QUEUES

The most widely used queueing models are based upon variations of the birth and death process. The equations describing the general birth-death process are easily developed by extending the arguments for the general Poisson process discussed in the last chapter to include departures or deaths. They are held in high esteem by practitioners because:

1. They can be adapted to fit a great many different system configurations.
2. They are easy to manipulate, especially for steady-state solutions.
3. They are robust in the sense that model results often closely match system performance even when some characteristics of the system seem to violate the basic assumptions of the model.

POSTULATES

The birth-death postulates include those of the pure birth process discussed in Chapter 4. The "births" represent customer arrivals which may be functionally related to the current state of the system or to the age of the process. In its most general form we could formulate the birth postulates as follows:

1. If the system is in state n at time t, the probability of a single arrival in the interval $(t, t + \Delta t)$ for small Δt is $\lambda_n(t)\, \Delta t$ where

$$\lim_{\Delta t \to 0} \frac{\lambda_n(t)\, \Delta t}{\Delta t} = \lambda_n(t)$$

2. The probability of more than a single arrival during a small interval of time Δt is $O(\Delta t)$ where

$$\lim_{\Delta t \to 0} \frac{O(\Delta t)}{\Delta t} = 0$$

Since we are now considering systems in which a customer departs the system after being served, i.e., a death occurs, we must also include the death postulates as follows:

3. If the system is in state n at time t, the probability of a single service in the interval $(t, t + \Delta t)$ for small Δt is $\mu_n(t)\, \Delta t$ where

$$\lim_{\Delta t \to 0} \frac{\mu_n(t)\, \Delta t}{\Delta t} = \mu_n(t)$$

4. The probability of more than a single service during a small interval of time Δt is $O(\Delta t)$ where

$$\lim_{\Delta t \to 0} \frac{O(\Delta t)}{\Delta t} = 0$$

In this chapter we will be concerned only with stationary processes in which the arrival and service rates do not change with time. For that reason we will temporarily drop the functional argument t in the birth and death postulates such that

$$\lambda_n(t) = \lambda_n \text{ for all } t \tag{5.1}$$

$$\mu_n(t) = \mu_n \text{ for all } t \tag{5.2}$$

TRANSITION MATRIX

The birth-death process exhibits the Markov property. The conditional probability of finding the system in state j at time $t + \Delta t$, given that it occupied state i at time t, depends only on the starting state and the transition time Δt. It is independent of the path used to reach state i. This permits us to construct a transition matrix for a simple Markov chain where the time between transitions is Δt.

In an effort to conserve space let us agree to ignore all terms $O(\Delta t)$. We can safely ignore those terms since for continuous processes we will ultimately be dividing all terms by Δt and by our postulates

$$\lim_{\Delta t \to 0} \frac{O(t)}{\Delta t} = 0$$

Let $p_{ij}(\Delta t)$ be the probability of moving from state i to state j during a time Δt. Beginning with state zero the system can move only to a higher order state or stay in state zero. It will remain in state zero if no customers arrive or if a single customer arrives and is served in $(t, t + \Delta t)$. The probability of these events is

$$p_{00}(\Delta t) = (1 - \lambda_0\, \Delta t) + \lambda_0\, \Delta t \mu_0\, \Delta t \tag{5.3}$$

The system will move from state zero to state one if a single customer arrives and none are served. Therefore

$$p_{01}(\Delta t) = \lambda_0\, \Delta t(1 - \mu_0\, \Delta t) \tag{5.4}$$

Multiple arrivals would be necessary to move the system from state zero to any state $j > 1$. But by the birth postulates the probability of such events for small Δt is zero. That is

$$p_{0j}(\Delta t) = 0, j > 1 \tag{5.5}$$

From higher order initial states, transitions can occur in either a positive or negative direction. The system can stay in state i, with no births or deaths, move from i to $i - 1$ with a death, or move from i to $i + 1$ when a birth in the interval $(t, t + \Delta t)$. The probabilities of these events are

$$p_{i,i-1}(\Delta t) = \mu_i \, \Delta t(1 - \lambda_i \, \Delta t) \tag{5.6}$$

$$p_{ii}(\Delta t) = \lambda_i \, \Delta t \mu_i \, \Delta t + (1 - \lambda_i \, \Delta t)(1 - \mu_i \, \Delta t) \tag{5.7}$$

$$p_{i, \, i+1}(\Delta t) = \lambda_i \, \Delta t(1 - \mu_i \, \Delta t) \tag{5.8}$$

Since multiple arrival or services are not admissible in our current system, $p_{ij}(\Delta t) = 0$ for $|i - j| > 1$. The complete transition matrix can now be written as

$$
\mathbf{T} = \begin{array}{c|ccc}
 & 0 & 1 & 2 \quad \cdots \\
\hline
0 & \lambda_0\mu_0\Delta t^2 + (1 - \lambda_0\Delta t) & \lambda_0\Delta t(1 - \mu_0\Delta t) & 0 \\[2ex]
1 & \mu_1\Delta t(1 - \lambda_1\Delta t) & \begin{array}{c}\lambda_1\mu_1\Delta t^2 \\ + (1 - \lambda_1\Delta t)(1 - \mu_1\Delta t)\end{array} & \lambda_1\Delta t(1 - \mu_1\Delta t) \\[3ex]
2 & 0 & \mu_2\Delta t(1 - \lambda_2\Delta t) & \begin{array}{c}\lambda_2\mu_2\Delta t^2 \\ + (1 - \lambda_2\Delta t)(1 - \mu_2\Delta t)\end{array} \\[3ex]
\vdots & & &
\end{array}
\tag{5.9}
$$

SOLUTION PROCEDURE

The solution for this continuous time Markov chain proceeds in exactly the same fashion demonstrated in the previous chapter. A probability vector at time $(t + \Delta t)$ is constructed by multiplying the vector at time t by the single stage transition matrix $\mathbf{T}$. In vector notation

$$\mathbf{P}(t + \Delta t) = \mathbf{P}(t)\mathbf{T} \tag{5.10}$$

This multiplication leads to the following set of equations

$$p_0(t + \Delta t) = p_0(t)[\lambda_0\mu_0 \, \Delta t^2 + (1 - \lambda_0 \, \Delta t)] + p_1(t)[\mu_1 \, \Delta t(1 - \lambda_1 \, \Delta t)]$$

$$p_n(t + \Delta t) = p_{n-1}(t)[\lambda_{n-1} \, \Delta t(1 - \mu_{n-1} \, \Delta t)]$$

$$+ p_n(t)[\lambda_n\mu_n \, \Delta t^2 + (1 - \lambda_n \, \Delta t)(1 - \mu_n \, \Delta t)]$$

$$+ p_{n+1}(t)[\mu_{n+1} \, \Delta t(1 - \lambda_{n+1} \, \Delta t)], \, n \geq 1 \tag{5.11}$$

By transferring $p_n(t)$ to the left side, dividing by Δt and taking the limit as Δt goes to zero, we are led to the set of differential-difference equations

$$\frac{dp_0(t)}{dt} = -\lambda_0 p_0(t) + \mu_1 p_1(t)$$

$$\frac{dp_n(t)}{dt} = -(\lambda_n + \mu_n)p_n(t) + \lambda_{n-1}p_{n-1}(t) + \mu_{n+1}p_{n+1}(t), \, n \geq 1 \tag{5.12}$$

This set of equations is probably the most important set in queueing theory from the viewpoint of easily accessed robust models. The reader should study it thoroughly as much of our future work stems directly from this model.

In the interest of completeness let us reintroduce time dependent arrival rates and service rates. The most general form of the birth-death equations then becomes the nonhomogenous set

$$\frac{dp_0(t)}{dt} = -\lambda_0(t)p_0(t) + \mu_1(t)p_1(t)$$

$$\frac{dp_n(t)}{dt} = -[\lambda_n(t) + \mu_n(t)]p_n(t) + \lambda_{n-1}(t)p_{n-1}(t)$$

$$+ \mu_{n+1}(t)p_{n+1}(t), \ n \geq 1 \tag{5.13}$$

In theory the two sets of equations (5.12) and (5.13) can be solved by the same techniques discussed in Chapter 4. Namely,

1. Define the transform

$$G(z, t) = \sum_{n=0}^{\infty} p_n(t)z^n$$

2. Reduce the infinite set of equations to a single differential equation in $G(z, t)$ by appropriate multiplication and addition.
3. Solve the differential equation for $G(z, t)$.
4. Invert $G(z, t)$ to obtain $p_n(t)$.

In practice these equations are extremely difficult to solve for most specifications of $\lambda_n(t)$ and $\mu_n(t)$. As we will demonstrate in future chapters, even when a formal closed form solution for $p_n(t)$ can be found, it may be so awkward to manipulate that it is of little practical value.

Let us hasten to add, however, that the picture is not as bleak as it may first appear. If one is willing to assume that most physical systems have finite capacity, the set of equations (5.13) is reduced to a finite number. We can then use numerical techniques for solving systems of differential equations on the digital computer to achieve high resolution approximations. The answers are presented in the form of tables of numbers which represent the probability distribution $p_n(t)$ rather than in compact equation form. But after all it is the numbers rather than the equation which we ultimately need to make system design decisions. We will explore this approach in considerable depth in a future chapter.

A second way to circumvent the difficulties posed by the set of equations (5.12) is to ignore the transient portions of the solution and go directly to the steady-state results. We describe a system as being in a transient state when its probabilistic behavior changes with time. For example, suppose that there are no customers present when a bank teller opens his window for business at 9:00 AM because the arrival process cannot function prior to that time. If we assume a constant arrival rate from that point forward, the distribution $p_n(t)$ would be a transient solution describing the state of the system say at 9:10 AM, 9:30 AM, etc. After the teller has been processing customers for a long time, say by 10:00 AM, the effect of the initial queue length zero may no longer be apparent. By this time the distribution $p_n(t)$ may stabilize such that $p_n(t) = p_n(t + \Delta t)$. When this occurs the system has reached steady state. For stationary queueing systems steady-state solutions then emphasize the long-run behavior while transient solutions focus on start-up conditions. For nonstationary systems such as those with time-varying arrival or service rates, there is no steady-state solution. In those cases only transient solutions have meaning.

The bulk of the remainder of this chapter emphasizes steady-state solutions for various configurations of the set of homogeneous equations (5.12).

TRANSITION RATES

Before attempting to solve our sets of differential-difference equations, let us pause to introduce an alternate technique for generating such equations. Reconsider the Chapman-Kolmogorov equation for discrete-time Markov processes given in equation (2.28). That is

$$\mathbf{P}(r, t) = \mathbf{P}(r, s)\mathbf{P}(s, t) \tag{5.14}$$

where the elements of the matrix $\mathbf{P}(r, t)$ represented the multistage transition probabilities for moving from state i to state j between stages r and t. What we now seek is a method for analyzing the continuous-time version of these equations.

Consider three successive points in time for a continuous-time Markov chain $r < s < t$. The same observation made for the discrete time case still holds. In moving from state i at time r the system must pass through some intermediate state k at time s before arriving in state j at time t. To distinguish the continuous from the discrete case let us adopt the matrix notation

$$\mathbf{T}(r, t) \stackrel{\Delta}{=} [p_{ij}(r, t)] \tag{5.15}$$

where the element $p_{ij}(r, t)$ is the conditional probability of moving from state i to state j in the time interval (r, t). The continuous-time Chapman-Kolmogorov equation is then

$$\mathbf{T}(r, t) = \mathbf{T}(r, s)\mathbf{T}(s, t) \tag{5.16}$$

Now consider the second transition in equation (5.16) to cover the interval of $(t, t + \Delta t)$. Rewriting, we have

$$\mathbf{T}(r, t + \Delta t) = \mathbf{T}(r, t)\mathbf{T}(t, t + \Delta t) \tag{5.17}$$

from which we can form the matrix difference equation

$$\mathbf{T}(r, t + \Delta t) - \mathbf{T}(r, t) = \mathbf{T}(r, t)\mathbf{T}(t, t + \Delta t) - \mathbf{T}(r, t) \tag{5.18}$$

Dividing both sides of this equation by Δt, then taking the limit as Δt approaches zero leads to

$$\frac{\partial \mathbf{T}(r, t)}{\partial t} = \mathbf{T}(r, t)\mathbf{R}(t) \tag{5.19}$$

where

$$\mathbf{R}(t) = \lim_{\Delta t \to 0} \frac{\mathbf{T}(t, t + \Delta t) - \mathbf{I}}{\Delta t} \tag{5.20}$$

Equation (5.19) is the forward Chapman-Kolmogorov equation for the continuous-time Markov chain. The matrix $\mathbf{R}(t)$ is known as the *transition rate matrix* or sometimes as the *infinitesimal generator* of the transition matrix $\mathbf{T}(r, t)$.

The elements of the matrix $\mathbf{R}(t)$ are

$$r_{ii}(t) = \lim_{\Delta t \to 0} \frac{p_{ii}(t, t + \Delta t) - 1}{\Delta t} \tag{5.21}$$

$$r_{ij}(t) = \lim_{\Delta t \to 0} \frac{p_{ij}(t, t + \Delta t)}{\Delta t}, i \neq j \tag{5.22}$$

The element $r_{ij}(t)$ represents the rate at which a process currently in state i will move to state j. The element $p_{ij}(t, t + \Delta t)$ for a fixed initial state i is a conditional probability and in general we know that

$$\sum_j p_{ij}(r, t) = 1 \tag{5.23}$$

for all r and t. Relating this information to the transition rates of equations (5.21) and (5.22), it is obvious that

$$\sum_j r_{ij}(t) = 0 \text{ for all } i \tag{5.24}$$

The rate matrix then has all row sums equal zero and all diagonal elements equal to the negative sum of all other elements in the same row.

If we expand equation (5.19) we obtain the individual elements

$$\frac{\partial p_{ij}(r, t)}{\partial t} = \sum_k p_{ik}(r, t)r_{kj}(t) \tag{5.25}$$

When the system begins operating in some initial state i at time r this system of equations can be solved subject to the boundary condition

$$p_{ij}(r, r) = \begin{cases} 1 & \text{if } j = i \\ 0 & \text{elsewhere} \end{cases} \tag{5.26}$$

Continuing the analogy with the discrete-time models of Chapter 2, let us write the state probability vector at time t as

$$\mathbf{V}(t) = [v_1(t), v_2(t), v_3(t) \dots] \tag{5.27}$$

where

$$v_i(t) = Pr[N(t) = i] \tag{5.28}$$

Beginning with an initial state distribution $\mathbf{V}(0)$, we have from the basic Markov chain argument

$$\mathbf{V}(t) = \mathbf{V}(0)\,\mathbf{T}(0, t) \tag{5.29}$$

Forming a matrix difference equation we have

$$\begin{aligned}
\mathbf{V}(t + \Delta t) - \mathbf{V}(t) &= \mathbf{V}(0)\,\mathbf{T}(0, t + \Delta t) - \mathbf{V}(t) \\
&= \mathbf{V}(0)\,\mathbf{T}(0, t)\,\mathbf{T}(t, t + \Delta t) - \mathbf{V}(t) \\
&= \mathbf{V}(t)[\mathbf{T}(t, t + \Delta t) - \mathbf{I}]
\end{aligned} \tag{5.30}$$

Dividing by Δt and taking the limit as Δt approaches zero we have

$$\frac{d\mathbf{V}(t)}{dt} = \mathbf{V}(t)\mathbf{R}(t) \tag{5.31}$$

The elements of this calculation can be expressed as

$$\frac{dv_j(t)}{dt} = \sum_k v_k(t)r_{kj}(t) \tag{5.32}$$

Both sets of equations (5.25) and (5.32) describe the same process. One may elect to solve for $p_{ij}(r, t)$ from equation (5.25) whenever it is necessary to emphasize the initial state of the system. When the initial state of the system is known only in a probability sense, given by the vector $\mathbf{V}(0)$, it is more desirable to use equation (5.32) to find the state probability distribution $[v_j(t)]$. If the system is known to start in state i, the initial state probability vector has elements

$$v_j(0) = \begin{cases} 1 & i = j \\ 0 & i \neq j \end{cases}$$

In this case solving for $v_j(t)$ is equivalent to solving for $p_{ij}(0, t)$.

BIRTH-DEATH RATE MATRIX

Let us now use the rate matrix technique to develop the birth-death homogeneous differential equations (5.12). The birth-death postulates led us to the transition probabilities over the interval $(t, t + \Delta t)$ expressed in the T-matrix (5.9). Relating those transition probabilities to the corresponding rate matrix elements through equation (5.21) we have

$$\begin{aligned}
r_{ii}(t) &= \lim_{\Delta t \to 0} \frac{p_{ii}(t, t + \Delta t) - 1}{\Delta t} \\
&= \lim_{\Delta t \to 0} \frac{\lambda_i \mu_i \Delta t^2 + (1 - \lambda_i \Delta t)(1 - \mu_i \Delta t) - 1}{\Delta t} = -(\lambda_i + \mu_i),\ i > 0 \\
r_{i,i+1}(t) &= \lim_{\Delta t \to 0} \frac{p_{i,i+1}(t, t + \Delta t)}{\Delta t} \\
&= \lim_{\Delta t \to 0} \frac{\lambda_i \Delta t(1 - \mu_i \Delta t)}{\Delta t} = \lambda_i
\end{aligned}$$

$$r_{i,i-1}(t) = \lim_{\Delta t \to 0} \frac{p_{i,i-1}(t, t + \Delta t)}{\Delta t}$$

$$= \lim_{\Delta t \to 0} \frac{\mu_i \, \Delta t(1 - \lambda_i \, \Delta t)}{\Delta t} = \mu_i, \ i > 0 \tag{5.33}$$

The corresponding transition rate matrix for the birth-death process then becomes

$$
\mathbf{R}(t) = \begin{array}{c c} & \begin{array}{cccccc} 0 & \quad 1 & \quad 2 & \quad 3 & \cdots \end{array} \\ \begin{array}{c} 0 \\ 1 \\ 2 \\ 3 \\ \vdots \end{array} & \left|\begin{array}{cccc} -\lambda_0 & \lambda_0 & 0 & 0 \\ \mu_1 & -(\lambda_1 + \mu_1) & \lambda_1 & 0 \\ 0 & \mu_2 & -(\lambda_2 + \mu_2) & \lambda_2 \quad \cdots \\ 0 & 0 & \mu_3 & -(\lambda_3 + \mu_3) \quad \cdots \end{array}\right. \end{array}
\tag{5.34}
$$

The differential equations can now be expressed as the matrix equation

$$\frac{d\mathbf{P}(t)}{dt} = \mathbf{P}(t)\mathbf{R}(t) \tag{5.35}$$

which leads directly to the set of equations (5.12) previous developed by other arguments.

Once the equivalence of the two methods for developing the necessary differential equations is understood, it is sometimes convenient to use system balance arguments to quickly construct those equations. For example, consider the transition rate flow diagram of Figure 5.1.

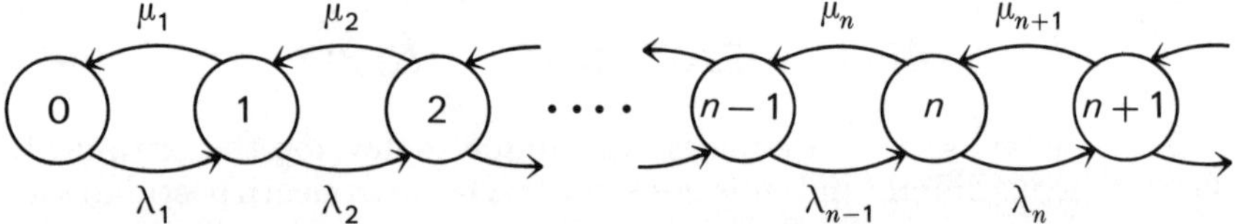

FIGURE 5-1   Flow Rate Diagram for the Birth-Death Process

The nodes of this diagram represent the states of the system. The arrows represent the rates at which probability "flows" from one state to another. To capture the dynamic nature of the system it is necessary to describe the input and output at each node. For the general node n we have

$$\text{Flow in} = \lambda_{n-1}p_{n-1}(t) + \mu_{n+1}p_{n+1}(t) \tag{5.36}$$

$$\text{Flow out} = \lambda_n p_n(t) + \mu_n p_n(t) \tag{5.37}$$

The net rate at which probability accumulates at this node is simply the difference between the flow in and the flow out. Therefore we can write

$$\frac{dp_n(t)}{dt} = \lambda_{n-1}p_{n-1}(t) + \mu_{n+1}p_{n+1}(t) - [\lambda_n p_n(t) + \mu_n p_n(t)] \quad (5.38)$$

which is the same general birth-death equation we have seen many times before.

STEADY-STATE SOLUTIONS

Most of the available queueing literature is concerned with the long-run behavior of systems. In such analyses the systems are assumed to have reached steady state in the sense that their probabilistic behavior becomes independent of time. This approach is often justified on the basis that most design problems are more concerned with a long-term average behavior than they are with temporary perturbations caused by start-up conditions or short interruptions in system operation. However, an even more important, if often unarticulated, reason for emphasizing steady-state solutions is that the transient solutions are simply too difficult to obtain. This author is of the opinion that transients should not be ignored in time-varying systems such as those encountered in the design of air traffic control procedures or mass transit systems. The effect of ignoring transients may be to vastly over-design the system in an attempt to counter peak loads or to invite system breakdown by assuming that all loads are nominal.

For example, suppose that the peak-hour arrival rate at an air terminal is eighty aircraft. Throughout the remainder of the day the arrival rate averages twenty per hour with very little variation. If a single runway can handle forty operations per hour it is tempting to say that two runways should be constructed to handle the peak load of eighty. However such a solution ignores the cost of idle facilities during the bulk of the day. What the analyst needs is a method of predicting the queue build-up and dissipation caused by the pulse load of eighty under single runway configuration. Steady-state analysis cannot be used to resolve this problem. Conversely if the system is running satisfactorily at a service rate of forty per hour one might need to know what delays would accumulate if the service rate were to be interrupted for a short period of time. Airline passengers in the New York area at the time of the infamous "Black Friday" have had firsthand experience with the consequences of service interruptions in a nearly saturated queueing system. In that instance, a thunderstorm which reduced the service rate to zero for a short period of time led to many hours of delay and many cancelled flights before the system again stabilized. Predicting the consequences of such perturbations in otherwise smoothly running systems requires more than steady-state analysis.

We will return to a discussion of transients in the next chapter. For now let us do as most analysts do and profess that we are really interested only in steady-state results.

To assume that the system has reached steady state is equivalent to stating that the state probability distribution is no longer changing with time. Mathematically this means that the derivative of $p_n(t)$ must be zero. Since the distribution is no longer time dependent let us drop the functional argument t so that $p_n(t)$ is replaced by p_n where

$$p_n = Pr(N = n)$$

Rewritting the birth-death equations (5.12) for steady-state conditions we have

$$0 = -\lambda_0 p_0 + \mu_1 p_1$$
$$0 = -(\lambda_n + \mu_n)p_n + \lambda_{n-1}p_{n-1} + \mu_{n+1}p_{n+1}, \ n \geq 1 \tag{5.39}$$

The above set of equations is most easily solved by iteration. Beginning with the first equation

$$p_1 = \frac{\lambda_0}{\mu_1} p_0 \tag{5.40}$$

Using that result in the second equation yields

$$\begin{aligned}
p_2 &= \frac{1}{\mu_2}(\lambda_1 + \mu_1)p_1 - \frac{\lambda_0}{\mu_2}p_0 \\
&= \frac{1}{\mu_2}(\lambda_1 + \mu_1)\frac{\lambda_0}{\mu_1}p_0 - \frac{\lambda_0}{\mu_2}p_0 \\
&= \frac{\lambda_0\lambda_1}{\mu_1\mu_2}p_0
\end{aligned} \tag{5.41}$$

and in general

$$p_n = \frac{\displaystyle\prod_{i=0}^{n-1}\lambda_i}{\displaystyle\prod_{i=1}^{n}\mu_i}p_0 \tag{5.42}$$

where the pi notation represents the product of indicated terms. Now that we have the general term p_n written as a function of p_0 we can invoke the definition of a probability distribution to determine p_0 and hence all p_n. Since

$$\sum_{n=0}^{\infty} p_n = 1 \tag{5.43}$$

it follows that

$$p_0 + \sum_{n=1}^{\infty} \frac{\displaystyle\prod_{i=0}^{n-1}\lambda_i}{\displaystyle\prod_{i=1}^{n}\mu_i}p_0 = 1 \tag{5.44}$$

from which

$$p_0 = \left[1 + \sum_{n=1}^{\infty} \frac{\displaystyle\prod_{i=0}^{n-1}\lambda_i}{\displaystyle\prod_{i=1}^{n}\mu_i}\right]^{-1} \tag{5.45}$$

Equations (5.42) and (5.45) then constitute the general steady-state solution for the birth-death process.

These equations are very robust. They can be used to model a great number of system configurations including infinite capacity, finite capacity, single-channel, multi-channel, infinite source, finite source, state-dependent service rate, and many other pure and mixed characteristics. All that is necessary is to define the proper set of arrival rate λ_n and service rate μ_n coefficients for the configuration being considered, then solve the resulting equations. We will demonstrate several such models in future sections.

SYSTEM IMPLICATIONS

Before launching into particular model development let us review the implications of modeling a system as a birth-death process. The birth-death postulates for the stationary case imply that the time between events, for both arrivals and departures, follows an exponential distribution. When n customers are present this means that the probability density function for the time until the arrival of the next customer is given by

$$a_n(t) = \lambda_n e^{-\lambda_n t} \tag{5.46}$$

The density function for time to departure of the next customer when the system is occupied is given by

$$s_n(t) = \mu_n e^{-\mu_n t} \tag{5.47}$$

The problem is to relate the properties of the exponential distribution to the physical behavior which might give rise to such properties.

The most important property of the exponential distribution is its lack of memory. This means that knowledge of the elapsed time since the last previous arrival or service is of no value in predicting the time at which the next event will occur. In terms of the arrival process such behavior is easy to understand. If one has a large population of customers each behaving independently of all others it is reasonable to assume that the length of time elapsing before the next arrival would be independent of what happens in all other intervals.

The forgetfulness property of the exponential distribution is more difficult to rationalize on the service side of the system. The forgetfulness property implies that the time already spent in servicing a customer is irrelevant for predicting the remaining amount of time he will be in service. In a banking operation, for example, this implies that the probability of a customer being discharged in a one-minute interval immediately after stepping to the teller's window is the same as the probability of his being discharged in the next minute even though he may have already been in service for five minutes. From the viewpoint of individual customers such a prospect is unappealing. It would not be a realistic assumption if every customer required the same service operations. However, if the system serves a variety of customers with widely varying service time demands, the exponential distribution may be quite reasonable as a model from the point of view of the service mechanism. If a particular customer has already been in service for some time it may be that he simply requires more extensive service than his fellow customers.

A second characteristic of the exponential distribution is that the density function is greatest at $t = 0$ and decreases monotonically for increasing t. This means that one must expect a higher frequency of short interevent intervals than long ones. This property is consistent with customers who arrive completely at random. It would not be a proper description for an arrival process in which customers pace their arrivals by time or by previous customer arrivals.

On the service side of the system a high frequency of short service times intersperced with occasional long services is characteristic of systems in which the service task differs among customers. The exponential distribution has been found to be particularly useful in modeling the duration of telephone calls where there are typically a large number of customers requiring reasonably short service and fewer requiring longer service. Such a description may also fit grocery check-out stations, hospital emergency room services, police radio calls, and many other facilities where service requested is usually of short duration but may occasionally be long.

The exponential distribution is not a good model for systems requiring nearly identical service for each customer. Nor is it a good model for systems requiring a fixed set-up time before each service can be performed. In both instances service times near zero would be essentially impossible to achieve. In these situations other service distributions provide more accurate descriptions. However, it is interesting to note that many of the usual performance measures such as the mean and variance for the number of customers in the system are not greatly influenced by the form of the service distribution over a wide range of functions as we shall demonstrate in future chapters. This observation permits us to use the standard birth-death model for system design even though we know that the service operation is not well approximated by the exponential distribution.

A third property of the exponential distribution which has physical significance is that the minimum of several exponentially distributed random variables is exponentially distributed. This permits us to use the birth-death equations for systems in which we have several classes of customers. Suppose that customers within a given class arrive in exponential fashion at rate a_i. The total system arrival rate across all classes is

$$\lambda = \sum_{i=1}^{m} a_i \tag{5.48}$$

Considering the interarrival time from the time of arrival of a random customer to the next customer of any type, this property suggests that time will be exponentially distributed with parameter λ. One can then ignore the distribution among customer classes and treat the entire arrival stream as a Poisson process.

This property of exponentially distributed minimums is useful on the service side to describe the behavior of multi-server systems. Suppose that one has m servers each currently providing service in exponential fashion at rate s. The time to next service completion from any member of the server population is exponentially distributed with parameter $\mu = ms$. This implies that we can model a busy multi-channel system as a single-server system with a modified service rate.

SYSTEM PERFORMANCE MEASURES

The measures of performance we will use in this chapter can all be related to the steady-state probability distribution p_n, where p_n is developed from the birth-death equations for particular specifications of $\{\lambda_n, \mu_n\}$.

The first performance measure of interest we could term *load level*. The question posed might ask: "What portion of time will the number of customers in the system exceed N?" This is easily obtainable from $\{p_n\}$ from

$$\widehat{P}_N = Pr[n > N] = \sum_{n=N+1}^{\infty} p_n = 1 - \sum_{n=0}^{N} p_n \qquad (5.49)$$

An alternative question of interest is: "Below what load limit will the system operate at least α-percent of the time?" Mathematically this says find an integer U_α such that

$$Pr[n \leq U_\alpha] \geq \alpha/100 > Pr[n \leq U_\alpha - 1] \qquad (5.50)$$

or

$$\sum_{n=0}^{U_\alpha - 1} p_n < \alpha/100 \leq \sum_{n=0}^{U_\alpha} p_n \qquad (5.51)$$

We will refer to U_α as the *alpha load limit*.

Other measures of performance involve averages or expected values. One such measure is the *offered load*. The offered load is defined as the mean number of arrivals occuring during a service time. If customers arrive at an average rate λ and the mean time to serve a customer is $\bar{t}$ the offered load is given by

$$a = \lambda \bar{t} \qquad (5.52)$$

Although this number is a dimensionless quantity it is often expressed in units called *Erlangs*.

The offered load is closely related to another measure of performance called *traffic intensity*. Traffic intensity can be viewed as the offered load per server. Thus for a c-channel parallel system where each channel has a mean service time of μ^{-1}, the traffic intensity is given by

$$\rho = a/c = \lambda/c\mu \qquad (5.53)$$

For more complex systems the definition of traffic intensity is not so straightforward. It may not be possible to meaningfully define it directly in terms of interarrival and service times. In such cases it must be related to the specific conditions of that queueing system.

In addition to the load offered to a system, we are also interested in the *carried load*. Carried load is that portion of the offered load which is not lost from the system. In other words carried load provides a measure of the customers seeking service who are ultimately satisfied. When the system has unlimited queue capacity and patient customers, the carried load is equal to the offered load. If $B(a)$ represents the probability that a customer in a system with an offered load a will be blocked or unable to obtain service, then the carried load is given by

$$a' = a[1 - B(a)] \tag{5.54}$$

In a standard multi-channel system under steady-state conditions the carried load is numerically equivalent to the mean number of busy servers.

The carried load is closely related to another standard measure of performance, the *utilization factor*. The utilization factor is the load carried per server under equilibrium conditions. For a c-channel system the utilization factor is given by

$$\rho' = \frac{a'}{c} \tag{5.55}$$

When a system has sufficient capacity that no calling customers are ever denied service and all channels have a mean service time of μ^{-1}, the utilization factor and traffic intensity are equivalent. That is

$$\rho' = \rho = \lambda/c\mu \tag{5.56}$$

The expected number of customers in the system can be calculated directly from the definition of mathematical expectation. The expected number in the system, including those being served, under steady-state conditions is given by

$$L = \sum_{n=0}^{\infty} np_n \tag{5.57}$$

The expected number of customers in the queue, exclusive of those actively engaged in service, can be calculated in similar fashion as

$$L_q = \sum_{n=c+1}^{\infty} (n - c) p_n \tag{5.58}$$

This expression assumes that no server in a c-channel system will remain idle when there are more than c customers present. When system specifications are more general we can determine the expected number in the queue by subtracting the expected number of busy servers from the expected number of customers in the system. That is

$$L_q = L - \sum_{i=0}^{c} i f_i \tag{5.59}$$

where f_i is the frequency with which i servers are occupied. For the special case of c identical channels in parallel

$$f_i = \begin{cases} p_i, & i < c \\ \sum_{j=c}^{\infty} p_j, & i = c \end{cases} \tag{5.60}$$

In this instance equation (5.59) reduces to (5.58) as follows:

$$L_q = L - \sum_{i=0}^{c-1} i p_i - c \sum_{i=c}^{\infty} p_i$$

$$= \sum_{n=0}^{\infty} n p_n - \sum_{n=0}^{c} n p_n - c \sum_{n=c+1}^{\infty} p_n$$

$$= \sum_{n=c+1}^{\infty} (n - c) p_n$$

In addition to the expected number in the system or in the queue one is often interested in the expected time it will take a customer to get through the different stages of the system. Fortunately there exists a simple relationship between the expected number of customers present under equilibrium conditions and their expected waiting times. Under surprisingly general conditions of arrival, queue, and service disciplines the following relationships hold

$$L = \lambda W \tag{5.61}$$
$$L_q = \lambda W_q \tag{5.62}$$

where W is the expected time required to process an arriving unit from the point of his arrival until his discharge from the system. W_q is the corresponding time spent in the queue while awaiting service. Equation (5.61) is referred to as *Little's formula* after the man who provided the first rigorous proof of that relationship (66). Later work by Jewell (53) provided a general set of sufficient conditions for this relationship to hold which go far beyond the standard Poisson input FCFS-type queues.

When some arriving units may not join the queue, because they are discouraged by the line length, the system is blocked or for other reasons, this strong relationship between expected line length and expected waiting time can still be preserved. In those cases we will define an *effective arrival rate* λ_e to include only those calling customers who actually gain admission to the system. The effective rate can be calculated by weighting the state dependent arrival rate by p_i and summing. That is

$$\lambda_e = \sum_{i=0}^{\infty} \lambda_i p_i \tag{5.63}$$

Once λ_e is known we can calculate the appropriate waiting times from

$$W = \frac{L}{\lambda_e} \tag{5.64}$$

and

$$W_q = \frac{L_q}{\lambda_e} \tag{5.65}$$

An intuitive explanation for these relationships goes as follows: Consider a customer who is just entering the service facility. As he takes an in-service position he glances backward to view the number of customers remaining in the queue. The average number he sees is L_q. But the time required to accumulate those customers must be equal to the average time our test customer spent in the queue. Since the arrival rate of entering customers is λ_e, it follows that $\lambda_e W_q$ must be equal to the average number in the queue.

Similar arguments follow for the total time in the system. Here we take the point of view of a departing customer who looks back to see an average of L customers left behind in the system. They must have accumulated during his through-put time. Therefore, $\lambda_e W$ must be equal to the average number in the system.

The various measures of performance we will use for the steady-state models of this chapter are summarized in Table 5-1.

TABLE 5-1

COMMON MEASURES OF PERFORMANCE FOR STEADY-STATE QUEUES

Symbol	Name	Definition
$\hat{P}_N$	load level	$\displaystyle\sum_{n=N+1}^{\infty} p_n$
U_α	alpha load limit	$\displaystyle\sum_{n=0}^{U_\alpha-1} p_n < \alpha/100 \le \sum_{n=0}^{U_\alpha} p_n$
a	offered load	$\lambda \bar{t}$ where $\bar{t}$ = average service time
a'	carried load	$a[1 - B(a)]$ where $B(a)$ = probability of blocking
ρ	traffic intensity	$a/c = \lambda/c\mu$ for multi-channel with identical channels
ρ'	utilization factor	a'/c
L	expected number in system	$\displaystyle\sum_{n=0}^{\infty} n\, p_n$
L_q	expected number in queue	$\displaystyle\sum_{n=c+1}^{\infty} (n - c)p_n$
W	expected time through system	L/λ
W_q	expected time in queue	L_q/λ
λ_e	effective arrival rate	$\displaystyle\sum_{i=0}^{\infty} \lambda_i p_i$

PROBLEMS

5.1 Consider each of the following as they relate to queueing models:

a) List the assumptions necessary to use the birth-death model to analyze a real system.

b) Distinguish among "transient," "stationary," and "steady-state."

c) What role does the Markov assumption play in the development of queueing models?

d) How would you change the state-variable definition in a system with a general service time distribution as opposed to exponential service time?

e) Suggest three general techniques one might employ for analyzing queueing systems.

5.2 Modify the basic birth-death postulates such that each arrival event signifies the arrival of two customers. The time between arrival events and the service time distribution remain exponential.

a) Construct a flow rate diagram for the process.

b) Form a transition rate matrix.

c) Develop an appropriate set of differential-difference equations.

5.3 Consider the problem of recruiting followers faced by the leader of a religious sect. He can handle a maximum of K followers. The rate at which he can attract new recruits is proportional to the number of followers currently in his group, i.e., $n\lambda$. However, outside influences are constantly at work to lure away members of his sect. The rate at which such appeals are successful is also proportional to the number of followers in the group, i.e., $n\mu$. He would like to be able to predict the size of his group.

a) Present arguments for and against modeling this as a birth-death process.

b) Assuming that the birth-death postulates are appropriate:

i) Construct a flow rate diagram for this system.

ii) Form a transition rate matrix.

iii) Develop an appropriate set of differential-difference equations.

iv) Suppose that he now has $K/2$ followers. Comment on the probability of ultimate success of his sect.

5.4 Consider a retail inventory control system in which demand occurs one unit at a time in exponential fashion at a rate λ per day. Replenishment time for all orders sent by the retailer to the wholesaler is exponentially distributed with a mean of $1/\mu$ days. The control policy states that when on-hand inventory eaches r units a replenishment order for Q will be placed. (Q is greater than r.) Customers are not willing to backorder.

a) Develop a flow rate diagram for this system.

b) Form a transition rate matrix.

c) Construct a proper set of differential-difference equations.

d) Find the steady-state solution.

5.5 Prove that the minimum of several exponentially distributed random variables is exponentially distributed.

5.6 Consider a service mechanism in which each service requires two distinct exponentially distributed service tasks with mean rates 2μ each. When a customer occupies either of these stages the service facility is blocked and new arrivals must queue until the customer in service is released from the second service operation. Arrivals occur in Poisson fashion at rate λ. The state of the system can then be defined by $p_{nj}(t)$ which gives the probability that n customers are present and the lead customer has j service tasks to complete.

a) Construct a flow rate diagram for this system.

b) Form a transition rate matrix.

c) Construct a proper set of differential-difference equations.

d) Suggest how you would find the steady-state probability for n in the system.

5.7 A computer system is either operating or down for repair. When operating there is a constant probability of breakdown $\lambda\,\Delta t$. Repair time is exponentially distributed with mean $1/\mu$. If the machine is currently down for repair, find the probability that it will be operating t hours from now.

5.8 Consider the problem of a group of people watching a window display. New observers arrive at a constant rate λ. Anyone currently in the crowd may leave at random at a constant rate μ per person.

a) Construct a flow rate diagram for this system.

b) Develop a set of differential-difference equations.

c) Find the equilibrium distribution for crowd size (if it exists).

5.9 Consider the problem of population growth in a rat colony. A single female produces male offspring at rate λ_m and female offspring at rate λ_f. Individual males die at rate μ_m; females die at rate μ_f. The state of the system is described by the number of males and females in the colony at time t.

a) Suppose that there are three females in the initial colony. Estimate the limiting probability of colony extinction. State all assumptions. (Hint: treat the number of females as a separate process.)

b) Develop a proper set of differential-difference equations to describe the total state of the colony at time t.

5.10 A time-sharing computer system services k remote terminals. If a terminal is using the central processor at time t, there is a constant probability $\mu\,\Delta t$ that its required processing will be completed at time $t + \Delta t$. An inactive terminal may call the CPU with probability $\lambda\,\Delta t$. The terminals operate independently of each other.

a) Construct a transition rate matrix.

b) Develop a set of differential-difference equations which describe the probability that n terminals will be in contact with the CPU at time t.

c) Find the steady-state probability that there will be n jobs in the CPU.

5.11 Reconsider the scenario in problem 5.3. Find the geometric transform for the probability $p_n(t)$.

5.12 Reconsider the scenario in problem 5.8. Find the geometric transform for $p_n(t)$.

6

PROTOTYPE STEADY-STATE MODELS

The last chapter presented a general solution for a class of problems known as birth and death processes. What we now must do is to relate this very powerful solution to commonly encountered system configurations. To accomplish this goal and to simultaneously illustrate how one can adapt the solution to new situations, we will develop a series of particular prototype models for a variety of system structures. These models are useful for their own sake whenever the system being studied matches the assumptions used in their formulation. However, it is often necessary to "bend and twist" them as reality intrudes. Hopefully the reader will learn such skills by carefully following the prototype developments.

SINGLE CHANNEL, UNLIMITED QUEUES

Let us next turn our attention to the (M/M/1) queueing system. This may be an appropriate model for systems such as a theater box office where a very large population of customers is seeking service in the form of ticket purchases from a single clerk. In terms of the birth-death parameters we have

$$\lambda_n = \lambda, \quad n \geq 0$$
$$\mu_n = \mu, \quad n \geq 1 \tag{6.1}$$

Applying equation (5.45) it follows that

$$p_0 = \left[1 + \sum_{n=1}^{\infty} \left(\frac{\prod_{i=0}^{n-1} \lambda_i}{\prod_{i=1}^{n} \mu_i} \right) \right]^{-1}$$

$$= \left[\sum_{n=0}^{\infty} \left(\frac{\lambda}{\mu} \right)^n \right]^{-1} \tag{6.2}$$

From the techniques of sum calculus we know that

$$\sum_{n=a}^{b} r^n = \frac{r^n}{r-1}\bigg|_{n=a}^{b+1}, \quad r \neq 1 \tag{6.3}$$

In this case the constant r is replaced by the ratio (λ/μ) and the upper limit b approaches infinity. If $\lambda/\mu \geq 1$, $p_0 = 0$ and there is no steady-state distribution. (That has intuitive appeal since an arrival rate which exceeds the maximum potential service rate can only lead to a queue which grows without bound.) If $\lambda/\mu < 1$, the sum can be evaluated as

$$\sum_{n=0}^{\infty} (\lambda/\mu)^n = \lim_{b \to \infty} \frac{(\lambda/\mu)^n}{\left(\dfrac{\lambda}{\mu} - 1\right)}\Bigg|_{n=0}^{b+1} = \frac{1}{1 - \lambda/\mu} \tag{6.4}$$

from which

$$p_0 = \left[\sum_{n=0}^{\infty} \left(\frac{\lambda}{\mu}\right)^n\right]^{-1} = 1 - \frac{\lambda}{\mu} \tag{6.5}$$

Finally, we can apply equation 5.42 to obtain the steady-state distribution

$$p_n = \frac{\displaystyle\prod_{i=0}^{n-1} \lambda_i}{\displaystyle\prod_{i=1}^{n} \mu_i} p_0$$

$$= \left(\frac{\lambda}{\mu}\right)^n \left(1 - \frac{\lambda}{\mu}\right)$$

$$= \rho^n (1 - \rho), \quad \rho < 1 \tag{6.6}$$

Thus the distribution for total customers present in the (M/M/1) model is seen to be geometric with a parameter $\rho = \lambda/\mu$.

With the distribution now in hand we can calculate the usual measures of performance. In the (M/M/1) system the offered load, carried load, traffic intensity, utilization factor, and expected number of customers engaged in service are all numerically equal. That is

$$a = a' = \rho = \rho' = (\lambda/\mu) \tag{6.7}$$

One could develop this expectation by arguing that there can be only zero or one customer in the service facility. The probability of zero is $f_0 = p_0 = (1 - \lambda/\mu)$. The probability of one in service is the probability that one or more customers are present in the system or $f_1 = 1 - p_0 = \lambda/\mu$. The expected number in service is then

$$S = \sum_{i=0}^{1} i f_i = \lambda/\mu \tag{6.8}$$

The expected number of customers in the system is

$$L = \sum_{n=0}^{\infty} np_n$$

$$= \sum_{n=0}^{\infty} n\rho^n(1 - \rho) \tag{6.9}$$

From the techniques of sum calculus we know that the following indefinite sum holds:

$$\sum nk^n = \frac{k^n[n(k - 1) - k]}{(k - 1)^2} \tag{6.10}$$

Therefore, replacing k with ρ we have

$$L = \lim_{b \to \infty} (1 - \rho) \left[\frac{\rho^n[n(\rho - 1) - \rho]}{(\rho - 1)^2} \right] \Bigg|_{n=0}^{b+1}$$

$$= \frac{\rho}{1 - \rho} = \frac{\lambda}{\mu - \lambda} \tag{6.11}$$

The expected number of customers in the queue can be obtained by direct calculation or by subtracting the expected number in service from the expected number in the system. It then follows that

$$L_q = L - S$$

$$= \frac{\rho}{1 - \rho} - \rho$$

$$= \frac{\rho^2}{1 - \rho} = \frac{\lambda^2}{\mu(\mu - \lambda)} \tag{6.12}$$

Expected waiting times are obtained by applying Little's formula. The expected total time to pass through the system is

$$W = L/\lambda$$

$$= \left(\frac{\lambda}{\mu - \lambda} \right) \Big/ \lambda$$

$$= \frac{1}{\mu - \lambda} = \frac{1}{\mu(1 - \rho)} \tag{6.13}$$

The expected delay time in queue is given by

$$W_q = L_q/\lambda$$

$$= \left(\frac{\lambda^2}{\mu(\mu - \lambda)} \right) \Big/ \lambda$$

$$= \frac{\lambda}{\mu(\mu - \lambda)} = \frac{\rho}{\mu(1 - \rho)} \tag{6.14}$$

Example.   Customers arrive at the express lane in a supermarket checkout area at a rate of fifteen per hour. Time studies reveal that the average time to check out a customer is two minutes. The analyst is willing to assume that the interarrival and service distributions can be approximated by exponential density functions. There is no problem with waiting space.

The appropriate model for the express lane is the (M/M/1) queueing model where $\lambda = 15$/hour and $\mu = 30$/hour. With those parameters, the average number of customers present will be

$$L = \frac{\lambda}{\mu - \lambda} = \frac{15}{30 - 15} = 1$$

The expected idle delay time experienced by a customer will be

$$W_q = \frac{\lambda}{\mu(\mu - \lambda)} = \frac{15}{30(30 - 15)}$$
$$= (1/30) \text{ hr.} = 2 \text{ minutes}$$

The total expected time for a customer to clear the system is

$$W = \frac{1}{\mu - \lambda} = (1/15) \text{ hr.} = 4 \text{ minutes}$$

MULTICHANNEL SYSTEMS

A logical extension to the single-channel model of the last section is the introduction of multiple servers. Initially we will confine our attention to the (M/M/c) system. The assumptions here are that:

1. Customers arrive in Poisson fashion at rate λ.
2. There are c identical servers or channel in parallel.
3. The service-time distribution for each channel is exponential with a mean of μ^{-1}.
4. Servers are noncooperative in the sense that one customer is never attended by more than a single server.
5. Arriving customers enter the first available open channel. When all channels are full they form a single queue with unlimited waiting space from which they are served in FCFS fashion.

This configuration could be diagrammed as in Figure 6-1. Such a model might accurately depict behavior at a tool crib with several clerks serving a large group of mechanics. It is also representative of the customer service counters for some airlines at new terminals such as the Dallas-Fort Worth terminal. Arriving passengers are asked to form a single queue from which they are served by the first available ticket agent.

The (M/M/c) model is a less accurate description of the typical grocery checkout or bank teller areas. In those systems customers commonly form multiple queues and may line-jump in an effort to expedite their service. However, from the point of view of total system performance the mathematical approximation of a single queue causes no difficulty. Attempting to account for multiple queues when all servers are identical is important only from the viewpoint of an individual customer. It has no

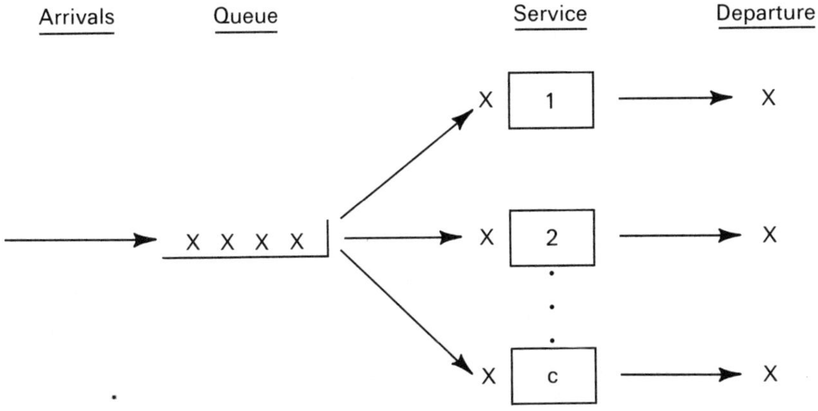

FIGURE 6-1   The (M/M/c) System

impact on gross performance measures such as average number of customers waiting and the utilization factor.

The steady-state probability distribution is easily constructed by
defining the set of λ_n and μ_n values. Under the Poisson arrival and
unlimited queue assumption, λ_n is a constant for all n. That is

$$\lambda_n = \lambda, \quad n = 0, 1, 2, \ldots \tag{6.15}$$

Service rate varies as a function of the number of customers present.
With one customer in the system only one server is active. The rate of
service is then μ. When two customers are present each is engaged in
service with a separate server. The total effective service rate is therefore
2μ and the corresponding departure time distribution is exponential with
a mean of $(2\mu)^{-1}$. The service rate continues to increase in linear fashion
with the number of customers present until all servers are busy. At that
point the service rate stabilizes at $c\mu$ where it remains as long as c or more
customers are present. To summarize

$$\mu_n = \begin{cases} n\mu, & 0 < n < c \\ c\mu, & n \geq c \end{cases} \tag{6.16}$$

From equation (5.45) we can calculate the probability that the system
will be empty.

$$p_0 = \left[1 + \sum_{n=1}^{\infty} \frac{\prod_{i=0}^{n-1} \lambda_i}{\prod_{i=1}^{n} \mu_i} \right]^{-1}$$

$$= \left[\sum_{n=0}^{c-1} \left(\frac{\lambda}{\mu}\right)^n \middle/ n! + \frac{(\lambda/\mu)^c}{c!} \sum_{n=c}^{\infty} \left(\frac{\lambda}{c\mu}\right)^{n-c} \right]^{-1} \tag{6.17}$$

Applying the techniques of sum calculus to the second summation in
equation (6.17) we have

$$\sum_{n=c}^{\infty} \left(\frac{\lambda}{c\mu}\right)^{n-c} = \sum_{j=0}^{\infty} \left(\frac{\lambda}{c\mu}\right)^{j}$$

$$= \frac{1}{1 - \dfrac{\lambda}{c\mu}} \quad \text{if } \lambda < c\mu \tag{6.18}$$

It then follows that

$$p_0 = \left[\sum_{n=0}^{c-1} \left(\frac{\lambda}{\mu}\right)^n \Big/ n! + \frac{(\lambda/\mu)^c}{c!\left(1 - \dfrac{\lambda}{c\mu}\right)} \right]^{-1} \tag{6.19}$$

and

$$p_n = \frac{\displaystyle\prod_{i=0}^{n-1}\lambda_i}{\displaystyle\prod_{i=1}^{n}\mu_i} p_0$$

$$= \begin{cases} \dfrac{(\lambda/\mu)^n}{n!}p_0, & 0 \le n \le c \\[3ex] \dfrac{(\lambda/\mu)^n}{c!c^{n-c}}p_0, & n > c \end{cases} \tag{6.20}$$

if $\lambda < c\mu$.

Again a steady-state solution exists only if the traffic intensity $\rho = \lambda/c\mu$ is less than unity. This means that the arrival rate must be less than the maximum potential service rate as is intuitively obvious. If $\rho \ge 1$ all p_n values are zero which indicates that there is no steady-state solution. Physically this implies that the queue will in the limit grow without bound. However, it should be noted that $\rho \ge 1$ does not exclude the possibility of calculating a time-varying probability distribution $p_n(t)$ as will be demonstrated in the next chapter.

With the distribution $\{p_n\}$ in hand we can now calculate the usual measures of performance. Since no customers are lost due to blocking, the offered load and carried load are identical.

$$a = a' = \lambda \bar{t} = \lambda/\mu \tag{6.21}$$

Similarly, the traffic intensity and utilization factor are also identical.

$$\rho = \rho' = \lambda/c\mu \tag{6.22}$$

The expected number of customers in the queue is calculated by using the indefinite sum identity in equation (6.10) as

$$L_q = \sum_{n=c}^{\infty} (n - c)\, p_n$$

$$= \sum_{n=c}^{\infty} (n - c)\, \frac{(\lambda/\mu)^c}{c!}\, (\lambda/c\mu)^{n-c} p_0$$

$$= \frac{(\lambda/\mu)^c}{c!} p_0 \sum_{n=c}^{\infty} (n - c)(\lambda/c\mu)^{n-c}$$

$$= \frac{(\lambda/\mu)^c}{c!} p_0 \sum_{j=0}^{\infty} j(\lambda/c\mu)^j$$

$$= \frac{(\lambda/\mu)^c}{c!} p_0 \left\{ \frac{(\lambda/c\mu)^j [j(\lambda/c\mu - 1) - \lambda/c\mu]}{(\lambda/c\mu - 1)^2} \bigg|_{j=0}^{\infty} \right\}$$

$$= \frac{(\lambda/\mu)^c \rho p_0}{c!(1 - \rho)^2} \tag{6.23}$$

where $\rho = \lambda/c\mu < 1$.

The mean number of busy servers, and consequently the expected number of customers actively engaged in service, is numerically equal to the carried load $a' = \lambda/\mu$. It therefore follows that the total number of customers in the system can be calculated as the sum of the expected number in the queue and the expected number in service.

$$L = L_q + \frac{\lambda}{\mu} \tag{6.24}$$

The mean waiting times are calculated directly from Little's formula.

$$W_q = L_q/\lambda \tag{6.25}$$

and

$$W = L/\lambda = \frac{L_q}{\lambda} + \frac{1}{\mu}$$

$$= W_q + \frac{1}{\mu} \tag{6.26}$$

Although the (M/M/c) model appears awkward to use for hand calculations, it is easily solved by the digital computer. A general purpose solution program which gives numerical values for this and several other models is contained in Appendix B.

Example.   The customer service counter for Alpha Airways is currently staffed by two agents. Arriving passengers are asked to form a single queue from which they will be served on a FCFS basis. The service counter records show that they serve an average of twenty-four customers per hour. The mean service time per customer has been measured at two minutes. Management is considering the possibility of reducing the staff to a single agent at the counter. The labor union representing the agents has countered that proposal by suggesting that in the interest of improved customer service the staff should really be expanded to three agents. The problem facing the analyst is to compare the three alternatives. Both sides are willing to assume that arrivals follow a Poisson process and that service times are exponentially distributed.

The appropriate model to use for the current two-agent configuration is (M/M/2) with $\lambda = 24$ per hour, $\mu = 30$ per hour. The fraction of time that both agents are expected to be idle is equal to the probability that no customers are present. From equation (6.19) that fraction is

$$p_0 = \left[\sum_{n=0}^{1} \frac{\left(\frac{24}{30}\right)^n}{n!} + \frac{\left(\frac{24}{30}\right)^2}{2\left(1 - \frac{24}{60}\right)} \right]^{-1}$$

$$= [1 + 0.8 + 0.5333]^{-1} = 0.43$$

One agent will be idle whenever one customer is present. The fraction of time this will occur is

$$p_1 = \left(\frac{24}{30}\right) p_0 = 0.344$$

Both agents will be busy whenever two or more customers are present. That is

$$Pr\ (Busy) = \sum_{n=2}^{\infty} p_n = 1 - p_0 - p_1 = 0.226$$

The expected number of customers in the queue is given by equation (6.23)

$$L_q = \frac{\left(\frac{24}{30}\right)^2 \left(\frac{24}{60}\right)(0.43)}{2\left(1 - \frac{24}{60}\right)} = 0.153$$

The expected number in the system is then

$$L = L_q + \frac{\lambda}{\mu} = 0.153 + 0.8 = 0.953$$

The corresponding waiting times are

$$W_q = L_q/\lambda = \frac{0.153}{24}$$
$$= 0.006375\ hr. = 0.3825\ min.$$
$$W = L/\lambda = \frac{0.953}{24}$$
$$= 0.397\ hr. = 2.3825\ min.$$

Idle agent-hours per eight-hour shift can be calculated from

$$I = (c - \lambda/\mu)(8)$$
$$= 1.2(8) = 9.6\ man\ hours/shift$$

Similar measures of performance can be calculated for the single agent and the three agent staffing policies by replacing c with 1 and 3, respectively. The results of those calculations are summarized below.

	Current System 2 agents	Management Proposal 1 agent	Union Proposal 3 agents
p_0	0.43	0.2	0.44
W_q	0.3825 min.	8.0 min.	0.1043 min.
W	2.3825 min.	10.0 min.	2.1043 min.
Idle agent hours	9.6 hrs.	1.6 hrs.	17.6 hrs.

A cursory examination of the table reveals that management's proposal will reduce the idle agent hours per shift by 83 percent at the expense of increasing the waiting time per customer by 320 percent. The union proposal will improve customer service by reducing the total wait time by 12 percent at the expense of increasing agent idle hours by 83 percent. The analyst can now retire from the scene while the negotiating teams ponder the impact of their alternatives.

TRUNCATED SYSTEMS

Although it may sometimes be mathematically convenient to assume that a queueing system has infinite capacity, as a practical matter very few physical systems can really meet that specification. The rationalization used is to argue that even though we recognize that few systems really have infinite capacity, their capacity may be sufficiently large relative to the load range of interest that capacity restrictions can be safely ignored. Our interest in this section concerns those systems for which infinite capacity is not a safe assumption.

Examples of systems with limited queue capacity surround us in daily life. Telephone switching systems have only a limited number of lines available for calls. Hospital emergency rooms have only a small number of beds to accommodate incoming patients. Airline terminals have a limited space in which to park aircraft for loading and unloading. In each case customers who arrive when the system is saturated either take their business elsewhere or must be accommodated by special means. What we now seek are models to help analyze and design such constrained systems.

When a system reaches its capacity or truncation point the effect is to reduce the arrival rate to zero until such time as a customer is served to again make queue space available. In terms of the birth-death parameters this means that for a truncation point $n = k$

$$\lambda_n = \begin{cases} \lambda, & n < k \\ 0, & n \geq k \end{cases} \qquad (6.27)$$

The proper specification for μ_n will depend upon whether one is analyzing a single or multiple channel system.

M/M/1/k

Consider first the single server system (M/M/1/k). Following the usual calculations from equations (5.42) and (5.45), we have

$$p_0 = \left[1 + \sum_{n=1}^{\infty} \frac{\prod_{i=0}^{n-1} \lambda_i}{\prod_{i=1}^{n} \mu_i} \right]^{-1}$$

$$= \left[\sum_{n=0}^{k} \left(\frac{\lambda}{\mu} \right)^n \right]^{-1}$$

$$= \frac{1 - \left(\frac{\lambda}{\mu} \right)}{1 - \left(\frac{\lambda}{\mu} \right)^{k+1}} = \frac{1 - \rho}{1 - \rho^{k+1}}, \qquad \rho \neq 1 \tag{6.28}$$

and

$$p_n = \frac{\prod_{i=0}^{n-1} \lambda_i}{\prod_{i=1}^{n} \mu_i} p_0$$

$$= \left(\frac{\lambda}{\mu} \right)^n p_0 = \frac{\rho^n (1 - \rho)}{1 - \rho^{k+1}}, \qquad n = 0, 1, \ldots, k \tag{6.29}$$

It is important to note here that it is no longer necessary for λ to be less than μ for the steady-state solution to exist. The sum in the expression for p_0 is a finite sum which will always have a finite value as long as λ is finite. The intuitive reason why the system can have a steady-state solution even in the face of an arrival rate which exceeds the service rate is that some of the arriving customers will be denied service due to being blocked by a saturated queue. The *effective arrival rate,* defined as that portion of the total arrival stream which does get service, will, in fact, be less than the service rate. Customers are denied service when the system is in state k. It therefore follows that the effective arrival rate is

$$\lambda_e = \lambda(1 - p_k) \tag{6.30}$$

Because the current system is truncated, the offered load, defined as the number of arrivals per service time, and carried load are no longer equal. Since the system is single channel, the offered load and traffic intensity, defined as the offered load per channel, are, however, equal.

$$a = \rho = \lambda/\mu \tag{6.31}$$

The carried load, being that portion of the offered load which is served, and the utilization factor, being the carried load per channel, are given by

$$a' = \rho' = \lambda_e/\mu$$

$$= \frac{\lambda}{\mu}(1 - p_k) = \rho(1 - p_k) \tag{6.32}$$

The average number of customers in the system is calculated in straightforward fashion as

$$L = \sum_{n=0}^{k} np_n$$

$$= \frac{1 - \rho}{1 - \rho^{k+1}} \sum_{n=0}^{k} n\rho^n$$

$$= \frac{1 - \rho}{1 - \rho^{k+1}} \left[\frac{\rho[n(\rho - 1) - \rho]}{(\rho - 1)^2} \bigg|_{n=0}^{k+1} \right]$$

$$= \frac{\rho - (k + 1)\rho^{k+1} + k\rho^{k+2}}{(1 - \rho)(1 - \rho^{k+1})} \tag{6.33}$$

The average number in the queue can be obtained by subtracting the average number in service (which is numerically equal to the carried load) from the average number in the system. That is

$$L_q = L - a'$$

$$= L - \rho(1 - p_k) \tag{6.34}$$

The corresponding expected waiting times are obtained from Little's formula. The only modification is that the proper arrival rate to use is the effective rate λ_e. It follows that

$$W_q = L_q/\lambda_e$$

$$= \frac{L_q}{\lambda(1 - p_k)} \tag{6.35}$$

and

$$W = L/\lambda_e$$

$$= \frac{L}{\lambda(1 - p_k)} \tag{6.36}$$

Note that the total time in the system could also be calculated as the sum of time in queue and time in service as

$$W = W_q + \frac{1}{\mu} \tag{6.37}$$

M/M/c/c

Let us next consider a special type of truncated multi-channel system in which the truncation point is equal to the number of channels. This system has no waiting room. When all channels are busy arriving customers are denied service and must seek satisfaction elsewhere. Certain types of telephone traffic systems with no provision for delaying blocked calls exhibit such behavior.

The specification of λ_n and μ_n combines the earlier truncation and multi-channel examples.

$$\lambda_n = \begin{cases} \lambda, & 0 \le n < c \\ 0, & n \ge c \end{cases} \tag{6.38}$$

$$\mu_n = \begin{cases} n\mu, & 1 \le n \le c \\ 0, & \text{elsewhere} \end{cases} \tag{6.39}$$

Using these parameters in equations (5.42) and (5.45) it follows that

$$p_0 = \left[1 + \sum_{n=1}^{c} \frac{\prod_{i=0}^{n-1} \lambda_i}{\prod_{i=1}^{n} \mu_i} \right]^{-1}$$

$$= \left[\sum_{n=0}^{c} \frac{(\lambda/\mu)^n}{n!} \right]^{-1} \tag{6.40}$$

and

$$p_n = \frac{\prod_{i=0}^{n-1} \lambda_i}{\prod_{i=1}^{n} \mu_i} p_0$$

$$= \frac{(\lambda/\mu)^n/n!}{\sum_{j=0}^{c} (\lambda/\mu)^j/j!}$$

$$= \frac{(c\rho)^n/n!}{\sum_{j=0}^{c} (c\rho)^j/j!} \tag{6.41}$$

where $\rho = \lambda/c\mu$.

The computational burden is eased somewhat for this model by observing that equation (6.41) has components which closely resemble functions of the Poisson distribution. By multiplying the numerator and denominator by $e^{-c\rho}$ we are led to

$$p_n = \frac{e^{-c\rho}(c\rho)^n/n!}{\sum_{j=0}^{c} e^{-c\rho}(c\rho)^j/j!} \tag{6.42}$$

This indicates that the steady-state distribution for the (M/M/c/c) system is simply a truncated Poisson distribution with parameter $c\rho = \lambda/\mu$. Its values are readily obtainable from tables of the Poisson distribution.

The system is saturated whenever all channels are busy. That is

$$p_c = \frac{e^{-c\rho}(c\rho)^c/c!}{\sum_{j=0}^{c} e^{-c\rho}(c\rho)^j/j!} \tag{6.43}$$

This very famous formula is known as *Erlang's loss formula*. In its original context of telephone traffic p_c is the portion of incoming calls which will receive a busy signal.

M/M/c/k

The next obvious extension is to consider a multichannel truncated system with waiting space such that $k > c$. Following our usual procedure

$$\lambda_n = \begin{cases} \lambda, & 0 \leq n < k \\ 0, & n \geq k \end{cases} \tag{6.44}$$

$$\mu_n = \begin{cases} n\mu, & 1 \leq n \leq c \\ c\mu, & c < n \leq k \end{cases} \tag{6.45}$$

from which

$$p_0 = \left[1 + \sum_{n=1}^{k} \frac{\prod_{1=0}^{n-1} \lambda_i}{\prod_{i=1}^{n} \mu_i} \right]^{-1}$$

$$= \left[\sum_{n=0}^{c-1} (\lambda/\mu)^n/n! + \frac{(\lambda/\mu)^c}{c!} \sum_{n=c}^{k} (\lambda/c\mu)^{n-c} \right]^{-1}$$

$$= \left[\sum_{n=0}^{c-1} (\lambda/\mu)^n/n! + \frac{(\lambda/\mu)^c}{c!} \left(\frac{1 - \rho^{k-c+1}}{1 - \rho} \right) \right]^{-1} \tag{6.46}$$

and

$$p_n = \begin{cases} \dfrac{(\lambda/\mu)^n}{n!} p_0, & 0 \leq n \leq c \\[2ex] \dfrac{(\lambda/\mu)^n}{c!c^{n-c}} p_0, & c < n \leq k \end{cases} \tag{6.47}$$

Note that the expression for p_n is identical in form to the standard multichannel model of equation (6.20). They differ only in the p_0 term.

It is again true that λ can exceed $c\mu$ because of the effect of truncation. The effective arrival rate, representing those calling customers who actually receive service is calculated as

$$\lambda_e = \lambda(1 - p_k) \tag{6.48}$$

The effective arrival rate is less than the service rate, $\lambda_e < c\mu$, under steady-state conditions.

From this information we can determine the offered load, carried load, traffic intensity and utilization factor as

$$a = \lambda/\mu \tag{6.49}$$

$$a' = (\lambda/\mu)(1 - p_k) \tag{6.50}$$

$$\rho = \lambda/c\mu \tag{6.51}$$

$$\rho' = (\lambda/c\mu)(1 - p_k) = \rho(1 - p_k) \tag{6.52}$$

Expected queue length is calculated from the definition and the use of sum calculus as

$$L_q = \sum_{n=c}^{k} (n - c)p_n$$

$$= \frac{(\lambda/\mu)^c}{c!} p_0 \sum_{n=c}^{k} (n - c)(\lambda/c\mu)^{n-c}$$

$$= \frac{(\lambda/\mu)^c}{c!} p_0 \left\{ \left. \frac{(\lambda/c\mu)^j [\, j(\lambda/c\mu - 1) - \lambda/c\mu]}{(\lambda/c\mu - 1)^2} \right|_{j=0}^{k-c+1} \right\}$$

$$= \frac{(\lambda/\mu)^c \rho p_0}{c!(1 - \rho)^2} \left\{ 1 - \left[(k - c)(1 - \rho) + 1 \right] \rho^{k-c} \right\} \qquad (6.53)$$

Since the carried load is equivalent to the mean number of busy servers it follows that we can calculate the expected number in the system as

$$L = L_q + a'$$

$$= L_q + \frac{\lambda}{\mu}(1 - p_k) \qquad (6.54)$$

With the effective arrival rate, expected number in the queue and expected number in the system now known to us, it is a simple matter to establish the mean waiting times. From Little's formula it follows that

$$W_q = L_q/\lambda_e$$

$$= \frac{L_q}{\lambda(1 - p_k)} \qquad (6.55)$$

and

$$W = L/\lambda_e$$

$$= L_q + 1/\mu \qquad (6.56)$$

Example. A consultant has been hired to analyze the needs for loading dock facilities at an air freight terminal. The present terminal has four docks on the main concourse. Any aircraft which arrive when all docks are full are assigned to less desirable docks on the back concourse. At the present time nearly 50 percent of the arriving flights are diverted to the back concourse. The average service time per aircraft is two hours on the front concourse and three hours on the back.

Two proposals are to be considered. The first proposal is to expand the main concourse by adding docks so that at least 80 percent of the arriving aircraft can be served there with the remainder being diverted to the back concourse. The second proposal is to keep the present number of docks on the main concourse but to add a holding area which can accomodate up to eight aircraft. Only when the holding area is full will aircraft be diverted to the back concourse.

Under current operations the system can be defined as the main concourse and treated as an (M/M/c/c) queue. From equation (6.43) and the data given

$$p_4 = \frac{(4\rho)^4 e^{-4\rho}/4!}{\displaystyle\sum_{j=0}^{4} e^{-4\rho}(4\rho)^j/j!} \approx 0.5$$

From the Poisson tables (or the Erlang loss tables) this value for p_4 is obtained when $4\rho = \lambda/\mu = 6.5$. The implied aircraft arrival rate is then

$$\lambda = 6.5\,\mu = 6.5/2 = 3.25 \text{ aircraft/hour}$$

Assuming that the expanded dock system will have the same offered load, what we seek is a value for c such that

$$p_c = \frac{(6.5)^c e^{-6.5}/c!}{\displaystyle\sum_{j=0}^{c} e^{-6.5}(6.5)^j/j!} \le 0.2$$

Again using the tables for parameter $c\rho = 6.5$, we find that

$$p_7 = 0.2174 \text{ and } p_8 = 0.1501$$

Therefore it is necessary to have a total of eight docks, i.e., four new ones on the main concourse to meet the desired 80 percent service objective.

The addition of a holding area for eight aircraft suggests a truncated system model with a total capacity of twelve, i.e., (M/M/4/12). From equations (6.46) and (6.47) and assuming the same offered load $a = \lambda/\mu = 6.5$ we have

$$p_0 = \left[\sum_{n=0}^{3} (6.5)^n/n! + \frac{(6.5)^4}{4!}\left(\frac{1 - (1.625)^9}{1 - 1.625}\right) \right]^{-1} \approx .0001$$

and

$$p_{12} = \frac{(6.5)^{12}}{4!\,4^8}\, p_0 = .3864$$

Under this system more than 38 percent of the arriving aircraft will still be diverted to the back concourse. The average queue in the front course holding area will be from equation (6.53)

$$L_q = \frac{(6.5)^4(1.625)(.0001)}{4!\,(1 - 1.625)^2}\left\{ 1 - \left[8(1 - 1.625) + 1\right](1.625)^8 \right\}$$

$$= 2.3627$$

Furthermore, the expected delay time awaiting service will be from equation (6.55)

$$W_q = L_q/\lambda(1 - p_{12})$$
$$= 2.3627/3.25(1 - .3864) = 1.185 \text{ hours}$$

When the two-hour service time is added, this means that the total expected processing time for those aircraft served in the front concourse will exceed back concourse service by 0.185 hours. Comparison of other performance measures is left as an exercise for the reader.

FINITE POPULATIONS

The next extension to the basic birth-death model concerns a limited number of customers in the calling population. A common application for this model is the machine servicing problem where one or more operators may service a group of m machines. A machine enters the service queue when it breaks down or otherwise requires attention. If there are n machines in the queue, there are $(m - n)$ potential customers remaining in the calling population.

The assumption made is that the elapsed time from completion of service on a particular machine until it next has need of service is exponentially distributed with parameter λ. In the machine repair context $1/\lambda$ is the mean time between failures for a particular machine. At any arbitrary point in time, the time to the next demand for service will be the minimum of the remaining running times for the $(m - n)$ machines not currently in the repair system. From the property of the minima of exponentially distributed random variables this distribution will also be exponential with parameter $\lambda_n = (m - n)\lambda$. An input stream with these characteristics is sometimes called *quasirandom input*.

(M/M/1/m/m)

Let us first consider a finite population system with a single server. Since the number of customers in the system can never exceed the total number of customers available in the calling population, this is also a limited capacity model where the capacity is equal to m.

The parameters for the steady-state birth-death equations are defined as

$$\lambda_n = \begin{cases} (m - n)\lambda & 0 \leq n < m \\ 0 & n \geq m \end{cases} \tag{6.57}$$

$$\mu_n = \begin{cases} \mu & 1 \leq n \leq m \\ 0 & n > m \end{cases} \tag{6.58}$$

Substituting these values in equations (5.42) and (5.45) we can determine the distribution for number of machines in the repair system as follows:

$$p_0 = \left[1 + \sum_{j=1}^{m} \frac{\displaystyle\prod_{i=0}^{j-1} \lambda_i}{\displaystyle\prod_{i=1}^{j} \mu_i} \right]^{-1}$$

$$= \left[\sum_{j=0}^{m} \frac{m!}{(m - j)!} \left(\frac{\lambda}{\mu}\right)^j \right]^{-1} \tag{6.59}$$

and

$$p_n = \frac{\displaystyle\prod_{i=0}^{n-1} \lambda_i}{\displaystyle\prod_{i=1}^{n} \mu_i} p_0$$

$$= \frac{m!}{(m - n)!} \left(\frac{\lambda}{\mu}\right)^n \bigg/ \sum_{j=0}^{m} \frac{m!}{(m - j)!} \left(\frac{\lambda}{\mu}\right)^j, \quad 1 \leq n \leq m \tag{6.60}$$

By judicious algebraic manipulation it is possible to express this distribution in a form easily located in standard Poisson tables. Let us cancel the term $(m!)$ in numerator and denominator then multiply both $e^{-\mu/\lambda}\left(\dfrac{\mu}{\lambda}\right)^m$. We can then write equation (6.60) as

$$p_n = \frac{e^{-\mu/\lambda}\left(\dfrac{\mu}{\lambda}\right)^{m-n}}{(m-n)!} \Bigg/ \sum_{j=0}^{m} \frac{e^{-\mu/\lambda}\left(\dfrac{\mu}{\lambda}\right)^{m-j}}{(m-j)!} \tag{6.61}$$

By shifting the origin it follows that

$$p_{m-i} = \frac{e^{-\mu/\lambda}\left(\dfrac{\mu}{\lambda}\right)^{i}\Big/i!}{\displaystyle\sum_{j=0}^{m} e^{-\mu/\lambda}\left(\dfrac{\mu}{\lambda}\right)^{j}\Big/j!} \tag{6.62}$$

which is simply a truncated Poisson distribution with parameter (μ/λ).

Some of the performance measures for this model are more difficult to calculate than with previously discussed birth-death systems. For example, the offered load a and carried load a', although equal for this model, cannot be specified without first calculating the effects of that load on the system. Furthermore, the delay time for an arriving unit as viewed from the position of that arriving unit is different from the average delay as viewed from the position of an outside observer of the system. However, the expected line length calculations do remain routine.

The expected number of customers in various portions of the system are related as follows:

Total customers = number in queue
$$\qquad\qquad + \text{number in repair} + \text{number running} \tag{6.63}$$

The expected number of total customers is simply the number of customers in the calling population, m. The number in repair in a single-channel system is either zero or one. The probability that one will be in repair is equal to the probability that one or more are currently broken or idle, i.e., $(1 - p_0)$. Therefore, the expected number in repair is

$$E(I) = \sum_{I=0}^{1} If(I)$$
$$= (1 - p_0) \tag{6.64}$$

The number of machines running is related to the number idle. If r are running, then $(m - r)$ must be idle, i.e., waiting to be served. From equation (6.62) the probability $g(r)$ that r will be running is given by the truncated Poisson distribution.

$$g(r) = p_{m-r} = \frac{e^{-\mu/\lambda}\left(\dfrac{\mu}{\lambda}\right)^{r}\Big/r!}{\displaystyle\sum_{j=0}^{m} e^{-\mu/\lambda}\left(\dfrac{\mu}{\lambda}\right)^{j}\Big/j!} \tag{6.65}$$

The expected number of machines running is then

$$E(r) = \sum_{r=0}^{m} rg(r)$$

$$= \frac{\sum_{r=0}^{m} re^{-\mu/\lambda}\left(\frac{\mu}{\lambda}\right)^{r}\Big/r!}{\sum_{j=0}^{m} e^{-\mu/\lambda}\left(\frac{\mu}{\lambda}\right)^{j}\Big/j!} \tag{6.66}$$

Let the Poisson distribution be represented by $p(r) = \dfrac{e^{-\tau}(\tau)^{r}}{r!}$ and

$P(r) = \sum\limits_{i=0}^{r} p(i)$. A compact form for $E(r)$ can be developed by noting the following relationship:

$$rp(r) = \frac{re^{-\tau}(\tau)^{r}}{r!} = \frac{\tau e^{-\tau}(\tau)^{r-1}}{(r-1)!} = \tau p(r-1) \tag{6.67}$$

where τ is the mean of the distribution. It follows that equation (6.66) can be written as

$$\sum_{r=0}^{m} \frac{re^{-\mu/\lambda}\left(\frac{\mu}{\lambda}\right)^{r}\Big/r!}{P(m)} = \frac{\mu}{\lambda}\sum_{r=1}^{m} p(r-1)/P(m)$$

$$= \frac{\mu}{\lambda}\left[\sum_{i=0}^{m} p(i) - p(m)\right]P(m)$$

$$= \frac{\mu}{\lambda}\left[1 - \frac{p(m)}{P(m)}\right] \tag{6.68}$$

but

$$\frac{p(m)}{P(m)} = p_0$$

from equation (6.62). Therefore we have

$$E(r) = \frac{\mu}{\lambda}(1 - p_0) \tag{6.69}$$

Returning to equation (6.63) we can now calculate the expected number in the queue and in the system as

$$L_q = m - (1 - p_0) - \frac{\mu}{\lambda}(1 - p_0)$$

$$= m - \left(1 + \frac{\mu}{\lambda}\right)\left(1 - p_0\right) \tag{6.70}$$

and

$$L = m - \frac{\mu}{\lambda}(1 - p_0) \tag{6.71}$$

The effective arrival rate can be determined from the product of the expected number of machines running and their constant breakdown rate.

$$\lambda_e = \lambda E(r) = \mu(1 - p_0) \tag{6.72}$$

Using this quantity and Little's formula we are now in a position to calculate the usual delay times.

$$W_q = L_q/\lambda_e$$

$$= \frac{m}{\mu(1 - p_0)} - \frac{1}{\mu}(1 + \frac{\mu}{\lambda}) \tag{6.73}$$

and

$$W = L/\lambda_e$$

$$= \frac{m}{\mu(1 - p_0)} - \frac{1}{\lambda} \tag{6.74}$$

The calculation of load factors requires more thought for the quasi-random input model than for those previously treated. In this context it is useful to consider the load offered by each machine currently operating. Let

$$\hat{a} = \lambda/\mu \tag{6.75}$$

be the offered load per potential customer. The total offered load under steady-state conditions is then the product of the load per source and the average number of machines running. That is

$$a = \hat{a}E(r)$$

$$= \frac{\lambda}{\mu} \left(\frac{\mu}{\lambda}\right)(1 - p_0) = (1 - p_0) \tag{6.76}$$

Since carried load and offered load are equal for this system, $(1 - p_0)$ is also the value of the carried load, traffic intensity, and utilization factor.

Arriving Customer Distribution

The probability distribution $\{p_n\}$ which we have developed for this and all other birth-death models in this chapter is sometimes called an outside observer's distribution. These are the state probabilities representing the portion of time the system occupies each state as witnessed by one who stands outside the system and records its behavior. The *arriving customers distribution* may, however, be something quite different. If one takes the viewpoint of an arriving customer, he is interested only in the state distribution at those instants at which a request for service takes place. For Poisson input streams these two distributions are identical. For quasi-random input, as well as other arrival patterns, they are not identical.

The arriving customer's distribution for quasi-random input bears a very simple relationship to the outside observer's distribution. In an m-customer birth-death queueing system the distribution seen by an arriving customer is the same as the outside observer's distribution for an

$(m - 1)$ customer system with the same parameters. (For a proof of this relationship see Cooper (23:82). Let $\{q_n(m)\}$ be the arriving customer's distribution and $\{p_n(m)\}$ be the outside observer's distribution for an m-customer system. Then

$$q_n(m) = p_n(m - 1) \tag{6.77}$$

Consider a three-machine system served by a single mechanic in which the mean time between failures on each machine is twelve hours and the mean time to service a broken machine is six hours. Based on a 24-hour day this implies that the parameters are

$$m = 3, \qquad \lambda = 2/\text{day}, \qquad \mu = 4/\text{day}$$

By applying equation (6.60) we can determine the outside observer's distribution to be

$$\begin{aligned}
p_0(3) &= .2105 & p_i(3) &= 0, \quad i > 3 \\
p_1(3) &= .3158 \\
p_2(3) &= .3158 \\
p_3(3) &= .1579
\end{aligned}$$

The arriving customer's distribution is from the same equations

$$\begin{aligned}
q_0(3) &= p_0(2) = 0.4 & q_i(3) &= 0, \quad i > 2 \\
q_1(3) &= p_1(2) = 0.4 \\
q_2(3) &= p_2(2) = 0.2
\end{aligned}$$

Note that there is no chance that an arriving customer will find three customers ahead of him even though from the outside observer's viewpoint three customers will be present nearly 16 percent of the time.

The interpretation of waiting time also requires modification. The expected time through the system from equation (6.74) is

$$W = \frac{m}{\mu(1 - p_0)} - \frac{1}{\lambda}$$

$$= \frac{3}{4(1 - .2105)} - \frac{1}{2} = 0.45 \text{ days}$$

This represents the time it will take the last machine in the queue to clear the system when the outside observer looks at the system at some random point in time.

From the viewpoint of an arriving customer, however, one is not interested in an observation at any random point in time but rather at the point in time at which the arrival event takes place. If the arriving customer finds no customers in front of him, his expected time to clear the system is simply his expected service time $1/\mu$. If n-customers are in front of him his expected time is $(n + 1)/\mu$. In terms of conditional expectations the expected times in the system from the viewpoint of an arriving customer are

$$W' = \sum_{n=0}^{m-1} E(w'|n)q_n(m)$$

$$= \sum_{n=0}^{m-1} \frac{(n+1)}{\mu} q_n(m) \tag{6.78}$$

and

$$W'_q = \sum_{n=1}^{m-1} \frac{n}{\mu} q_n(m) \tag{6.79}$$

Applying equation (6.78) to the three-machine problem under consideration we have an arriving customer's expected time in the system of

$$W' = \sum_{n=0}^{m-1} \frac{(n+1)}{\mu} q_n(m)$$

$$= \frac{1}{4}(0.4) + \frac{2}{4}(0.4) + \frac{3}{4}(0.2)$$

$$= 0.45 \text{ days}$$

This value is the same as that obtained from the outside observer's viewpoint. However, if one simply applies equation (6.74) to the two-machine case, as intimated by the relationship between the q and p distributions, the expected time in the system would be estimated as 0.33 days. Since there is obviously a difference in the two values, the analyst should be very careful about defining the performance measures he is using to compare systems.

(M/M/c/m/m)

It is now a simple matter to extend the limited population model to cover systems with multiple servers. The parameters are now

$$\lambda_n = \begin{cases} (m-n)\lambda & 0 \le n < m \\ 0 & n \ge m \end{cases} \tag{6.80}$$

$$\mu_n = \begin{cases} n\mu & 1 \le n \le c \\ c\mu & c < n \le m \\ 0 & n > m \end{cases} \tag{6.81}$$

Substituting these values in equations (5.42) and (5.45) we are led to

$$p_0 = \left[\sum_{j=0}^{c-1} \frac{m!}{(m-j)!j!} \left(\frac{\lambda}{\mu}\right)^j + \sum_{j=c}^{m} \frac{m!}{(m-j)!c!c^{j-c}} \left(\frac{\lambda}{\mu}\right)^j \right]^{-1} \tag{6.82}$$

and

$$p_n = \begin{cases} \dfrac{m!}{(m-n)!n!} \left(\dfrac{\lambda}{\mu}\right)^n p_0, & 0 < n \le c \\[4mm] \dfrac{m!}{(m-n)!c!c^{n-c}} \left(\dfrac{\lambda}{\mu}\right)^n p_0, & c < n \le m \end{cases} \tag{6.83}$$

The expected number in the system and expected number in the queue are calculated directly from the definitions.

$$L_q = \sum_{n=c}^{m} (n - c)p_n \tag{6.84}$$

and

$$L = \sum_{n=0}^{m} np_n$$

$$= m - \sum_{r=0}^{m} rp_{m-r} \tag{6.85}$$

The outside observer waiting times are obtained from Little's formula and the effective arrival rate. The effective arrival rate is the product of the failure rate of individual machines and the expected number of machines running.

$$\lambda_e = \lambda E(r) = \lambda \sum_{r=0}^{m} rp_{m-r} \tag{6.86}$$

The corresponding waiting times are then

$$W_q = L_q/\lambda_e$$

$$= \sum_{n=c}^{m} (n - c)p_n / \lambda \sum_{r=0}^{m} rp_{m-r} \tag{6.87}$$

and

$$W = L/\lambda_e$$

$$= \frac{m}{\lambda \sum_{r=0}^{m} rp_{m-r}} - \frac{1}{\lambda} \tag{6.88}$$

As with the single channel case, the arriving customer's distribution is equivalent to an $(m - 1)$-customer system as viewed by an outside observer. That is

$$q_n(m) = p_n(m - 1) \tag{6.89}$$

The corresponding waiting times for an arriving customer are

$$W'_q = \sum_{n=c}^{m-1} \frac{n - c + 1}{c\mu} q_n(m) \tag{6.90}$$

and

$$W' = W'_q + \frac{1}{\mu} \tag{6.91}$$

The computational burden is greater for the multi-channel, finite population model because there does not appear to be a neat closed-form expression for $\{p_n\}$. However, all of the sums are finite and should pose no insurmountable difficulty, given access to a modest amount of computing equipment. In addition, numerical values for this model have been extensively tabled by Peck and Hazelwood (74) so that many problems involving finite populations can be solved by table look-up.

STATE-DEPENDENT RATES

One final illustration of the versatility of the steady-state birth and death model concerns altering the service and arrival rates as functions of the number of customers in the queueing system. We have already examined a special case of state-dependent service rates every time we constructed a multi-channel model. In every multi-channel model the service rate increased linearly with the number of customers present up to the level $c\mu$ with c customers in the system and remained at that level for larger numbers. We similarly treated state-dependent arrival rates when we considered the finite population models. What we now seek are methods for generalizing such state dependence to account for nonlinear service and arrival rate functions.

Pressure Coefficients

Conway and Maxwell introduced a single channel model with state-dependent service expressed in terms of a "pressure coefficient" (21). Hillier, Conway, and Maxwell later expanded this approach to include multiple servers (44). The models arose from the observation that a large backlog of work often causes servers to work faster than when there is little backlog. The doctor may spend less time in friendly conversation with his patients when his waiting room is full. The gas station attendant may offer fewer courtesies in the form of oil checks and windshield washes when a line of cars is waiting at the gas pumps. In both cases the effect is to increase the service rate as the queue size increases.

Let

μ = expected service rate when there is only one customer
    in the system

γ = pressure coefficient which relates the service rate to
    system state ($\gamma \geq 0$)

The model for state dependent service rate and Poisson input can then be developed from equations (5.42) and (5.45) by specifying the parameters

$$\lambda_n = \lambda, \qquad n = 0, 1, 2, \ldots \qquad (6.92)$$

$$\mu_n = n^\gamma \mu, \qquad n = 1, 2, \ldots \qquad (6.93)$$

Note that when $\gamma = 1$ we have the same effect as an infinite number of servers in a multi-channel system. If the queue is truncated $\gamma = 1$ would lead us to the Erlang loss formula. When $\gamma = 0$ we have the standard single-channel model. An intermediate value such as $\gamma = 0.5$ would imply a service rate proportional to the square root of the number of customers in the system. The state distribution is then specified as

$$p_0 = \left[1 + \sum_{n=1}^{\infty} \frac{\prod_{i=0}^{n-1} \lambda_i}{\prod_{i=1}^{n} \mu_i} \right]^{-1}$$

$$= \left[\sum_{n=0}^{\infty} \frac{(\lambda/\mu)^n}{(n!)^{\gamma}} \right]^{-1} \tag{6.94}$$

and

$$p_n = \frac{(\lambda/\mu)^n}{(n!)^{\gamma}} p_0, \qquad n = 1, 2, \ldots \tag{6.95}$$

There does not appear to be a closed-form expression for p_n for general values of γ. However, it is possible to obtain numerical approximations for the distribution, and subsequently all the usual measures of performance, by summing a large but finite number of terms on the computer. The general purpose program in Appendix B can be used to obtain such an approximation.

This model can also be expanded to include state dependent arrival rates. Long queues may tend to slow the arrival stream by discouraging potential customers from entering. They may seek service elsewhere or postpone needed service in hopes of finding later periods of lighter traffic. The theoretical model posed to capture this effect suggests that

$$\lambda_n = \frac{\lambda}{(n + 1)^{\beta}} \tag{6.96}$$

where β is analogous to the pressure coefficient γ. If one combines this specification with $\mu_n = \mu$, the steady-state birth-death equations appear in exactly the same form as equation (6.95) except that γ is replaced by β.

The above observation suggests an even more general model in which both arrival and service rates are influenced by queue length.

Let

$$\lambda_n = \frac{\lambda}{(n + 1)^{\beta}} \tag{6.97}$$

and

$$\mu_n = n^{\alpha}\mu \tag{6.98}$$

Then

$$p_0 = \left[\sum_{n=0}^{\infty} \frac{(\lambda/\mu)^n}{(n!)^{\alpha+\beta}} \right]^{-1} \tag{6.99}$$

and

$$p_n = \frac{(\lambda/\mu)^n}{(n!)^{\alpha+\beta}} p_0, \qquad n \geq 1 \tag{6.100}$$

Therefore, one only needs to tabulate p_n for various values of pressure coefficient γ to have general results which are applicable for any desired combination of state dependent arrivals and service such that $\gamma = \alpha + \beta$.

This model has also been expanded to multiple servers where the pressure coefficients apply to the number of customers per server. The parameter specifications now become

$$\lambda_n = \begin{cases} \lambda, & 0 \leq n \leq c - 1 \\ \left(\dfrac{c}{n+1}\right)^{\beta}\lambda, & n \geq c \end{cases} \tag{6.101}$$

$$\mu_n = \begin{cases} n\mu, & 1 \leq n \leq c \\ \left(\dfrac{n}{c}\right)^{\alpha}c\mu, & n > c \end{cases} \tag{6.102}$$

The corresponding probability statements are

$$p_0 = \left[\sum_{n=0}^{c-1} \frac{(\lambda/\mu)^n}{n!} + \frac{(\lambda/\mu)^c \sum\limits_{n=c}^{\infty} (\lambda/\mu)^{n-c}}{c!\,(n!/c!)^{\gamma}c^{(1-\gamma)(n-c)}} \right]^{-1} \tag{6.103}$$

$$p_n = \begin{cases} \dfrac{(\lambda/\mu)^n}{n!}p_0, & 0 < n \leq c \\[4mm] \dfrac{(\lambda/\mu)^n\,p_0}{c!(n!/c!)^{\gamma}\,c^{(1-\gamma)(n-c)}}, & n > c \end{cases} \tag{6.104}$$

Example.   A study of radio communications traffic in a metropolitan police department suggests that the rate at which a records check is conducted is a function of the number of requests for information the radio operator currently has outstanding. Radio calls occur at random at the rate of fifteen per hour. When only a single request is outstanding the total transaction time, including the radio conversation and the record search averages four minutes. When five calls are active the transaction time is reduced to two minutes per record search. The analyst for the department feels that the model given by equation (6.95) is adequate for an initial description.

From the problem statement $\lambda = \mu = 15$ per hour. The pressure coefficient γ can be calculated from the boundary conditions, namely

$$\mu_5 = 5^{\gamma}\mu$$

or

$$30 = 5^{\gamma}(15)$$

By taking logs it follows that

$$\gamma \ln 5 = \ln 2$$

from which

$$\gamma \approx 0.43$$

The probability distribution can then be expressed as

$$p_n = \frac{p_0}{(n!)^{0.43}}$$

where

$$p_0 = \left[\sum_{n=0}^{\infty} (n!)^{-0.43} \right]^{-1}$$

The determination of particular numerical values is left as an exercise for the reader.

Channel Control

Another variation of state-dependent rates is often observed in customer service systems such as banks and grocery stores where the number of channels available for service is changed as a function of the queues observed by the manager. For example, a grocery may have four checkout counters which may be manned at different levels depending on the number of customers awaiting checkout. Management may decide that only one counter will be open if five or less customers are present. When six to ten customers are present he may open a second counter and so forth until the number of customers in the system exceeds fifteen. At that point all available counters will be open for service. The intent is to keep the average backlog per counter somewhere in the neighborhood of five.

Let the decision variable controlling the opening and closing of channels be the integer x. The service rate parameter as a function of number in the system can then be expressed as

$$\mu_n = \begin{cases} \mu & , & 1 \le n \le x \\ 2\mu & , & x < n \le 2x \\ \vdots & & \\ (c-1)\mu , & (c-2)x < n \le (c-1)x \\ c\mu & , & (c-1)x < n \end{cases} \tag{6.105}$$

If we assume a Poisson arrival stream, $\lambda_n = \lambda$ for all n, we are in a position to write the usual steady-state birth-death equations.

Let

$$K(n, i) = \frac{(\lambda/\mu)^n}{(i!)^x (i+1)^{n-ix}} \tag{6.106}$$

Then

$$p_n = \begin{cases} K(n, i) p_0 , & 0 \le n \le (i+1)x; \ i = 0, 1, 2, \ldots, c-1 \\ \dfrac{(\lambda/\mu)^n}{(c!)^x c^{n-cx}} p_0 , & n > cx \end{cases} \tag{6.107}$$

and

$$p_0 = \left[\sum_{i=0}^{c-1} \sum_{n=ix}^{(i+1)x} K(n, i) + \frac{(\lambda \mu)^{cx}}{(c!)^x (1 - \lambda/c\mu)} \right]^{-1} \tag{6.108}$$

Although this expression is somewhat awkward to work with, it does involve finite sums and can therefore be solved exactly. Furthermore the expected number of customers in the system can also be expressed in reasonably compact form as

$$L = \sum_{n=0}^{cx-1} np_n + \frac{p_0(\lambda/\mu)^{cx}}{(c!)^x} \left[\frac{\lambda/c\mu}{(1 - \lambda/c\mu)^2} + \frac{cx}{(1 - \lambda/c\mu)} \right] \tag{6.109}$$

The number of customers actively engaged in service is

$$s = \begin{cases} 1, & 1 \le n \le x \\ 2, & x < n \le 2x \\ \vdots & \\ c, & (c-1)x < n \end{cases}$$

The expected number in service is therefore

$$E(s) = \sum_{i=1}^{c-1} \sum_{n=(i-1)x+1}^{ix} ip_n + c \sum_{n=(c-1)x+1}^{cx} p_n$$

$$+ c \sum_{n=cx+1}^{\infty} \frac{(\lambda/\mu)^n p_0}{(c!)^x c^{n-cx}}$$

$$= \sum_{i=1}^{c-1} \sum_{n=(i-1)x+1}^{ix} ip_n + c \sum_{n=(c-1)x+1}^{cx} p_n$$

$$+ \frac{c(\lambda/\mu)^{cx}}{(c!)^x} \left[\frac{1}{(1 - \lambda/c\mu)} - 1 \right] p_0 \qquad (6.110)$$

Since the expected number in the queue is equal to the difference between the expected number in the system and the expected number in service we have

$$L_q = L - E(s) \qquad (6.111)$$

If we assume a single queue in front of this multi-channel system, we can again use Little's formula to calculate the average waiting times as

$$W_q = L_q/\lambda \qquad (6.112)$$

and

$$W = L/\lambda \qquad (6.113)$$

Example. Student registration at a large university is conducted in an office with six service windows. Actual data taken from this system yield an average time between arrivals for registrants of 24.53 seconds. The service time averages 129.05 seconds. Both the interarrival and service distributions have been checked by standard statistical tests and can be safely approximated by exponential distributions. The registrar uses his clerical personnel both to serve students and to perform filing tasks. For that reason he does not want to keep all the service windows manned throughout the day. It is his practice to open and close windows as the queue of students ebbs and flows. The problem is to analyze the effect of his staffing policy as a function of potential decision levels he might employ in making his service rate changes.

The parameters suggested for modeling this system are

$$c = 6 \text{ clerks available for duty}$$
$$\lambda = 2.448 \text{ arrivals/minute}$$
$$\mu = 0.465 \text{ services/minute}$$
$$\lambda/\mu = 5.26$$

Values for the decision variable have been treated for $x = [1, 2, 4, 6]$. Expected number in the system, expected waiting time and the expected number of idle clerks have been calculated for each of the x values. Those values are listed in Table 6-1.

TABLE 6-1
REGISTRATION SYSTEM PERFORMANCE

	$x = 1$	$x = 2$	$x = 4$	$x = 6$
L	10.12	14.05	22.55	31.42
W (minutes)	4.13	5.74	9.21	12.83
E (idle clerks)	0.74	0.74	0.74	0.74

One of the more interesting observations from this analysis is that the expected number of idle clerks, and hence the expected number available for auxillary tasks, is constant regardless of the x-value selected. However, the expected number of students present and their time in the system is dramatically influenced by the window staffing policy. This seems to suggest that if student service is the goal of the system, the registrar should consider leaving all service windows open all of the time. He does not gain clerical time by juggling the staff. We should hasten to add, however, that he may gain larger and hence more usable chunks of time by doing so.

SUMMARY

The models of this chapter have all been developed from the steady-state birth-death equations. The goal has been to give the reader some inkling of the great variety of physical systems which can be modeled from these equations. They are probably the most powerful results in queueing theory.

Although most of the models in this chapter can be reduced to compact algebraic expressions, that is not the case with many systems one might wish to model. However, with the wide availability of computer equipment it is not difficult to obtain numerical approximations for any variation of the birth-death process one wishes to analyze. All that is necessary is to properly specify the parameters λ_n and μ_n, employ equation (5.42) and (5.45) to develop the state distribution, then calculate whatever performance measures are desired. There is a general purpose computer program in the appendix to facilitate that calculation.

PROBLEMS

6.1 Consider the general equation

$$p_n = \frac{\displaystyle\prod_{i=0}^{n-1} \lambda_i}{\displaystyle\prod_{i=j}^{n} \mu_i} p_0, \qquad n = 1, 2, \ldots$$

a) Define each term in the model.

b) List the assumptions necessary to use this model in the analysis of a real system.

c) Demonstrate the application of this model to a system of your own design.

6.2 A commune receives new members at the rate of fifteen per month. As long as anyone is present they lose members at the rate of twenty per month.

a) Find the steady-state probability distribution for commune size.

b) Calculate the average size of the commune.

c) Find the 60 percent "load limit" for the commune.

d) What portion of the time will the size of the commune be greater than ten?

6.3 The service department at a large automobile dealer wants to analyze their parts-counter operation. Mechanics arrive at the parts counter with orders for parts in Poisson fashion at a rate of twenty per hour. The time to fill a single mechanic's order is estimated to be exponentially distributed with a mean of five minutes. The parts counter can accomodate anywhere from one to four clerks. Mechanics are paid $8.50 per hour and clerks are paid $6.00 per hour.

a) Find the optimum number of clerks to assign to the parts counter. State all assumptions.

b) Under optimum conditions, find the average time a mechanic will be delayed at the parts counter.

c) What is the average length of the queue?

d) How many minutes of clerk idle time will the system accumulate per day?

6.4 A university computer center is being asked to analyze their use of card readers. Under present policies there are four readers dispersed about the campus. Each reader processes an average of thirty jobs per hour. Individual job input time is estimated to average 1.2 minutes.

a) Analyze the current system. Calculate average customer delay time.

b) What will be the average delay time if the four readers are placed in a central location?

c) What ratio of reader cost per hour to customer delay cost per hour would justify reducing the proposed central facility from four to three readers?

6.5 An airline maintenance center presently operates with seven crews for emergency maintenance. Aircraft arrive for maintenance in Poisson fashion at the rate of five per day. Required maintenance appears to be exponentially distributed with a mean of one day per aircraft. The cost of aircraft downtime is estimated to be $400 per hour. Additional maintenance crews (up to a total of ten due to space restrictions) can be hired for $100 per hour.

a) Calculate the utilization factor for the current facility.

b) How many crews, if any, should be added?

6.6 An employment counselor maintains an office which can accomodate only four applicants. When one applicant is present the counselor requires an average of four minutes to complete an interview. When more than one applicant is present the average interview lasts three minutes. Applicants arrive at a rate of fifteen per hour unless two or more are in the system. At that point new arrivals are sufficiently discouraged to drop the rate to ten per hour.

a) Find the expected number of applicants in the office.

b) What portion of time will the counselor be busy?

c) If the true underlying demand for counseling is fifteen per hour, what portion of those needing it will be denied service?

6.7 Consider the problem faced by a crew of two mechanics who must maintain a fleet of four aircraft. The mean between time failures on each aircraft is estimated to be eight hours. When a single aircraft is down, both mechanics work on it with an expected repair time of two hours. When more than one is down, the mechanics work separately with the expected repair time for each becoming 2.5 hours. State all assumptions concerning your model before answering the following questions.

a) Develop the p. d. f. for number of aircraft operating.

b) Find the expected downtime per aircraft failure.

c) Find the probability that an arriving aircraft will receive immediate service.

6.8 Use the computer programs in Appendix B to solve the following problem.

Student registration at a large university is conducted in an office with six service windows. The service time at each window averages 2.2 minutes. The arrival process is such that the student arrival rate is functionally related to the number of students already in the registrar's office. That relationship is

$$\lambda_n = \frac{6.9}{(n + 1)^{0.2}} \text{ per minute}$$

The registrar exercises channel control by opening a new window whenever x Mod(n) customers are present, i.e., if $x = 3$, a single window would be available unless four customers were waiting whereupon a second window would open until seven were waiting, and so forth up to the maximum of six available windows. The maximum capacity of the system is 100 students.

a) Compare system performance measures for $x = 1, 5,$ and 10.

b) What strategy would you recommend if

    i) Student delay time costs $10 per student hour, denying service to a student costs $40 and maintaining an open window costs $100 per hour per window.

    ii) Student delays are free but the goal is to have 50 students in the system no more than ten percent of the time.

6.9 Use the computer programs in Appendix B to solve the following problem.

A company maintenance department has a crew of ten mechanics to maintain a group of sixty automatic screw machines. The mean time between breakdowns on a machine is estimated to be four hours. A single mechanic can repair a machine in an average of one hour. However, when available a crew of two will be dispatched for repair. The crew can accomplish the repair in an average of forty minutes. When only twenty machines are operable an outside contractor is hired to repair additional breakdowns as they occur.

a) Write an expression for the model to be used. State all assumptions.

b) Find the usual measures of performance for the current system.

c) Suppose that the contractor provides instant repairs by installing a substitute machine while the breakdown is repaired at a cost of $300 per breakdown. How much per hour could the company afford to pay to expand their own repair crew to twelve mechanics?

6.10 Consider the problem faced by a gas station operator who must design his facilities to handle a potential demand of 120 cars per hour. He can install up to five islands with two pumps each. The cost of pumps is amortized over one year and is estimated to be $10,000 per pump. State codes require that he have at least one attendant per island. With one attendant per island, the average service time for a single customer is six minutes. With two attendants per island, the servers cooperate by operating with two servers per car as long as there is a single car at an island. Service time then averages four minutes. As soon as additional vehicles arrive they revert to single man operation with each man taking an average of six minutes to serve a car. Wages plus fringe benefits amount to $10 per attendant per hour for eight-hour shifts. All hours over eight per day must be paid at double rate. Staggered shifts are not permitted. The difference between retail and wholesale price of gas sold per customer averages $1.50 per sale. Local zoning regulations specify that 1) only ten cars, beyond those at the pumps, can occupy the station; 2) the station can be open no more than ten hours per day; and 3) the station can operate no more than 300 days per year.
Determine the number of gas islands, number of attendants, hours of operation per day, days per year, and expected annual profit for an optimum design. What is the average customer sojourn time?

6.11 Consider an ordering system in which customers arrive at a rate λ (Poisson fashion) to place orders for a custom-built trailer. Production time for each order follows a negative exponential distribution with mean time of five days. Service time for any customer in the system is independent of the number in the system and orders are permitted to cross.

a) Show how this can be modeled as a queueing system with infinite many servers. Develop the probability distribution for customer orders in the system.

b) Suppose that the system is truncated at fifty orders. Determine the critical rate λ for which an order processing system with capacity fifty will be able to accomodate 90 percent of the orders.

6.12 Consider the maintenance operation for bus repair at a university. The shop foreman has three mechanics which he can assign, if necessary, to repairing breakdowns among the school's twenty buses. However, since he also has other tasks which he wants the mechanics to perform, he is reluctant to commit them all at once to the repair task. His policy is to provide a single mechanic for buses as long as there are no more than three buses in the shop. When more than three but less than seven buses are present, two mechanics are assigned. When seven or more buses are down, all mechanics are required to work on bus repair. Furthermore, shop space is limited so that no more than ten buses at a time can be accommodated. Breakdowns occurring when the shop is full are repaired by a local auto dealer who provides a loaner vehicle for the university's use while a bus is in his shop. The dealer has ten buses in his stock. An analysis of transportation department records indicates that the average time between failures on a bus is eight hours. Shop records indicate that the mean repair time per breakdown is two man hours.

a) Construct a model for the repair shop activities which will reveal a probability distribution for shop load.

b) How many man hours per day are available for using bus mechanics on other tasks?

c) What portion of time will the university be using "loaner" vehicles?

d) Find the average fleet size available for operation by the university.

e) Find the proportion of the total fleet under the control of the university which will be university owned.

6.13 A motor freight terminal has eight covered loading docks at the main warehouse. Any trucks which arrive when all docks are full are assigned to open docks near the old rail spur. Arrival rates and service times have been studied and can be approximated by Poisson processes. Recent data indicate that 30 percent of the arriving trucks must unload at the open docks due to saturation at the main warehouse. How many covered docks should be added at the main warehouse so that at least 90 percent of the shipments can be unloaded under cover?

6.14 In many queueing systems it may be recognized that the customer population is really finite, but for computational convenience it is treated as an infinite population. Consider for example a machine repair problem with two mechanics with individual mean service times of one hour per repair serving a population of m machines each of which has a mean time between failures of twelve hours.

a) Plot the average queue length and p_0 versus population size for both the finite source and the infinite source approximation when $m = 6, 12, 18, 24$.

b) Perform the same calculations for a mean service time of two hours.

c) What "principles of approximation" can you propose on the basis of your plots?

6.15 In analyzing some queueing systems it may be convenient to ignore waiting space limitations for steady-state solutions and contrariwise to introduce such limitations when none exist for transient solutions. Consider a single channel system with capacity k in which $\lambda = 1$, $\mu = 2$.

a) Plot the average waiting time and p_0 versus k for both the truncated model and its corresponding infinite approximation when $k = 1, 5, 10, 15$. (Treat the infinite model as if $\lambda = \lambda_e$.)

b) Repeat these calculations for a mean service rate of $\mu = 1.1$.

c) What "principles of approximation" can you propose on the basis of your plots?

7

TRANSIENT SOLUTIONS

The purpose of this chapter is to extend our analysis into the area of transient behavior of systems. Historically, many authors and practitioners have dismissed transient solutions as having very little practical importance. For the most part they have confined their attention to studying relaxation times to justify the use of steady-state solutions and occasionally to analyzing how long it would take for the queue to become unmanageable if the arrival rate would suddenly increase to a value much greater than the service rate.[1] There is relatively little literature which discusses closed-form solutions for systems in which arrival rates and service rates vary systematically over time.

IMPORTANCE

It is the premise of this author that the transient behavior of many systems is the most important characteristic of those systems. Consider, for example, the study of air traffic congestion at a major airline terminal. Arrival rates are determined by airline schedules and may vary from near zero in the 1:00 AM to 4:00 AM period to well over 100 per hour during the evening rush. Service rates also vary over time. A local thunderstorm can briefly cut the service rate to zero. Ice on the runway may also drastically curtail service for brief periods. To base decisions concerning air traffic control at such a terminal on steady-state results is to invite chaos. There is no steady state.

The reasons one does not often find transient analysis used in such systems are that 1) very few transient solutions exist and 2) those solutions which are available are extremely awkward to manipulate. For those reasons it is a common practice to resort to simulation studies in an attempt to analyze time-varying behavior. However, simulation is an experimental tool with all of the statistical pitfalls of decision making

---

[1] Relaxation time, sometimes called the time constant, is a concept borrowed from electrical engineering. It is a measure of the time it takes the system to approach steady state.

from experimental data. In addition it is very expensive to obtain sufficiently large samples of simulation results to have confidence in the conclusions drawn.

What we hope to offer in this chapter and the next is a set of tools which can provide time-varying solutions for a wide range of realistic problems. The solutions are often obtained by numerical solution to systems of differential equations and by iterative calculation of state probabilities using basic Markov chain arguments. In these cases the output is not an equation but rather a table of numbers which enumerates the numerical values of the state distribution. But after all it is that set of numbers which is most important in our analysis and design decisions.

CLOSED-FORM SOLUTIONS

Before we discuss numerical techniques for solution let us pause to illustrate some of the characteristics of existing closed-form solutions. We will examine a simple two-state system (M/M/1/1), the classical (M/M/1) queue, and the single channel truncated (M/M/1/k) queue. In so doing we hope to demonstrate how such solutions can be obtained and the limitations of the resulting equations.

(M/M/1/1) Transients

The (M/M/1/1) system is one with a Poisson arrival stream, exponential service, and no waiting room. The system has only two states; either it is busy or it is empty. Physical systems with no storage capacity such as a single telephone instrument could be treated with this model.

As with all variations of the birth-death model, the solution to the (M/M/1/1) system begins with a set of differential-difference equations. The transition rates are defined as

$$\lambda_n = \begin{cases} \lambda, & n = 0 \\ 0, & \text{elsewhere} \end{cases}$$

$$\mu_n = \begin{cases} \mu, & n = 1 \\ 0, & \text{elsewhere} \end{cases}$$

This leads to a modified version of equation (5.12), namely

$$\frac{dp_0(t)}{dt} = -\lambda p_0(t) + \mu p_1(t) \tag{7.1a}$$

$$\frac{dp_1(t)}{dt} = -\mu p_1(t) + \lambda p_0(t) \tag{7.1b}$$

This set of equations is the exception to the general rule of computational difficulties encountered in finding transient solutions. The computation is made easy by observing that, since the system only has two states, the probability of either one can easily be expressed as a function of the other. Since the total probability at any time t must be unity, it follows that

$$p_0(t) = 1 - p_1(t) \qquad (7.2)$$

and

$$p_1(t) = 1 - p_0(t) \qquad (7.3)$$

By substituting the expression for $p_1(t)$ in equation (7.1a), we have a simple first order linear differential equation with constant coefficients.

$$\frac{dp_0(t)}{dt} = -\lambda p_0(t) + \mu[1 - p_0(t)] \qquad (7.4)$$

We can rewrite equation (7.1b) in similar fashion as

$$\frac{dp_1(t)}{dt} = -\mu p_1(t) + \lambda[1 - p_1(t)] \qquad (7.5)$$

Although these equations can be solved by a variety of techniques, let us use Laplace transforms. Taking transforms of both sides of equation (7.4), we are led to the subsidiary equation

$$s\mathcal{L}(p_0(t)) - p_0(0) = -\lambda\mathcal{L}(p_0(t)) + \frac{\mu}{s} - \mu\mathcal{L}(p_0(t)) \qquad (7.6)$$

Solving for $\mathcal{L}(p_0(t))$ we have

$$\mathcal{L}(p_0(t)) = \frac{\mu}{s(s + \lambda + \mu)} + \frac{p_0(0)}{s + \lambda + \mu} \qquad (7.7)$$

which by partial fraction expansion becomes

$$\mathcal{L}(p_0(t)) = \frac{\mu}{(\lambda + \mu)s} - \frac{\mu}{(\lambda + \mu)(s + \lambda + \mu)} + \frac{p_0(0)}{s + \lambda + \mu} \qquad (7.8)$$

Finally, inverting term by term yields

$$p_0(t) = \frac{\mu}{\lambda + \mu}[1 - e^{-(\lambda+\mu)t}] + p_0(0)\, e^{-(\lambda+\mu)t} \qquad (7.9)$$

The complete solution depends upon the particular value for $p_0(0)$ specified as a boundary condition.

One can apply exactly the same transform procedure to equation (7.5) or simply use $p_1(t) = 1 - p_0(t)$ to obtain

$$p_1(t) = \frac{\lambda}{\lambda + \mu}[1 - e^{-(\lambda+\mu)t}] + p_1(0)\, e^{-(\lambda+\mu)t} \qquad (7.10)$$

The steady-state solution can be obtained by setting the derivatives equal to zero in equations (7.4) and (7.5) or by taking the limit as t approaches infinity in equations (7.9) and (7.10). By either technique, we have

$$p_0 = \frac{\mu}{\lambda + \mu} \qquad (7.11)$$

$$p_1 = \frac{\lambda}{\lambda + \mu} \qquad (7.12)$$

The transient character of the solution is best illustrated by a particular example. Suppose that a telephone is not busy at $t = 0$. The normal calling rate for this telephone is ten calls per hour. The mean duration for a telephone conversation is four minutes. The parameters for this problem are then

$$\lambda = 10, \ \mu = 15, \ p_0(0) = 1$$

We are interested in the probability that the line will still be available after an elapsed time t. From equation (7.9) we have

$$p_0(t) = \frac{15}{10 + 15}[1 - e^{-25t}] + e^{-25t}$$
$$= 0.6 + 0.4\, e^{-25t}$$

where t is measured in hours. Converting to a time scale in minutes we have

$$p_0(t) = 0.6 + 0.4\, e^{-0.417t}$$
and
$$p_1(t) = 0.4 - 0.4\, e^{-0.417t}$$

These functions are plotted in Figure 7.1. From the graph it is obvious that the values for $p_0(t)$ and $p_1(t)$ will asymptote their steady-state values as t becomes large. Since the expressions for $p_0(t)$ and $p_1(t)$ each contain a constant, which is the steady-state value of that term, and a transient portion, one might ask how long the system must run before initial

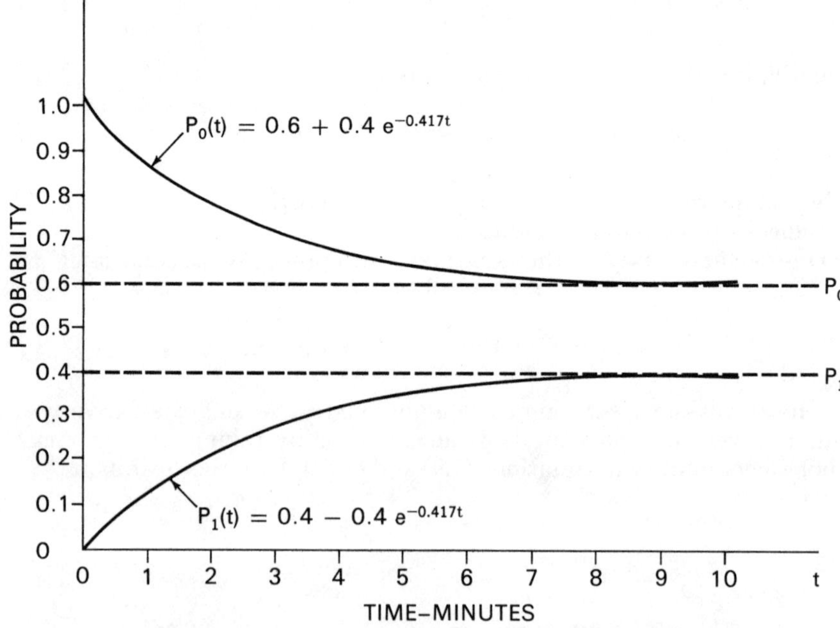

FIGURE 7-1 Transient Solution for Two-state System

conditions can be ignored and a steady-state solution employed. Suppose we arbitrarily define that point as the point in time at which the transient component is only 10 percent as large as its initial value. For this example this means find a value t so that

$$0.4\, e^{-4.17t} = 0.04$$

Solving we have

$$t = 5.53 \text{ minutes}$$

This says that, after slightly more than five and one-half minutes, knowing that the system was originally idle has little influence on predicting future state probabilities. If the system is to be analyzed over an eight-hour day, the influence of the transient occurs over such a short portion of the total time horizon that it can be safely ignored. If, on the other hand, we wanted to know how it behaved over the first four minutes of the day it would not be wise to use the steady-state results.

The effect of the offered load on the time to reach steady state can be easily calculated. Suppose that the call rate doubles to $\lambda = 20$ while the service rate remains constant at $\mu = 15$ per hour. The revised transient element is

$$\frac{\mu}{\lambda + \mu}\, e^{-(\lambda+\mu)t} = 0.4286\, e^{-0.583t}$$

The time necessary to reduce this element by 90 percent is $t = 3.95$ minutes. In this case the higher call rate means that we will reach steady state (by our definition) 1.58 minutes sooner. Similarly if we reduce the call rate to 5 per hour the time goes up to 6.9 minutes or an increase of 1.37 minutes to reach steady state.

Another measure sometimes used to indicate the time it takes a system to approach steady state is the so-called *relaxation time*. Relaxation time is a concept borrowed from the field of electrical engineering. The relaxation time for our simple two-state system is $1/(\lambda + \mu)$. This is the time it takes the transient to die down to $(1/e)$ of its initial value. The relaxation time is also sometimes called the *time constant*. The time constant T is a measure of the duration of a term of the form $e^{-t/T}$. For the case of $\lambda = 10$, $\mu = 15$, the relaxation time is $(1/25)$ hours or 2.4 minutes as compared with out 90 percent criterion of 5.53 minutes.

(M/M/1) Transients

The (M/M/1) single server unlimited queue is the most easily solved and most easily manipulated model of all the steady-state models treated in Chapter 6. However, transient solutions are an entirely different matter. The difficulties encountered for this most simple of systems in, first, finding the transient solution and, second, in using that solution once it is available are sufficiently formidable to discourage all but the most curious and highly skilled mathematicians from further explorations of transients.

The solution begins with the now familiar infinite set of differential-difference equations for the birth-death process in which $\lambda_n = \lambda$ and $\mu_n = \mu$ for all n. That is

$$\frac{dp_0(t)}{dt} = -\lambda p_0(t) + \mu p_1(t)$$

$$\frac{dp_n(t)}{dt} = -(\lambda + \mu)\, p_n(t) + \lambda p_{n-1}(t) + \mu p_{n+1}(t),\ n \geq 1 \qquad (7.13)$$

The difficulties encountered in solving this set of equations stem largely from the fact that the equation is different for $n = 0$ than it is away from the boundary. Several techniques for finding a solution to this particular problem have been used over the years, including spectral analysis, combinatorial methods, and transform techniques. In keeping with our earlier practice for solving similar sets of equations for the pure birth process we will examine the transform technique as first suggested by Bailey (4).

Recall from Chapter 4 that we defined the geometric transform (or generating function) for $p_n(t)$ as

$$G(z, t) = \sum_{n=0}^{\infty} p_n(t)z^n \qquad (7.14)$$

Since $G(z, t)$ is a function of both the transform variable z and the time variable t we also noted that

$$\frac{\partial G(z, t)}{\partial t} = \sum_{n=0}^{\infty} \frac{dp_n(t)}{dt} z^n$$

The solution procedure begins by reducing the infinite set of differential-difference equations to a single equation in the geometric transform by multiplying the nth equation in the set (7.13) by z^n and adding. That is

$$\frac{dp_0(t)}{dt} z^0 = -\lambda p_0(t)z^0 + \mu p_1(t)\, z^0$$

$$\frac{dp_n(t)}{dt} z^n = -(\lambda + \mu)p_n(t)z^n + \lambda p_{n-1}(t)\, z^n$$

$$+ \mu p_{n+1}(t)\, z^n,\ n \geq 1$$

$$\sum_{n=0}^{\infty} \frac{dp_n(t)}{dt} z^n = -\lambda \sum_{n=0}^{\infty} p_n(t)z^n - \mu \sum_{n=1}^{\infty} p_n(t)z^n$$

$$+ \lambda \sum_{n=1}^{\infty} p_{n-1}(t)z^n + \mu \sum_{n=0}^{\infty} p_{n+1}(t)z^n \qquad (7.15)$$

By multiplying and dividing the various sums in this expression by appropriate powers of z, it can be rewritten as a function of the transform $G(z, t)$ as

$$\sum_{n=0}^{\infty} \frac{dp_n(t)}{dt} z^n = -\lambda \sum_{n=0}^{\infty} p_n(t) z^n - \mu\left[\sum_{n=0}^{\infty} p_n(t) z^n - p_0(t)\right]$$

$$+ \lambda z \sum_{n=0}^{\infty} p_n(t) z^n$$

$$+ \frac{\mu}{z}\left[\sum_{n=0}^{\infty} p_n(t) z^n - p_0(t)\right]$$

which reduces to

$$\frac{\partial G(z, t)}{\partial t} = -\lambda G(z, t) - \mu[G(z, t) - p_0(t)] + \lambda z G(z, t)$$

$$+ \frac{\mu}{z}[G(z, t) - p_0(t)]$$

$$= G(z, t)[\lambda(z - 1) - \mu(1 - 1/z)] + \mu p_0(t)[1 - 1/z]$$

$$= \frac{z - 1}{z}[G(z, t)(\lambda z - \mu) + \mu p_0(t)] \tag{7.16}$$

If one begins operating with a known number of customers i present, the boundary condition for this equation is

$$G(z, 0) = \sum_{n=0}^{\infty} p_n(0) z^n = z^i \tag{7.17}$$

Equation (7.16) can now be treated as a differential equation in $G(z, t)$ to be reduced to an algebraic equation in the Laplace transform where

$$\mathcal{L}(G(z, t)) = \int_0^{\infty} e^{-st} G(z, t)dt \tag{7.18}$$

and

$$\mathcal{L}\left(\frac{\partial G(z, t)}{\partial t}\right) = s\mathcal{L}(G(z, t)) - G(z, 0)$$

$$= s\mathcal{L}(G(z, t)) - z^i \tag{7.19}$$

Taking the Laplace transform of both sides of equation (7.16) we have

$$s\mathcal{L}(G(z, t)) - z^i = \left(\frac{z - 1}{z}\right)(\lambda z - \mu)\,\mathcal{L}(G(z, t)) + \frac{\mu(z - 1)}{z}\,\mathcal{L}(p_0(t))$$

$$\tag{7.20}$$

from which

$$\mathcal{L}(G(z, t)) = \frac{z^{i+1} + \mu(z - 1)\mathcal{L}(p_0(t))}{(\lambda + \mu + s)z - \mu - \lambda z^2} \tag{7.21}$$

The solution at this point becomes extremely complicated. One must first evaluate $\mathcal{L}(p_0(t))$. This is accomplished by matching zeros in the numerator and denominator of equation (7.21) and invoking Rouche's theorem from complex analysis.[2] Once $\mathcal{L}(p_0(t))$ is expressed as a function of the two transform variables z and s one must then invert the Laplace transform $\mathcal{L}(G(z, t))$ and the geometric transform $G(z, t)$. The calculations involved are extremely tedious and of little value for our immediate needs. Readers interested in the details of the calculation are referred to the original discussion by Bailey (4) or the excellent explanation by Gross and Harris (42). The end result is the function

$$p_n(t) = e^{-(\lambda+\mu)t} \left[\left(\frac{\mu}{\lambda}\right)^{\frac{i-n}{2}} I_{n-i}(2\sqrt{\lambda\mu}\,t) \right.$$

$$+ \left(\frac{\lambda}{\mu}\right)^{\frac{i-n+1}{2}} I_{n+i+1}(2\sqrt{\lambda\mu}\,t)$$

$$\left. + \left(1 - \frac{\lambda}{\mu}\right)\left(\frac{\lambda}{\mu}\right)^n \sum_{k=n+i+2}^{\infty} \left(\frac{\lambda}{\mu}\right)^{k/2} I_k(2\sqrt{\lambda\mu}\,t) \right] \qquad (7.22)$$

for a system which contains i customers at time zero. The function $I_n(x)$ is the modified Bessel function of the first kind defined as

$$I_n(x) = \sum_{k=0}^{\infty} \frac{(x/2)^{n+2k}}{k!(n+k)!}, \qquad n > -1 \qquad (7.23)$$

This is a formidable expression for an engineer in a hurry to work with, although it is possible to obtain numerical answers, given a proper set of tables and much patience. Figures 7.2 and 7.3 illustrate the behavior of this model for selected parameters. All the graphs assume a mean service rate of unity so that the time axis can be interpreted in general units of service time.

Figure 7.2 can be used to compare the transient behavior, relative to states zero and one, for a variety of utilization factors and initial conditions. For example, comparing 7.2-a, 7.2-c, and 7.2-e one can see that as the offered load or utilization factor increases from 0.2 through 0.8 the time necessary for the probabilities to approach their steady-state values increases markedly. By comparing 7.2-a to 7.2-b, 7.2-c to 7.2-d, and 7.2-e to 7.2-f one can see the influence of initial conditions. In the graphs on the left the system is initially empty and on the right the system initially contains one customer.

Figures 7.2-e and 7.2-f also suggest that there may be some oscillatory behavior in the probability functions, particularly at high load levels. The transient $p_n(t)$ may approach its steady-state value from one side, overshoot, then gradually stabilize from the opposite direction. Figure 7-3 offers a slightly more dramatic illustration of this behavior. In this case the system has an offered load greater than unity and an initial system load of two. The probability of finding the system empty rises very rapidly to a peak after slightly more than one time unit then decays slowly toward zero.

---

[2]Rouche's theorem is used to establish that only one zero of the denominator falls within the unit circle. A proof of this theorem may be found in complex variable books such as Churchill (17).

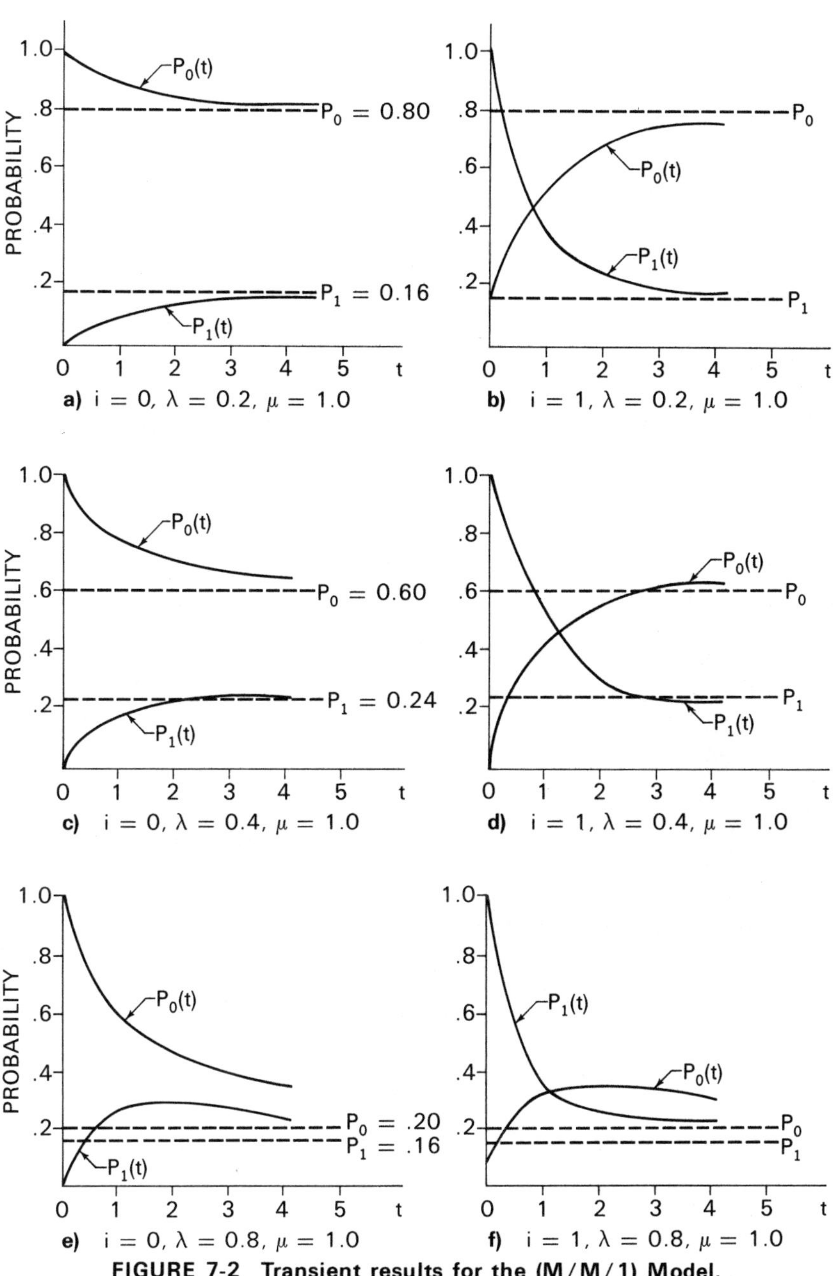

a) $i = 0$, $\lambda = 0.2$, $\mu = 1.0$

b) $i = 1$, $\lambda = 0.2$, $\mu = 1.0$

c) $i = 0$, $\lambda = 0.4$, $\mu = 1.0$

d) $i = 1$, $\lambda = 0.4$, $\mu = 1.0$

e) $i = 0$, $\lambda = 0.8$, $\mu = 1.0$

f) $i = 1$, $\lambda = 0.8$, $\mu = 1.0$

FIGURE 7-2 Transient results for the (M/M/1) Model.

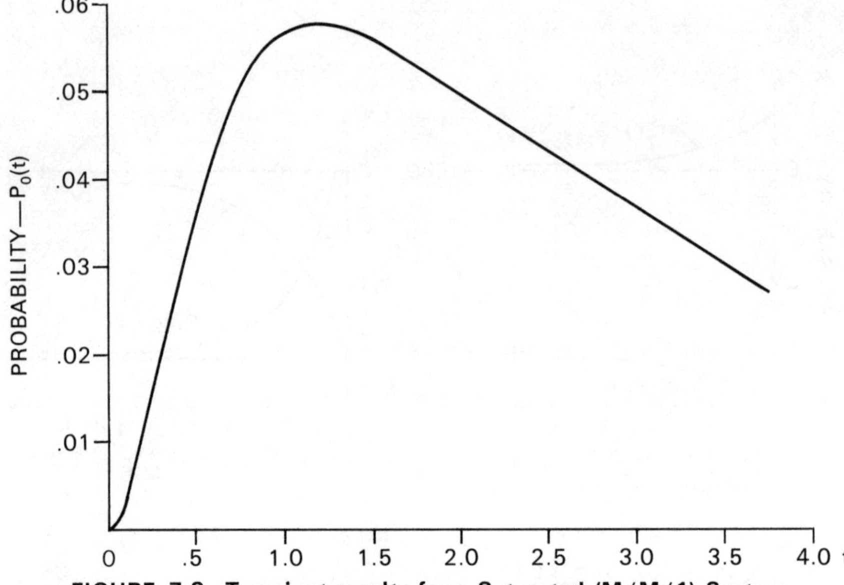

FIGURE 7-3 Transient results for a Saturated (M/M/1) System
$$i = 2, \ \lambda = 2, \ \mu = 1$$

(M/M/1/k) Transients

When the standard single channel Poisson system is truncated at some general capacity k, we are led to a finite system of linear differential equations with constant coefficients. That is

$$\frac{dp_0(t)}{dt} = -\lambda p_0(t) + \mu p_1(t)$$

$$\frac{dp_n(t)}{dt} = \lambda p_{n-1}(t) - (\lambda + \mu)\, p_n(t) + \mu p_{n+1}(t), \qquad 0 < n < k$$

$$\frac{dp_k(t)}{dt} = \lambda p_{k-1}(t) - \mu p_k(t) \tag{7.24}$$

This set of equations can also be written in terms of a probability vector and a rate matrix in the manner demonstrated in equation (5.35) as

$$\frac{d\mathbf{P}(t)}{dt} = \mathbf{P}(t)\, \mathbf{R}(t) \tag{7.25}$$

where $\mathbf{P}(t)$ contains the set of elements $\{p_n(t)\}$ and $\mathbf{R}(t)$ contains the set of coefficients $\{\lambda, \mu\}$. For example, a system truncated at $k = 2$ would be represented by the set of equations

$$\left(\frac{dp_0(t)}{dt}, \frac{dp_1(t)}{dt}, \frac{dp_2(t)}{dt} \right)$$

$$= (p_0(t), p_1(t), p_2(t)) \begin{pmatrix} -\lambda & \lambda & 0 \\ \mu & -(\lambda + \mu) & \lambda \\ 0 & \mu & -\mu \end{pmatrix} \tag{7.26}$$

Since we are dealing with a system of ordinary linear differential equations with constant coefficients there are a number of standard techniques we could use to find a solution. For example, see Kaplan (55:Ch. 6). Because of their type we would expect to find the general solution to these equations to be of the form

$$p_n(t) = \sum_{j=0}^{k} \beta_{nj}\, e^{\gamma_j t} \tag{7.27}$$

The coefficients β_{nj} depend upon the initial conditions of the system. The γ_j terms in the exponent are the eigenvalues of the rate matrix $\mathbf{R}(t)$. Because of the structure of $\mathbf{R}(t)$ one of the eigenvalues will always be zero, say γ_0. The β_{n0} term in the summation of equation (7.27) then corresponds to the steady-state value for $p_n(t)$. All other eigenvalues of $\mathbf{R}(t)$ are less than zero and represent transient behavior.

Laplace transforms offer an efficient solution procedure for particular configurations of the (M/M/1/k) model. Let us take transforms of both sides of equation (7.25). Recognizing that all of the elements of $\mathbf{R}(t)$ are constants, we have the subsidiary equation

$$s\mathcal{L}(\mathbf{P}(t)) - \mathbf{P}(0) = \mathcal{L}(\mathbf{P}(t))\mathbf{R} \tag{7.28}$$

Solving for the unknown vector of transforms $\mathcal{L}(\mathbf{P}(t))$, taking care to observe the rules of matrix algebra, yields

$$\mathcal{L}(\mathbf{P}(t)) = \mathbf{P}(0)[s\mathbf{I} - \mathbf{R}]^{-1} \tag{7.29}$$

where the negative exponent indicates the matrix inversion operation. The time dependent distribution $\mathbf{P}(t)$ is then obtained by inverting the transform.

To illustrate the technique suppose that we have a truncated system with the following set of parameters:

$$k = 2,\ \lambda = 4,\ \mu = 2,\ \mathbf{P}(0) = (0.2, 0.5, 0.3)$$

Following the format of equation (7.26) the rate matrix $\mathbf{R}$ can be written as

$$\mathbf{R} = \begin{pmatrix} -4 & 4 & 0 \\ 2 & -6 & 4 \\ 0 & 2 & -2 \end{pmatrix}$$

from which

$$(s\mathbf{I} - \mathbf{R}) = \begin{pmatrix} s+4 & -4 & 0 \\ -2 & s+6 & -4 \\ 0 & -2 & s+2 \end{pmatrix}$$

Inverting the matrix we have

$$(s\mathbf{I} - \mathbf{R})^{-1} = \frac{1}{s(s^2 + 12s + 28)}$$

$$\times \begin{bmatrix} s^2 + 8s + 4 & 4(s+2) & 16 \\ 2(s+2) & (s+4)(s+2) & 4(s+4) \\ 4 & 2(s+4) & (s+8)(s+2) \end{bmatrix}$$

The poles of this tansform, which are in fact the eigenvalues of the rate matrix, are located at $s = 0$, $s = -6 - \sqrt{8}$ and $s = -6 + \sqrt{8}$. By partial fraction expansion

$$(s\mathbf{I} - \mathbf{R})^{-1} = \frac{1}{s}\begin{bmatrix} \frac{1}{7} & \frac{2}{7} & \frac{4}{7} \\ \frac{1}{7} & \frac{2}{7} & \frac{4}{7} \\ \frac{1}{7} & \frac{2}{7} & \frac{4}{7} \end{bmatrix}$$

$$+ \frac{1}{s + 8.828}\begin{bmatrix} .2265 & -.5470 & .3204 \\ -.2735 & .6602 & -.3867 \\ .0801 & -.1934 & .1133 \end{bmatrix}$$

$$+ \frac{1}{s + 3.172}\begin{bmatrix} .6306 & .2612 & -.8918 \\ .1306 & .0541 & -.1847 \\ -.2230 & -.0923 & .3153 \end{bmatrix}$$

$$= \frac{1}{s}\mathbf{A} + \frac{1}{s + 8.828}\mathbf{B} + \frac{1}{s + 3.172}\mathbf{C}$$

Note that the matrices in this expansion have a familiar form. The matrix $\mathbf{A}$ associated with $(1/s)$ is a *stochastic matrix* representing the steady-state solution to this continuous-time Markov process while the matrices $\mathbf{B}$ and $\mathbf{C}$ are each *differential matrices* representing the transient part of the solution. Each transient element decays toward zero as it is multiplied by a term of the form e^{-at}. This solution for the (M/M/1/k) queueing system is in exactly the same form as our transform solutions for discrete Markov chains introduced in Chapter 3.

Inverting the transform we have

$$\mathcal{L}^{-1}((s\mathbf{I} - \mathbf{R})^{-1}) = \mathbf{A} + \mathbf{B}e^{-8.828t} + \mathbf{C}e^{-3.172t}$$

which, after accounting for initial conditions, yields

$$\mathbf{P}(t) = (0.2, 0.5, 0.3)[\mathbf{A} + \mathbf{B}e^{-8.828t} + \mathbf{C}e^{-3.172t}]$$

Expanding to find the individual state probabilities we have

$$p_0(t) = \frac{1}{7} - .06742\, e^{-8.828t} + .12452\, e^{-3.172t}$$

$$p_1(t) = \frac{2}{7} + .16268\, e^{-8.828t} + .05160\, e^{-3.172t}$$

$$p_2(t) = \frac{4}{7} - .09528\, e^{-8.828t} - .17612\, e^{-3.172t}$$

The reader is urged to plot a few points for these time-varying probabilities to better understand how the transients behave.

Solving the set of differential equations by transforms for every particular system configuration one might like to analyze would be a very tedious task. Fortunately there does exist a general solution which can be expressed as a function of the starting state of the system. This solution, using slightly different notation, is detailed by Morse (69:66). With r customers present at $t = 0$, the probability of finding n in the system at time t is

$$p_n^r(t) = p_n - \frac{2(\lambda/\mu)^{\frac{1}{2}(n-r)}}{k+1} \sum_{j=1}^{k} \left(\frac{\mu}{\gamma_j}\right)\left[\sin\left(\frac{jr\pi}{k+1}\right)\right.$$
$$\left. - \left(\frac{\lambda}{\mu}\right)^{\frac{1}{2}} \sin\left(\frac{j(r+1)\pi}{k+1}\right)\right]\left[\sin\left(\frac{jn\pi}{k+1}\right)\right.$$
$$\left. - \left(\frac{\lambda}{\mu}\right)^{\frac{1}{2}} \sin\left(\frac{j(n+1)\pi}{k+1}\right)\right] e^{\gamma_j t} \qquad (7.30)$$

where

$$p_n = \text{steady state value}$$

$$\gamma_j = 2\sqrt{\lambda\mu} \cos\left(\frac{j\pi}{k+1}\right) - (\lambda + \mu)$$

The γ_j terms are the eigenvalues of the rate matrix $\mathbf{R}(t)$. For example, in our previous numerical example

$$\gamma_1 = 2\sqrt{8} \cos\left(\frac{\pi}{3}\right) - (4+2) = -3.172$$

$$\gamma_2 = 2\sqrt{8} \cos\left(\frac{2\pi}{3}\right) - (4+2) = -8.828$$

which are consistent with the coefficients obtained by transform methods. The reader who is interested in details of the proof for this relationship as well as other similar transient results is urged to consult Morse (69:Ch. 6).

PROBLEMS

7.1 The queue for course change tickets has been observed to be thirty students at 8:00 AM The mean Poisson arrival rate is estimated to be sixty per hour in the morning and fifteen per hour in the afternoon. The single service window operates in exponential fashion with a mean of three minutes per student.

a) Estimate the mean line length at 10:00 AM.

b) What is the expected wait for a student who arrives at 10:00 AM?

c) If the queue at 12:00 noon is 80, how long will it take the system to reach equilibrium?

Hint: Use steady-state and saturated system approximations.

7.2 Consider a two-state system which is initially empty and subjected to an arrival rate of ten per hour for the first four minutes of operation. At that time the arrival rate drops to six per hour for eight minutes after which it jumps to twenty-five per hour for three minutes. Service rate remains constant at fifteen per hour. Plot the probability $p_0(t)$ for the first fifteen minutes of operation for this system. What can you say about steady-state approximations in this setting?

7.3 Consider a two state system in which $\lambda = 15$ and $\mu = 20$.

    a) Plot $p_0(t)$ and $p_1(t)$ for each of the following initial conditions:

        i) $p_0(0) = 0,\ p_1(0) = 1$

        ii) $p_0(0) = 0.8,\ p_1(0) = 0.2$

        iii) $p_0(0) = 0.5,\ p_1(0) = 0.5$

        iv) $p_0(0) = 0.2,\ p_1(0) = 0.8$

    b) Calculate the time it will take each of the systems above to reduce the transient element by 90 percent.

7.4 Invert the transform of equation (7.21) to prove that the transient solution for the (M/M/1) system is valid.

7.5 Show that the steady-state solution for the (M/M/1) system can be obtained from $\lim_{t \to \infty} p_n(t)$ where $p_n(t)$ is given by equation (7.22).

7.6 Consider an (M/M/1) system which is initially empty. If $\lambda = \mu = 1$, plot a curve for $p_0(t)$, for $t = 0, 1, \ldots, 10$. What is the limiting value for $p_0(t)$?

7.7 Consider a truncated system in which $k = 3$, $\lambda = 6$, $\mu = 2$, and $p_0(0) = 1$.

    a) Use transform techniques to find a solution for $p_n(t)$.

    b) Plot a curve for the expected number in the system as a function of t.

7.8 Consider the truncated system $k = 2$, $\lambda = 4$, $\mu = 2$, and $p_0(0) = 1$. (t is measured in hours)

    a) Find the transient solutions for $p_n(t)$.

    b) Suppose that the arrival rate drops to $\lambda = 2$ after the system has operated for three minutes. Find $p_n(t)$.

7.9 Prove that the general truncated model of equation (7.30) reduces to the special two-state system equations of (7.9) and (7.10).

7.10 Discuss the effects of using an approximate truncated model to analyze transients in an unlimited queue system. Support your hypothesis by comparing results from equation (7.30) with those from (7.22) for two extremes of traffic intensity and two truncation points.

7.11 Prepare a computer program which will calculate values for $p_n{}^r(t)$ suggested by equation (7.30). The program should be able to do the following:

a) Handle any size system up to $k = 50$ for arbitrary specification of λ and μ.

b) Print the time-varying probability distribution in increments of $t = 0.25$.

c) Calculate and print appropriate statistics at each reporting point (e.g., $E[n(t)]$).

d) Permit the experimenter to change arrival and service rates in step fashion over an extended run period. (For example, treat the time varying arrival problem in which $\lambda = 5$ for the first hour, 10 the second, 3 the third, etc. while the system runs continuously.)

8

TIME-VARYING INPUTS

We have demonstrated that closed-form solutions can be found, albeit with considerable computational burden, for transient behavior of some queueing systems as long as the arrival and service rates remain constant. We will now explore solution techniques for developing state probabilities when either the arrival rate or the service rate or both may vary over time. When the changes in rate occur with high frequency or with some continuous pattern, the system never reaches steady state in the usual sense of the word. Only transient solutions have meaning.

SYSTEM MODEL

We will limit our attentions to finite capacity systems which can be described by a finite set of differential-difference equations of the birth-death variety. In its most general form the system of equations with which we will work in this section is

$$\frac{dp_0(t)}{dt} = -\lambda_0(t)\, p_0(t) + \mu_1(t)\, p_1(t)$$

$$\frac{dp_n(t)}{dt} = \lambda_{n-1}(t)\, p_{n-1}(t) - [\lambda_n(t) + \mu_n(t)]\, p_n(t) + \mu_{n+1}(t)\, p_{n+1}(t),\ 0 < n < k$$

$$\frac{dp_k(t)}{dt} = \lambda_{k-1}(t)\, p_{k-1}(t) - \mu_k(t)\, p_k(t) \tag{8.1}$$

The variable coefficients in this set of differential equations pose special difficulties for most standard solution procedures. Even if $\lambda(t)$ and $\mu(t)$ are smooth well-behaved functions it may be impossible to find a compact expression for $p_n(t)$. When those rate functions are described by curves developed from empirical data, a complete formal solution is totally out of reach. We will circumvent these problems by following the approach suggested by Koopman (64).

Koopman suggests that one approximate the time-varying arrival and service-rate functions by a series of step functions. The width of the step should be selected on the basis of the rate of change in $\lambda(t)$ and $\mu(t)$. If those functions are changing very slowly one can tolerate a fairly long step size. If they are changing rapidly, shorter steps are necessary to achieve a reasonable approximation. Within the time period covered by a single step the arrival and service rates are assumed to be constant. The set of equations (8.1) then become the familiar finite set of differential equations with constant coefficients.

Figure 8-1 illustrates how one might approximate time-varying arrival rate and service rate functions for a single-channel system by a series of step functions. The arrival rate function is typically developed by recording the average number of arrivals per hour of the day over a several

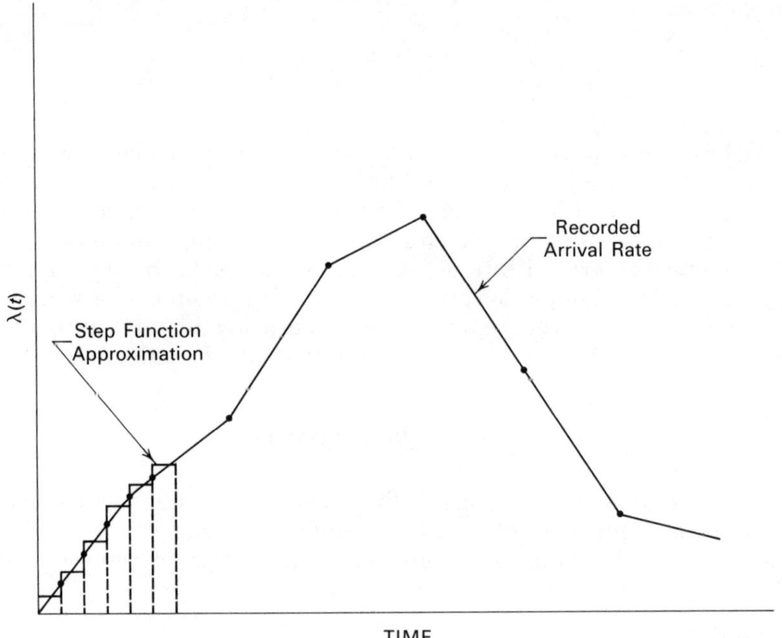

FIGURE 8-1　Step Function Approximation to a Time-Varying Arrival Rate

day period. The service rate function may be an experimental variable designed to answer questions such as what happens if the service facility becomes temporarily inoperative during a traffic rush period. The hourly averages are connected by straight lines. If our calculation interval is, say, fifteen minutes, each hour would then be divided into four steps with the arrival and service rates assumed to be equal to the average of the right and left hand limits contained within each step. These values provide the constant coefficients for the set of differential equations.

The solution technique is to:

1. Begin with some known boundary condition, e.g., no customers present, at time zero.
2. Assume a constant arrival and service rate over the first time interval T_1.

3. Solve the system of equations (8.1) for $p_n(t)$. This solution is valid over the interval $[0, T_1]$.
4. At the end of the first time interval the state of the system is described by the probability distribution $p_n(T_1)$. These become the boundary conditions for the next interval.
5. Select new constant arrival and service rates and resolve the set of equations using $p_n(T_1)$ as the boundary conditions.
6. Repeat this solution procedure over the number of intervals defining the planning horizon.

Step 3, the solution of the set of equations within each interval, can be accomplished by several different techniques. For example, in a single-channel system one could determine the conditional state probabilities $p_n{}^r(t)$ using equation (7.30) for each interval with its uniquely defined values for λ and μ. The variable t would then measure elapsed time from the left-hand end point of that interval. The unconditional distribution in that interval is obtained by weighting $p_n{}^r(t)$ by the state probability distribution evaluated at the end of the previous time interval. That is for interval j and an assumed constant interval width T,

$$p_n(t) = \sum_{r=0}^{k} p_r[(j-1)T] p_n{}^r[t - (j-1)T],$$

$$(j-1)T \leq t \leq jT, \qquad j > 0 \tag{8.2}$$

A second alternative is to find the solution within each time interval using numerical methods for solving systems of linear differential equations with constant coefficients. There are several commercial computer packages available for executing such calculations. For small scale (M/M/1/k) systems the numerical methods may not be as computationally efficient as the use of equations (7.30) and (8.2). However, they are more general and permit ready access to much more complicated models including multi-channel and feedback configurations where there are no known closed-form solutions. For that reason we will emphasize numerical methods for the balance of this chapter.

NUMERICAL METHODS

There are a number of standard numerical methods available for solving simultaneous ordinary initial value differential equations. Two which are available in the IBM Scientific Subroutine Package, SSP (48) involve the fourth-order *Runge-Kutta* method and a fourth-order *predictor-corrector* method.

The predictor-corrector method provides the user with estimates of the error bound of the solution while the Runge-Kutta method does not. However, in comparing the output from the two routines over several prototype time-varying queueing problems, Hartman found that the accuracy of the two methods for the problems tested was identical (43). Furthermore the predictor-corrector routine required approximately 50 percent longer execution time than the Runge-Kutta routine for the same problem. Consequently he recommended that the Runge-Kutta method be used to solve the equations for the time-varying queueing model. That is the method used in the programs contained in Appendix C.

The fourth-order Runge-Kutta method evaluates the dependent variable in a differential equation at the end of a small step h of the independent variable. The method approximates the first four terms of the Taylor Series expansion of the dependent variable. The equations we must solve form a system of first-order ordinary differential equations of the type

$$\frac{dy_i}{dt} = y_i{}' = f_i[t, y_1(t), y_2(t), \ldots y_n(t)] \tag{8.3}$$

with initial conditions $y_i(t_0) = y_{i0}$. What we seek are the values $y_i(t_0 + h)$ where h is an increment in the independent variable t and $i = 1, 2, \ldots n$. The variable h is an arbitrary but small increment in t. The output is a set of approximations $\bar{y}_i(t_0 + h)$ to the true values $y_i(t_0 + h)$. The Runge-Kutta method is an algorithm designed to approximate the Taylor Series solutions

$$y_i(t_0 + h) = y_i(t_0) + hy_i{}'(t_0) + \frac{h^2}{2} y_i{}''(t_0) + \ldots \tag{8.4}$$

Runge-Kutta does not require evaluation of derivatives beyond the first but rather obtains approximations at the expense of several evaluations of the first derivatives. The classical fourth-order method requires four evaluations of the first derivative to obtain a Taylor Series approximation through terms of order h^4. Let

$$\bar{y}_n = \bar{y}(t_0 + nh) \text{ and } (t_0 + nh) = t_n$$

then for the single variable problem the fourth-order Runge-Kutta method calculates

$$y_{n+1} = y_n + \frac{1}{6} (k_1 + 2k_2 + 2k_3 + k_4) \tag{8.5}$$

where

$$k_1 = h \cdot f(t_n, \bar{y}_n)$$
$$k_2 = h \cdot f(t_n + h/2, \bar{y}_n + k_1/2)$$
$$k_3 = h \cdot f(t_n + h/2, \bar{y}_n + k_2/2)$$
$$k_4 = h \cdot f(t_n + h, \bar{y}_n + k_3)$$

A complete discussion of this method is given by Conte (20:220) and by Ralston and Wilf (77). For our purposes it suffices to know that the method is well proven and is available in a standard computer software package.

In the programs presented in Appendix C the set of equations (8.1) is solved by Runge-Kutta within each interval. An interval is defined by the width of the step used to approximate the arrival and service rates by constants as shown in Figure 8-1. Starting at the beginning of an interval

T_j the equations are solved in increments of length h until the end of the interval is reached. The values at the end of the interval became the initial conditions for the next interval. Operating statistics are calculated and printed at the end of each interval of length T_j. For any given problem there is an optimum increment h. The trade-off is between execution time and accuracy. After experimenting with several values, Hartman found that a step size of $h = .03$ time units provided convergence for his problems. Larger values of h also provided convergence but very little decrease in execution time. Consequently, he used $h = .03$ in his programs. Using this technique, execution time for problems involving a 24-hour test period in a system with a maximum capacity of twenty-five typically run about thirty seconds on the IBM 360/75.

PERFORMANCE MEASURES

The measures of performance for the truncated time-varying queue offered in the program in Appendix C are all related to the state-probability distribution. The probability distribution $p_n(t)$ is printed at the end of each time interval T_j. From that distribution the program calculates the expected number in the queue, the standard deviation for number in the queue, and the expected total customer minutes delay time in that interval. The program also calculates an approximation to the expected number of rejects, i.e., customers who cannot gain access to the system because of saturation. The mean and standard deviation for queue length follow directly from the definitions and require no special definition. However, the expected total delay time and expected rejects are not so straightforward.

The expected total delay time within a computation interval is calculated as the product of the average number of customers in the queue during that interval and the length of the interval. If one assumes that the expected number in the queue varies in linear fashion over the interval, the average number present can be approximated from

$$E(Q) = \frac{E(Q_{j-1}) + E(Q_j)}{2} \tag{8.6}$$

where $E(Q_j)$ is the expected number in queue at the end of interval j, as calculated from $p_n(t)$. The total expected holding or delay time within an interval of length T is then

$$E(\text{HOLD}) = E(Q) \cdot T$$

$$= \left[\frac{E(Q_{j-1}) + E(Q_j)}{2} \right] T \tag{8.7}$$

These holding times are summed over the entire execution period for the convenience of the analyst.

$$E(\textit{THOLD}) = \sum_{j=1}^{N} E(\text{HOLD}(j)) \tag{8.8}$$

where $E(\text{HOLD}(j))$ is the expected value of customer-minutes delay in period j.

The cumulative number of rejects, i.e., customers denied service, is approximated by multiplying $p_k(t)$, the probability of saturation, by $\lambda(t)$, the arrival rate and the time during which the system is saturated. This calculation takes place at each step in the Runge-Kutta approximation rather than just at the end of a major subdivision.

STANDARD PATTERNS

The numerical solution routine can be used to study single and multi-channel system response to a variety of prototype driving functions, including sinusoid arrival and service rate patterns, input surges and service breakdowns. Some typical results for single-channel service are graphed in Figure 8-2.

Figure 8-2 illustrates the effect of periodic demands on the expected number of customers in the system. These data were obtained at fifteen-minute intervals for a single channel system with a truncation point of 26. The service rate was assumed to be constant at a rate of twenty customers per hour. Demand rate was assumed to be periodic ranging from zero to twenty per hour over a six-hour cycle. Several observations can be made from these curves.

1. The expected number of customers in the system is periodic with the same period as the demand rate function.
2. The expected number of customers present peaks at a later time than the maximum point on the arrival rate curve. In this case the lag is forty-five minutes.
3. The expected number of customers reaches a minimum value after the arrival rate has reached zero but with less lag than present at the maximum points. In this case the lag is fifteen minutes.

Runs made at other service rate levels indicate that the peak-to-peak lag between arrival rate and expected number in the system increases as the service rate decreases. For example, when $\mu(t) = 10$ the lag time is greater than one hour and fifteen minutes.

There is also an interesting periodic pattern in the variance for number of customers in the system. For those sample runs tested, the variance has the same period as the other curves, but reaches a peak when $E(N_t)$ is at an average or nonextreme value and reaches a minimum when $E(N_t)$ is at an extreme point. Results for other combinations of time-varying demand and service rate functions are easily obtained from the computer programs in Appendix C.

PERIODIC STEADY STATE

We have already noted that there is no steady-state solution in the usual sense of the word for systems which have time-varying demand and

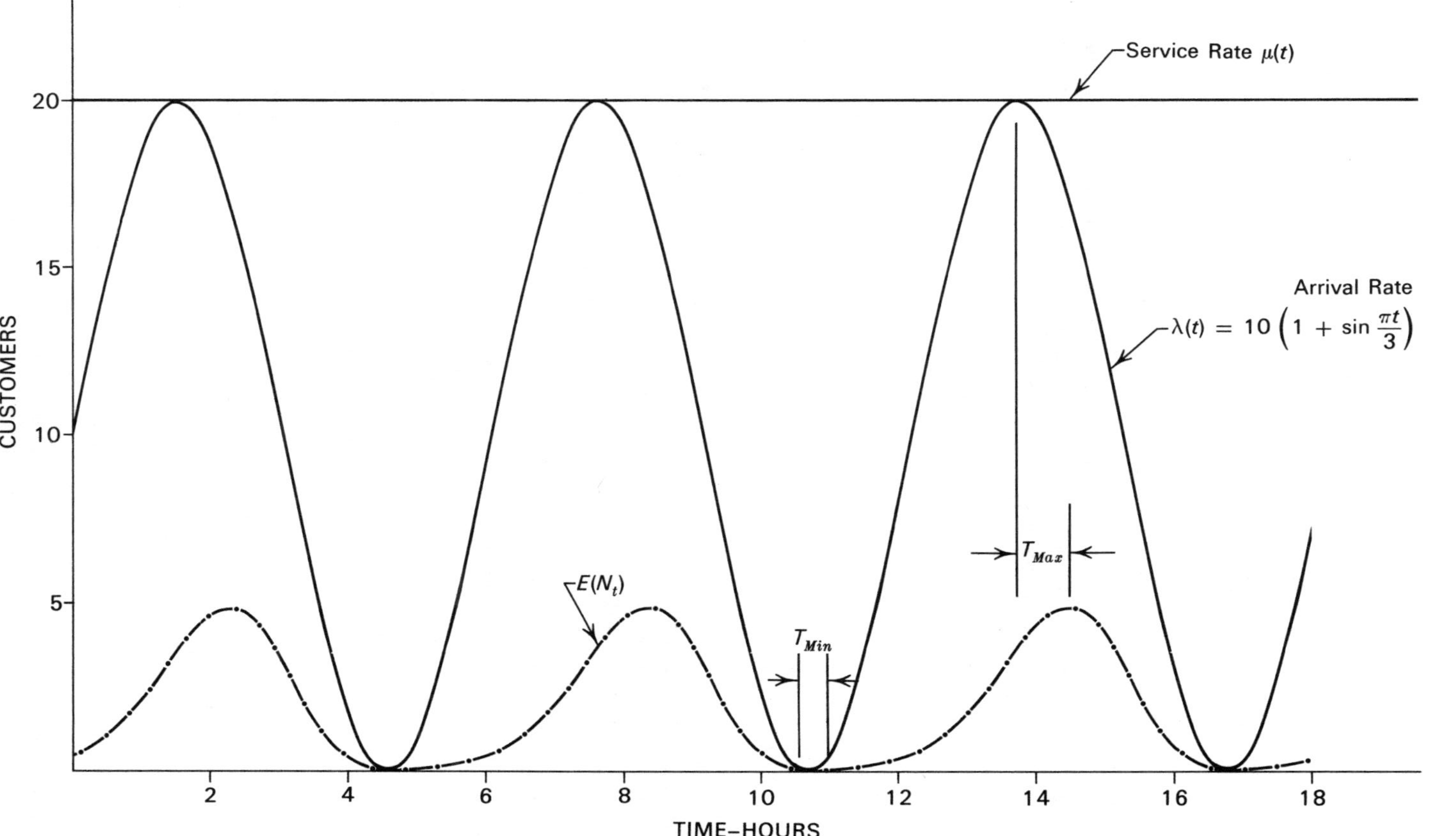

FIGURE 8-2   Time Lags with Periodic Demand

service rate functions. However, the results shown in Figure 8-2 do suggest that there may be a "periodic steady state" for those systems with periodic driving functions. For a system to have a periodic steady state with period T means that the probability distribution $p_n(t)$ exactly repeats itself every T time units after the system has run for a long time.

Koopman noted this periodicity in the state distribution for air traffic in major terminal areas (64). His data on aircraft arrivals were diurnal in the sense that the arrival rate for aircraft was nearly zero at midnight and followed the same average daily pattern of peaks and valleys. The probability distribution he obtained from numerical solutions to the truncated birth-death equations also appeared to be periodic with the same 24-hour period as the input data. Koopman formally proved by classical methods that such a periodic steady state does exist. His results for the single channel system can be stated as follows:

1. When the arrival rate and service rate functions are periodic such that $\lambda(t + T) = \lambda(t)$ and $\mu(t + T) = \mu(t)$, there is a unique periodic solution to the system of stochastic equations (8.1) such that $p_n(t + T) = p_n(t)$.
2. Every probability distribution which satisfies equations (8.1) together with the conditions $\lambda(t + T) = \lambda(t)$ and $\mu(t + T) = \mu(t)$, approaches the periodic solution at a geometric rate as t increases indefinitely.

There are several ways one could go about constructing a periodic solution. The most straightforward method is simply to repeat the periodic input data over several periods and carry forward $p_n(t)$ without reinitializing until it satisfactorily reproduces each of its values at both ends of an interval $[sT, (s + 1)T]$. At that point the distribution

$$\hat{p}_n(t) = p_n(\tau + t), \qquad sT \leq \tau \leq (s + 1)T, \qquad 0 \leq t \leq T \qquad (8.9)$$

can be assumed to be the periodic steady-state distribution.

An alternative approach is to first calculate the conditional probability of finding the system in state n at time T for every possible starting state. That is

$$Pr[n \text{ present at } t = T | i \text{ present at } t = 0) = p_n^{\,i}(T) \qquad (8.10)$$

These probabilities can be obtained by solving equation (7.30) or by running the numerical solution programs $(k + 1)$ times, one for each possible starting state, all evaluated at $t = T$. The resulting set of values $\{p_n^{\,i}(T)\}$ form a one-stage transition matrix for a simple Markov chain, i.e., $p_n^{\,i}(T)$ is the probability that the system will move from i to n in time T.

The periodic steady-state values for $\{\hat{p}_n(0)\}$ can be obtained from elementary matrix operations first introduced in Chapter 2. Let the vector $\mathbf{C}$ contain the elements $\{p_n(0)\}$ and the matrix $\mathbf{A}$ contain the elements $\{p_n^{\,i}(T)\}$. Then the unknown elements of $\mathbf{C}$ can be obtained from the basic steady-state equations

$$C = CA \tag{8.11}$$

and

$$\sum_{n=0}^{k} c_n = \sum_{n=0}^{k} \hat{p}_n(0) = 1$$

With the periodic steady-state distribution now known at time $t = 0$, it is a simple matter to use the values $\hat{p}_n(0)$ as initial conditions and resolve for $p_n(t)$ from equations (8.1). The resulting solution over the interval $[0, T]$ will be the desired periodic steady state $\hat{p}_n(t)$.

CONSTANT SERVICE TIME

Modified Poisson arrival patterns appear to be reasonable approximations for many real-world queues. However, it is not so clear that an exponential service law will often be realistic. As an alternative to exponential service let us now consider systems with constant service times. In these systems service time within an interval has no variance but rather requires a fixed amount of time h for every customer served in that interval.

This is the model introduced in the air traffic control example of Chapter 3. Time-varying arrival and service-rate functions are again approximated by a series of step functions. If the empirical data suggests that the service rate in an interval T_j is $\mu(T_j)$, the implied constant service time is

$$h_j = 1/\mu(T_j) \tag{8.12}$$

To ease the problem formulation let us treat the system much as a conveyor belt would function. The service system continues to cycle every h time units whether or not customers are present. This causes problems only when a customer arrives at an otherwise empty system midway in a service epoch. Under the assumption being made here he cannot gain access to the service facility until the beginning of the next service epoch. When this type of assumed behavior is contrary to the realities of the operating system, the model results will be at best approximations to system performance. The assumption has very little distorting effect, however, when h or $p_0(t)$ are small.

Whenever there is not an integral number of service epochs in an interval T_j we will assume that the length of the epoch overlapping two intervals will be determined by the service rate in the interval in which the epoch started. This assumption will cause slight distortion in results only during periods in which service rates change dramatically.

The model is a simple Markov chain in which transitions take place over time h. The time between transitions is constant within the interval covered by one step in our step function approximation to the time-varying service rate but may be a different constant in different intervals. The single stage transition matrix for a single channel system from the beginning of one service epoch to the beginning of the next is described as

$$\mathbf{A}(h\,|\,\lambda,\mu) = \begin{array}{c|ccccc} & 0 & 1 & 2 & & k \\ \hline 0 & f(0) & f(1) & f(2) & \cdots & \sum_{i=k}^{\infty} f(i) \\[2ex] 1 & f(0) & f(1) & f(2) & \cdots & \sum_{i=k}^{\infty} f(i) \\[2ex] 2 & 0 & f(0) & f(1) & \cdots & \sum_{i=k-1}^{\infty} f(i) \\[2ex] 3 & 0 & 0 & f(0) & \cdots & \sum_{i=k-2}^{\infty} f(i) \\ \vdots & & & & & \\ k & 0 & 0 & 0 & \cdots & \sum_{i=1}^{\infty} f(i) \end{array}$$

$$(8.13)$$

where

$$f(i) = \frac{(\lambda/\mu)^i \, e^{-\lambda/\mu}}{i!}$$

The time-varying probability distribution can now be calculated in iterative fashion by repeated matrix multiplication.

$$\mathbf{P}(t + h) = \mathbf{P}(t)\mathbf{A} \tag{8.14}$$

where

$$\mathbf{P}(t) = (p_0(t), p_1(t) \ldots p_k(t))$$

At the end of a step interval, say at time $t = sT$, the matrix $\mathbf{A}$ is changed to reflect the new constant arrival and service rates. The vector existing at that time $\mathbf{P}(sT)$ is printed out, then used as input to the next set of iterations. The resulting time-varying probability distribution can be used exactly as we used the results of the exponential service model to calculate any desired measures of performance.

The constant service time model programmed for the convenience of the reader is contained in the Appendix. That program also contains the capability of treating bulk service problems where service takes place in batches of size KI. We will have more to say about the bulk service problem in future chapters. For purposes of this chapter $KI = 1$. The program will calculate the entries in the matrix $\mathbf{A}$, carry out the Markov iteration, calculate appropriate statistics and print the results. All that is required of the user is a specification for system capacity and the values for $\lambda(t)$ and $\mu(t)$.

ROBUST MODEL

One might argue that neither the constant service time nor exponential service time assumptions are very realistic. More probably a real-world service distribution will fall somewhere between these two extremes. In fact the common queueing theory practice of fitting an Erlang or a gamma distribution to service time data describes such an intermediate distribution. The exponential distribution is the low-order member of the gamma family with shape parameter $k = 1$. The constant service time distribution is the high-order member of the gamma family with shape parameter $k = \infty$. The gamma distribution is given by

$$f(t) = \frac{(\mu k)^k}{(k - 1)!}\, t^{k-1}\, e^{-k\mu t} \tag{8.15}$$

The mean and variance for this distribution are

$$E(t) = 1/\mu \tag{8.16}$$

and

$$Var(t) = 1/k\mu^2 \tag{8.17}$$

There is further reason to argue that the exponential and constant service time assumptions bound most other reasonable service distributions. Consider a single server queue with a Poisson arrival system and exponential service time with mean "c" to be compared with a similar system in which service time is a constant equal to "c." From an information theory point of view we can define the *entropy* of a distribution $f(x)$ as

$$H(x) = -\int_0^\infty f(x) \ln f(x)\, dx \tag{8.18}$$

Reza (79:78–84) has proven that if we try to find a function $f(x)$ which maximizes entropy subject to

$$\int_0^\infty f(x)\, dx = 1 \tag{8.19}$$

and

$$\int_0^\infty xf(x)dx = c \tag{8.20}$$

the distribution will be the exponential

$$f(x) = (1/c)e^{-x/c} \tag{8.21}$$

We can treat the constant service as a density

$$g(x) = \lim_{b \to a} \frac{1}{b - a} \tag{8.22}$$

where

$$c = \frac{b + a}{2} \tag{8.23}$$

Replacing $f(x)$ by $g(x)$ in equation (8.18) it is obvious that constant service will give a value of $H(x) \rightarrow -\infty$ and therefore minimizes entropy.

What this discussion of entropy leads to is the observation that random or exponential service maximizes uncertainty while constant service minimizes uncertainty. The thought is that most service distributions of practical significance should fall between these two extremes.

Both Koopman (64) and Hartman (43) have compared results for real system time-varying demand data run against both constant and exponential service distribution assumptions. The somewhat surprising, but very encouraging result, is that the envelope formed for most performance measures by these two models is extremely tight. Examples of the narrow range between exponential and constant service time results for the expected number and standard deviation of customers in a system subjected to sinusoid input are shown in Figures 8-3a and 8-3b.

Since the performance measure bands are so narrow and since we have argued that most service time distributions should fall between the extremes of constant and exponential service, it seems reasonable to use the model which is most easily manipulated with little regard to the actual form of the service distribution. For that reason it appears that the exponential service distribution model, which is a very robust tool and is easily solved by numerical methods, is the proper choice for analysis and design of most time-varying queueing systems.

MULTIPLE CHANNELS

The numerical solution routines contained in Appendix C can be used to analyze both single and multi-channel systems. A commonly encountered problem which can be easily handled by these programs is to evaluate the effect of combining or splitting service facilities.

To illustrate one application consider the time-varying demand rate shown in Figure 8-4. This curve is similar to the data Koopman reported as generated from an FAA study of landings and departures at La-Guardia airport during a one-month period in 1968 (76). The initial analysis supposes that single-channel, exponentially distributed service is available at a constant rate of 45 operations per hour. The demand rate varies from near zero to a peak load of 66 per hour during the evening rush period. Even though the total expected number of potential services in a 24-hour period exceeds the total expected demand, there are periods during which demand far exceeds service capabilities and other periods in which one possesses an excess service capacity. In such an environment it is obvious that steady-state solutions have little to offer.

The solid line in Figure 8-5 shows how the expected number of aircraft in the system will vary over time if only a single channel (runway in this context) is available and the maximum capacity is 25. As was true in our previous analysis of simple sinusoid demand curves, there is a lag between the time at which peak demand occurs and the time at which the average number of aircraft present will reach its maximum. However, the general shape of the expected value curve is similar to that of the demand rate.

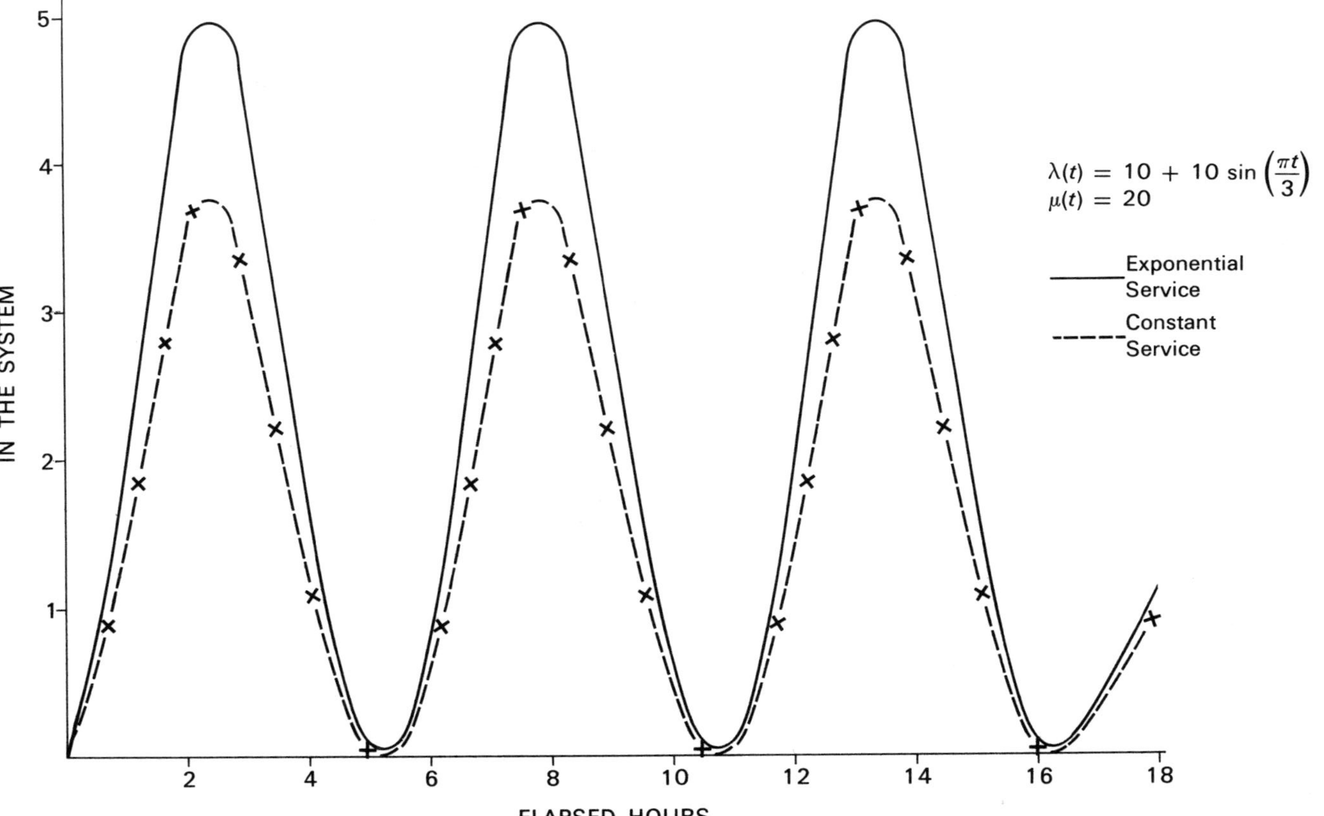

FIGURE 8-3a Performance Envelope for Means

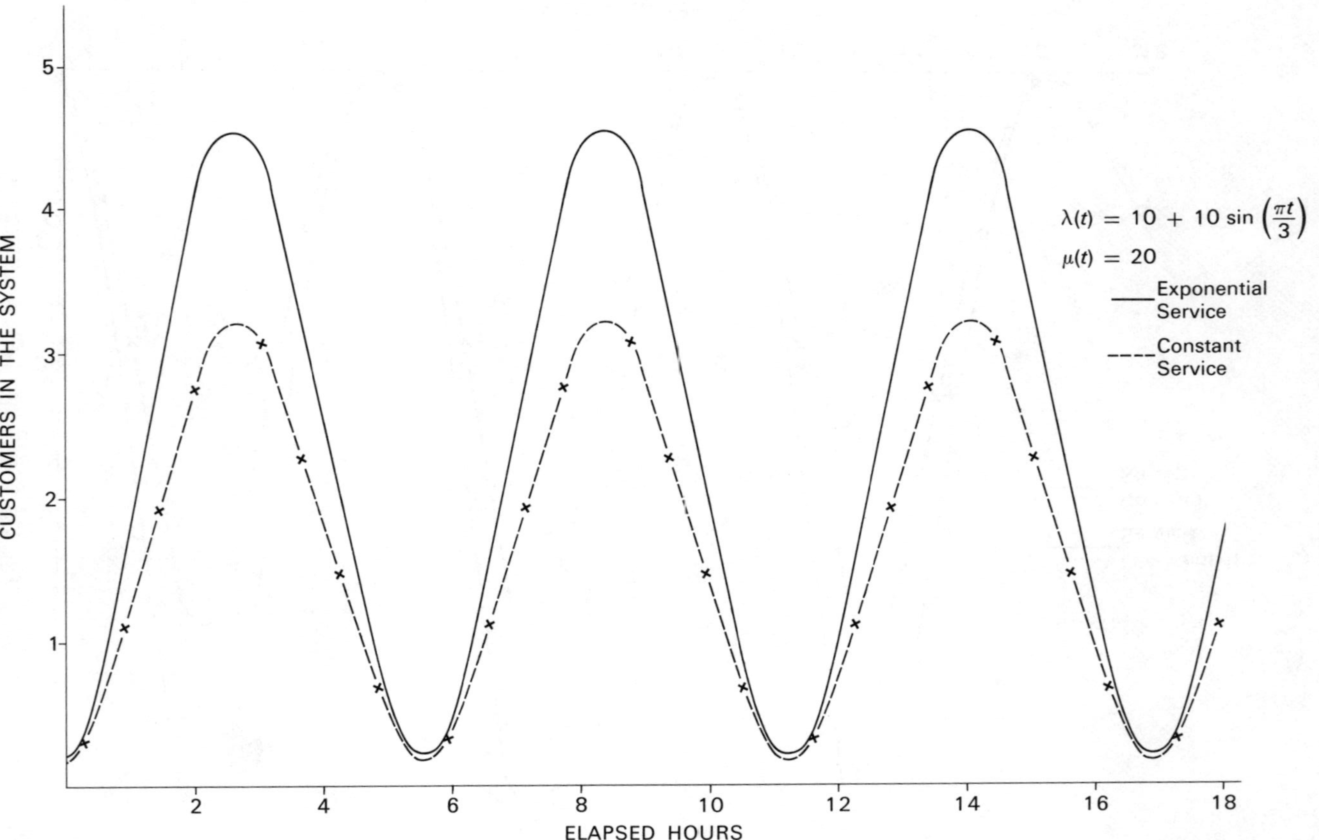

FIGURE 8-3b Performance Envelope for Standard Deviation

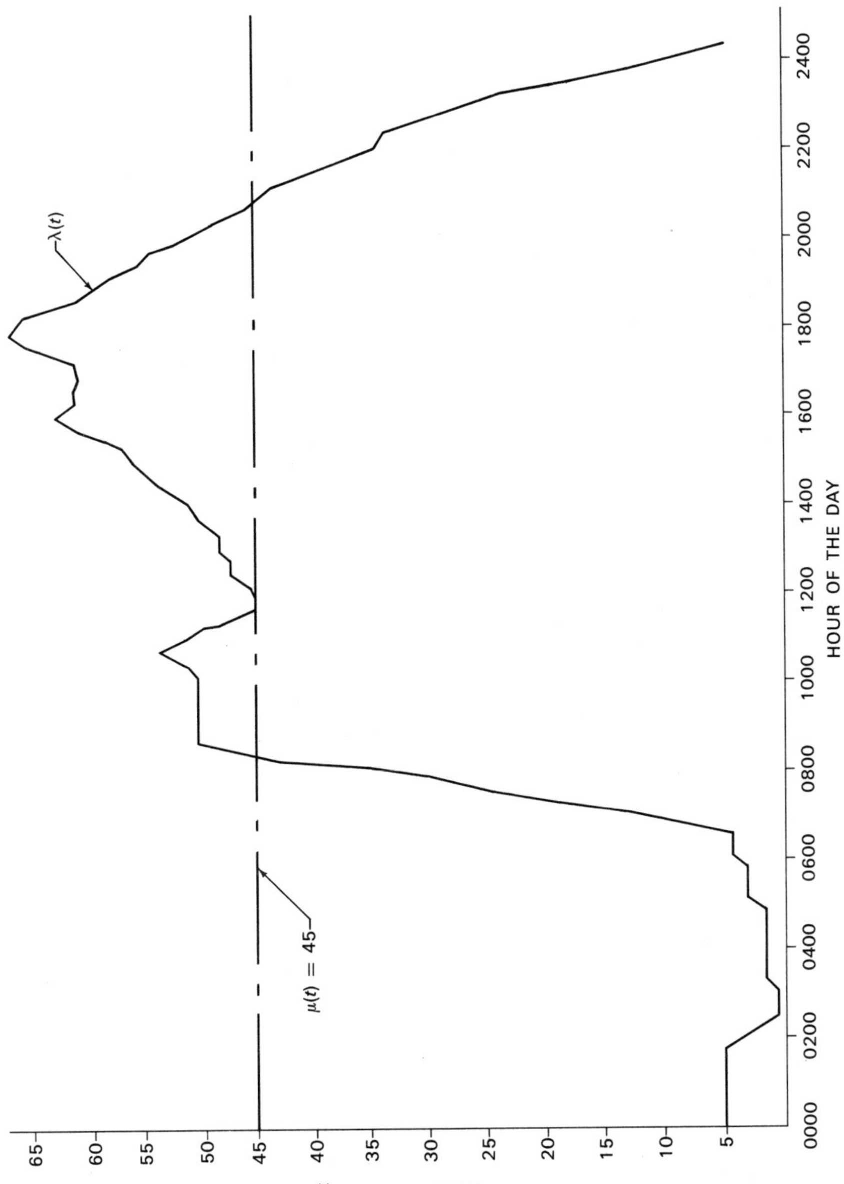

FIGURE 8-4 Demand Pattern for a Large Air Terminal

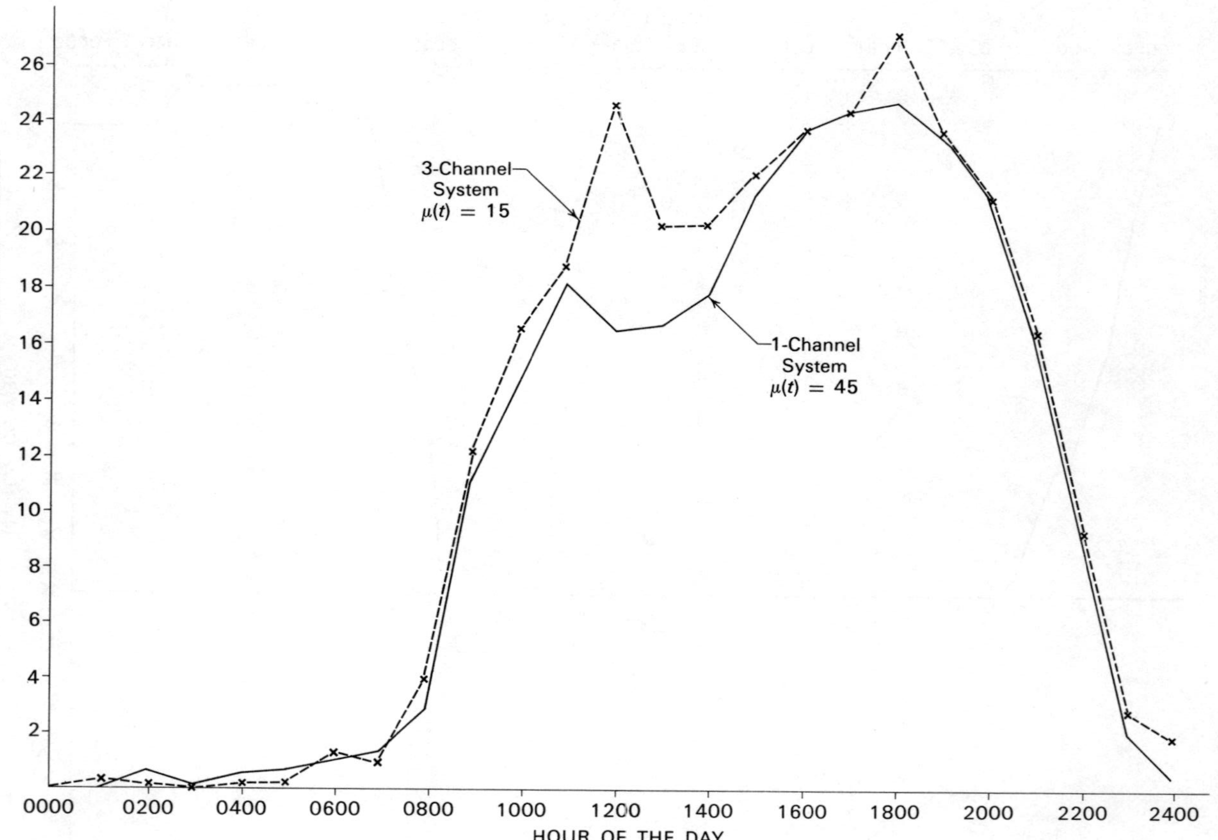

FIGURE 8.5 Multichannel Comparison

The dashed line in Figure 8-5 illustrates the effect on expected number present one might find if the system were redesigned to provide three separate channels, each with a service rate equal to one-third of the single-channel system. In the modified system the maximum potential service rate is the same as it was for the single-channel system, 45 per hour. However, aircraft will be served at that maximum rate only when there are three or more aircraft in the system. When only one is present the rate is 15 per hour and when two are present it is 30 per hour as opposed to the single-channel configuration in which the rate is always 45 per hour.

Comparing the two curves we can see that the three-channel system tends to provide poorer service, particularly at high load levels. Furthermore, the lag time between peak demand and peak expected number of aircraft present is longer for the three-channel system than it is for the single-channel system. It is also true that the three-channel design is more likely to become saturated with the result that aircraft must be diverted to other facilities. In fact the expected number present at 1800 hours (6:00 PM) is very nearly equal to the absolute capacity of 25. The expected total number of aircraft which must be diverted to other facilities will exceed 100 by 6:00 PM in the multi-channel system but will never reach that level in the equivalent single-channel design. If the average number of customers present and the average number rejected are the sole measures of performance, it is obvious that a centralized facility in this context is preferred.

MODEL VERIFICATION

One method which may be useful in establishing the worth of numerical methods such as those advocated in this chapter is to compare the results of the time-varying queueing model to those obtained by other techniques when both are driven by the same input data. That is the tack we will pursue. The following paragraphs contain material adapted from the M.S. Thesis written by Anderson H. Walters under the direction of the author.[1]

The basis for comparison is a 1970 simulation study reported as a part of the Summer Predoctoral Fellowship Program in Engineering Systems Design conducted by NASA and West Virginia University (84). The objective of that effort was to study through simulation how best to permit maximum flow of aircraft into and out of a terminal area safely and economically so that delays are minimized. Since the simulation study considered many more variables than one can capture in our basic limited capacity queueing model, the first task was to extract a subproblem which could be fit into our framework. The results for this admittedly more abstract system could then be compared with the corresponding portion of the simulated system.

---

[1] Adapted with permission of the author from Anderson H. Walters, "The Application of an Analytical Model to the Analysis of Air Traffic in the Terminal Area," M.S. Thesis, The Ohio State University, 1976.

Demand data can be generated from traffic counts made from time to time at terminals with FAA control towers. The NASA-West Virginia study used data generated at Atlanta, Georgia, on July 9, 1970. These data were felt to reflect a representative day of traffic movement at that terminal. The traffic count was made by aircraft category where the category system, defined by maximum aircraft takeoff weight, is that used by the Airport Capacity Handbook (2). Categories 1 and 2 correspond to single engine and light twin engine aircraft commonly flown by general aviation. Categories 3, 4, and 5 cover weights ranging from 33, 000 pounds, which correspond to light jet-aircraft of the Lear Jet variety, through 260,000 pounds, which correspond to commercial carriers such as the Boeing 707. Category 6 represents the wide-body and heavy jets such as the Boeing 747.

The first task was to abstract from the available demand data a composite demand rate $\lambda(t)$ which could be used in the time-varying queueing model. Category 6 was easily eliminated since no arrivals of aircraft of that class were recorded on July 9, 1970. Operating procedures in effect at Atlanta at that time routinely separated general aviation aircraft from commercial carriers by conducting most general aviation operations from a separate runway. For that reason it was decided to combine categories 3, 4, and 5 to represent the arrival stream to the major runway. The resulting arrival counts, used as the step function input to our queueing model, are plotted in Figure 8-6.

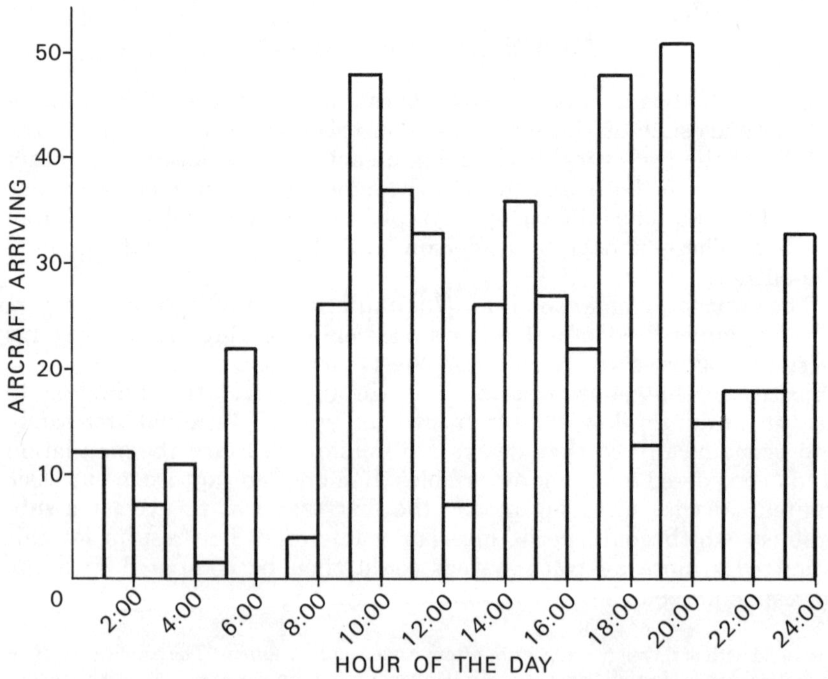

FIGURE 8-6 Hourly Arrivals of Large Aircraft at Atlanta, July 9, 1970

Note that this terminal is characterized by extreme variability in demand over a 24-hour period and hence not easily analyzed by steady-state models. Simulation or time-dependent queueing models are essential to obtain a real understanding of how the system operates.

The maximum queue capacity is determined by the number of holding stack spaces available and the number of aircraft which can be safely monitored on radar by a crew of air traffic controllers. The New York area, one of the largest, can handle approximately sixty aircraft in holding patterns. Most other terminals can handle considerably less. For that reason a capacity of twenty-five was chosen for this study. That number proved to be adequate since the expected number of aircraft denied service due to saturation calculated by our model was less than one over the 24-hour period.

The final ingredient necessary to run the model is an estimate of the service-rate capabilities. Since the commercial traffic at Atlanta is normally confined to a single runway, a single-channel model is appropriate. From FAA data it is possible to establish typical performance profiles for each category of aircraft in terms of the expected approach, transition, and climb speeds. For example, for category 5 aircraft the expected final approach speed is 150 knots, the initial approach speed is 170 knots, transition speed is 200 knots, and climb speed is 290 knots. The FAA Terminal Air Traffic Control Handbook 7110.61, 1 January 1976 specifies that the minimum separation distance between two "large" aircraft in trail, considered in this study to be categories 3, 4, and 5, shall be three miles. This separation is a safety measure required for wake turbulence avoidance and normally expected system accuracy. The separation distance on final approach is the determining factor for the maximum potential service rate.

The time necessary for an aircraft to fly the required three miles can be calculated as 1.36, 1.20, and 1.04 minutes for categories 3, 4, and 5, respectively, at assumed final approach speeds of 115, 130, and 150 knots.

Observation of operations at major terminals and personal conversations with air traffic controllers led to the conclusion by Walters and the author that the regulatory minimum separation is rarely met. Rather it is a common practice to increase separation distances to provide allowances for unforeseen error and controller confidence that he can safely maintain the traffic flow. For that reason it was decided to add fifteen seconds to the required time separations as a realistic assumption for error allowance.

Finally it was necessary to combine service rates for categories 3, 4, and 5 to obtain a single composite service rate to use in the model. This was accomplished by first counting the number of arrivals of each class in the "representative" 24-hour period then converting these to percentages. The conclusion was that about 3 percent of the total demand in the arrival stream of concern was category 3, 31 percent was category 4, and 66 percent was category 5. The expected service time for a randomly arriving aircraft was then determined by weighting each of the calculated separation times plus error allowance by the percentage of aircraft falling in that category. That is

$$\bar{t} = .03(1.36 + .25) + .31(1.20 + .25) + .66(1.04 + .25)$$
$$= 1.3492 \text{ minutes}$$

which converts into an hourly service rate of

$$\mu(t) = \frac{60}{t} = 44.47 \text{ per hour.}$$

For purposes of modeling this number was rounded to $\mu(t) = 45$ per hour.

In summary, the model developed for comparison with the NASA-West Virginia simulation was a single-channel, time-varying queueing model with a constant service rate $\mu(t) = 45$, a capacity of $k = 25$ and a step function demand rate defined by Figure 8-9. A program similar to that in Appendix C was then used to generate the time-varying probability distribution and corresponding measures of performance. Some results of that analysis are plotted in Figures 8-7, 8-8, and 8-9. From those graphs it is easy to see which hours of the day will see the most traffic congestion and at what times arriving aircraft are most likely to be delayed.

To gain some insight into the power of the analytic solution, certain performance measures were compared with the much more complete simulation model. Since the objective of the simulation study was stated in terms of delay time, the total expected holding time was of prime importance. The simulation model was run for ten simulated days covering the hours of 0800 to 1800 each day. Multiplying the average simulated delay time per aircraft by the number of simulated movements over that period revealed a total expected delay time per 10-hour day of 1532 minutes. The expected holding time from the analytic model over the period of 0800 to 1800 hours was 1534.5 minutes.

The difference between the two model results appears insignificant. The time-varying queueing model can approximate the output of the simulation very closely. Furthermore, there appears to be a significant savings in computation time. The simulation model consisting of over 2500 cards reportedly required 80 seconds on a CDC 6600 computer to simulate ten 10-hour days. The queueing model consisted of 197 cards and required 5.61 seconds on an IBM 370 to obtain analytic results for the full 24-hour day. From these observations it can be concluded that numerical techniques for time-varying queueing models can be profitably employed in the analysis of complex systems often thought amenable only to simulation.

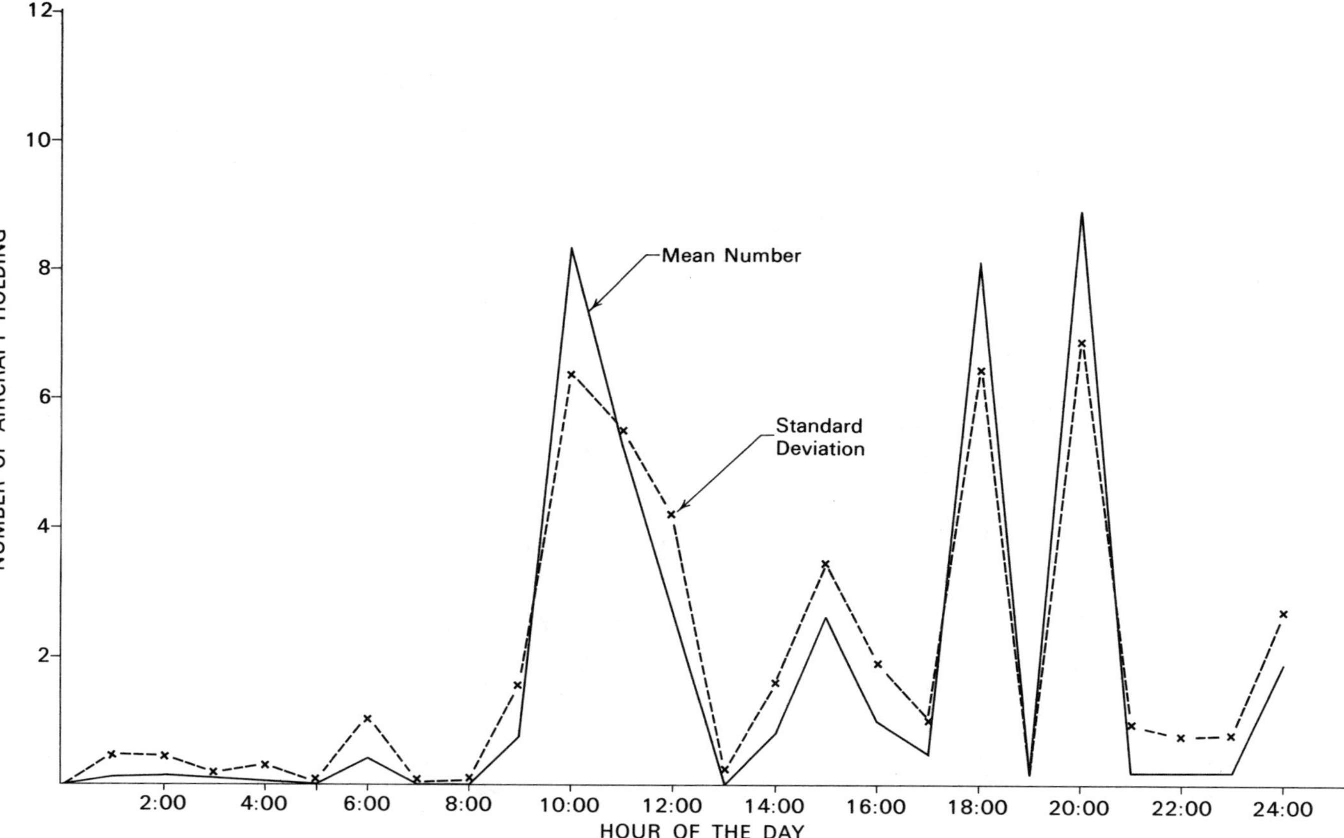

FIGURE 8-7 Mean and Standard Deviation for Number of Aircraft Holding

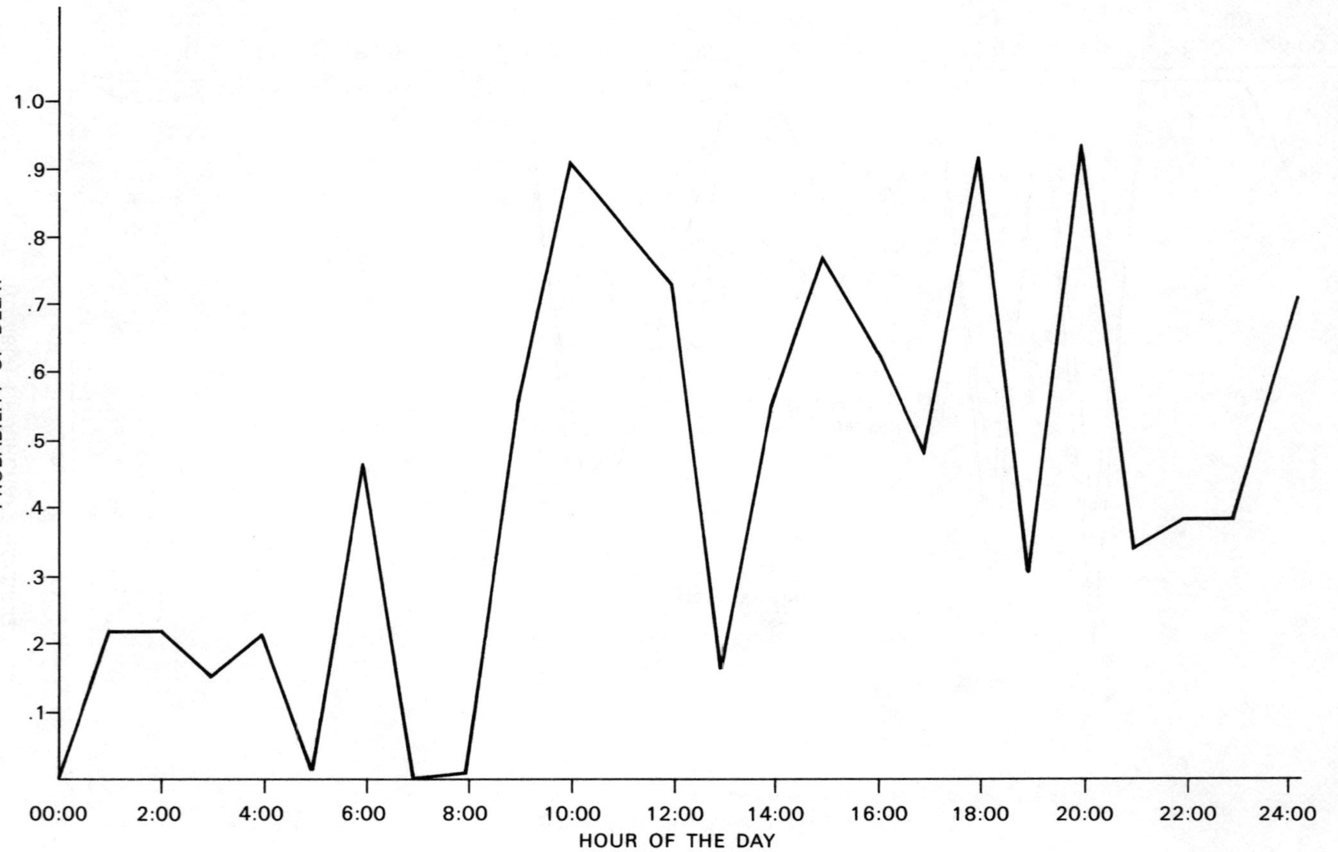

FIGURE 8-8 Probability of Delay

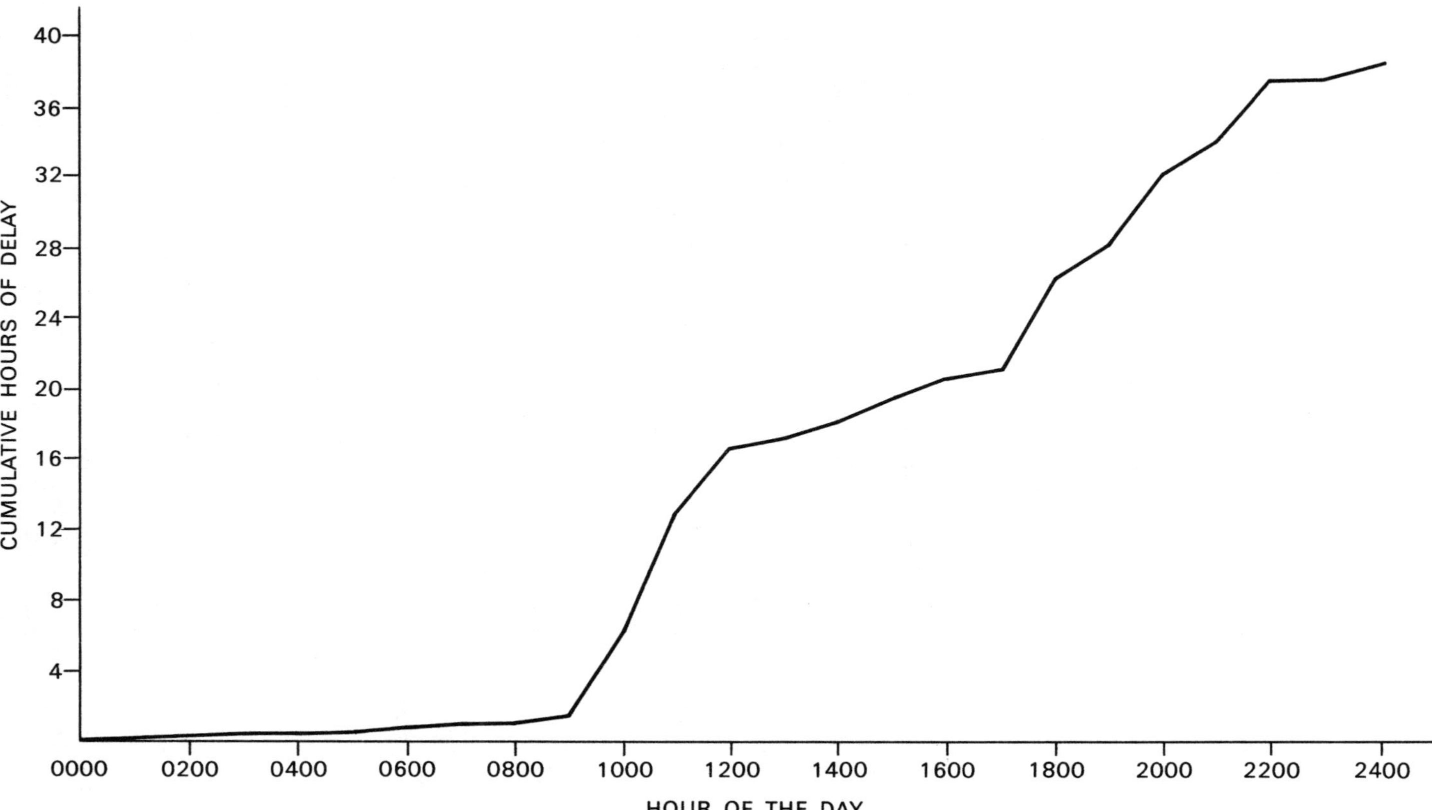

FIGURE 8-9 Aircraft-Hours of Delay

PROBLEMS

8.1 Pick a particular system from your own experience which exhibits time-varying queue behavior. Design a data collection program and conduct preliminary investigations of the applicability of the time-varying solution programs in Appendix C for analyzing your test system.

8.2 Consider a system with capacity 5 which is subjected to time-varying Poisson arrivals and exponential service according to the following pattern:

Δt	$\lambda(t)$	$\mu(t)$
0–2	5	2
2–4	4	4
4–6	2	3
6–8	2	4

a) Use the equations from Chapter 7 and the relationship suggested in Equation 8.2 to find $p_0(t)$ and $p_5(t)$ at the end of each one-hour period from [0, 8].

b) Perform the same calculation using the programs in Appendix C.

8.3 Analyze the behavior of a single channel system with capacity 25 which is subjected to periodic demand and service rate functions of the following types:

a) Service rate = 20/hour and arrival rate = $10\left(1 + \sin\frac{\pi t}{3}\right)$ where t is measured in hours.

b) Service rate = 10/hour and arrival rate = $10\left(1 + \sin\frac{\pi t}{3}\right)$.

c) Service rate = arrival rate = $10\left(1 + \sin\frac{\pi t}{3}\right)$.

d) Service rate = $11 + 10\sin\left(\frac{\pi t}{3}\right)$ and arrival rate = $10\left(1 + \sin\frac{\pi t}{3}\right)$.

e) Service rate = $10\left(1 - \sin\frac{\pi t}{3}\right)$ and arrival rate = $10\left(1 + \sin\frac{\pi t}{3}\right)$.

In each case plot the expected number in the system, the variance and number of rejects. Explain any relationships, e.g., phase shifts and periodic response, which appear to offer general insights into time-varying system behavior.

8.4 Consider the arrival pattern for aircraft in a terminal area as depicted in Figure 8-4. Perform a sensitivity analysis on the effect of changing the single-channel service rate from 15 to 25 to 45 to 55 per hour. Assume a fixed capacity of 25 aircraft and exponentially distributed service times.

8.5 Perform the same sensitivity analysis indicated in problem 8.4 for a deterministic service time system. Compare these results with the exponential service times.

8.6 Compare the results of Figure 8-5 with a two-channel system in which the service rate per channel is 22.5/hour. Suggest a proper timing for opening and closing the second channel so that total system performance is comparable to the single-channel system operating at a service rate of 45/hour.

8.7 Consider the arrival pattern for aircraft in a terminal area as depicted in Figure 8-6. Assume that service times are exponentially distributed at a rate of 30 per hour except for periods of severe weather interruptions and that the system has a capacity of 25 aircraft. Discuss the system impact of dropping the service rate to 15 per hour during the 5:00 to 6:00 PM period. Compare this result for a similar interruption in the 12:00 to 1:00 PM period.

8.8 Consider a customer arrival pattern similar to that in Figure 8-6. Suppose that service rate is fixed at 40 per hour. System capacity can be set at 20, 30, or 50 units for costs of $15,000, $30,000, or $60,000 per week, respectively. Each customer rejected costs the system $35. Each minute a customer is delayed costs the system $5. Determine the optimum capacity level.

8.9 Consider a customer arrival rate pattern similar to that in Figure 8.4. Suppose that one can staff the service facility with 1, 2, 3, or 4 servers in four-hour time blocks, e.g., we could have a two-channel system from 0000 to 0400, a three-channel system for 0400 to 0800, etc. The service rate per channel is 15 per hour. The system capacity is twenty customers. Each channel while operating costs $1000 per hour. Each customer rejected incurs a special handling cost of $50 and each minute a customer is delayed costs $6. Determine an optimal staffing policy.

8.10 A traffic light at the entrance to a plant parking lot is timed so that the green portion of the cycle lasts two minutes and the red portion three minutes. Under "full queue" conditions, twenty cars can get through the light in a single green period. Between midnight and 6:30 AM demand averages five cars per hour. Between 6:30 and 7:00 AM it increases from a rate of five to three hundred per hour. Thereafter it drops immediately to thirty per hour where it remains until 4:00 PM. In the interval of 4:00 to 4:15 the demand rate is eight hundred per hour whereafter it again drops to five. Waiting space is limited to fifty cars. When the queue is full, cars denied service go to an alternate access gate.

a) Analyze the current behavior of this system.

b) Suggest system design changes under the assumption that cross-street traffic is constant at a rate of one hundred vehicles per hour.

8.11 Prove that the probability distribution which maximizes entropy, subject to the condition that the expected value be c, is the exponential distribution. Prove that the constant service-time distribution minimizes entropy. What do these results have to do with the analysis of general service-time queues? What can you say about transient results when the service-time distribution has a ratio $(\sigma/\mu) > 1$?

8.12 Consider the set of differential-difference equations (8.1).

    a) Write the special case for single-channel systems.

    b) Write the special case for *KI* channel systems.

    c) Write the special case for a system in which channels open depend upon customers present, e.g., if $n < x$, one channel is operative, if $x \leq n < 2x$, two channels are open, etc.

    d) Modify the program in Appendix C to solve the model in part (*c*) above.

8.13 Write the Fortran statements necessary to change the program of Appendix C so that the experimenter can specify any desired initial probability vector rather than the current practice of assuming that $p_0(0) = 1$. Check your program's functioning by executing test runs with $p_0(0) = 1$ and $p_5(0) = 1$ for a 5-capacity system with $\lambda(t) = 10$, $\mu(t) = 15$.

8.14 Use the program developed in problem 8.13 to analyze the time for an initially saturated two-channel system to reach steady state for capacities of 20, 10, and 5 when $\lambda(t) = 40$, $\mu(t) = 22$. Use known steady-state results as your benchmark.

8.15 Change the constant service-time program of Appendix D so that the experimenter can specify any desired initial probability vector rather than the current practice of assuming that $p_0(0) = 1$. Check your program's functioning by executing test runs with $p_0(0) = 1$ and $p_{10}(0) = 1$ for a 10-capacity system with $\lambda(t) = 12$, $\mu(t) = 10$.

9

IMBEDDED MARKOV CHAINS

Most of the models so far introduced in this textbook have been predicated on either the birth-death postulates or a constant service time assumption. What we now want to explore are techniques which can be used to model systems which may not satisfy the basic birth-death postulates. In this chapter we will consider systems with general service time distributions, e.g., (M/G/1), and general arrival distributions e.g., (GI/M/1). The principal method we will use to model such systems is that of imbedded Markov chains. We will later use similar arguments for batch arrivals and bulk service.

IMBEDDING PROCESS

Consider the problem posed by attempting to model a metal machining system with Poisson arrivals. The service process requires a random set-up time to put the work piece in the chuck, select the tools needed, and make the proper adjustments for speeds and feeds. The actual machining and inspection time is a variable whose value depends upon the quality of the material being used and the amount of metal removal required. The analyst estimates the machining and inspection time for a three-inch bushing at fifteen minutes and feels that the set-up time is uniformly distributed in the interval of five to fifteen minutes. His total service distribution for this production run is

$$s(t) = 0.1, \qquad 20 \leq t \leq 30 \tag{9.1}$$

His task is to develop a probability distribution for the number of bushings awaiting machining at some random point in time.

The first observation to make is that this system is not a simple Markov process. In our previous modeling efforts we assumed that knowledge of the number of customers present at time t was all that was needed to predict the number which would be present at time $t + \Delta t$. In the present system this is no longer true. For example, if we examine the system five minutes after service starts on the work piece currently occupying the machine, the probability that it will be completed in the next second is zero. If we examine it 25 minutes after service starts, the probability of completion in the next second is the conditional probability of service time falling between 25 minutes and 25 plus one second, given that it exceeds 25 minutes. Predicting system behavior then requires knowing not only how many customers are present but also how long the lead customer has been in service. No continuous distribution other than the exponential exhibits the convenient property of forgetfulness necessary to treat this as a simple Markov process. What we need for this case and indeed for the general-service distribution case is a way to make the system look as though it is Markovian when in fact it is not. The technique for accomplishing this is the imbedded Markov chain first introduced by Kendall (56, 57).

The technique used to analyse this particular class of stochastic processes is to work with an enumerable Markov chain formed by directing attention to so-called regeneration points. A regeneration point is a point in time at which the process is renewed in the sense that past history becomes irrelevant for predicting future behavior. In the (M/G/1) queue, for example, the epochs at which individual customers depart from the system form a sequence of regeneration points. In a sense one might view such an analysis as taking snapshots of the system at a sequence of discrete times, defined by departure events, rather than accounting for all changes on a continuous time scale. Let N_t represent the number of customers present at any arbitrary point in time. What we now see is a sequence of values for the number in the system only at those instants immediately following service. That is

$$\{N_{T_i}\} = \{N_{T_1}, N_{T_2} \text{---}\} \tag{9.2}$$

where N_{T_i} represents the number in the system at the time the ith service is completed. This discrete parameter process which has been extracted from the continuous parameter process is referred to as an *imbedded Markov chain*.

The reason for the success of this technique when applied to the (M/G/1) queue is that the queue size at a departure epoch is statistically independent from the number of customers arriving during the next service interval. This is true as long as the input is Poisson. For more general arrival distributions, queue size and arrivals in a service epoch are not statistically independent.

The same argument can be made for the (GI/M/1) queue. Here we define the regeneration points as those instants just prior to the arrival of a customer. Because of the forgetfulness property of the exponential service distribution, the number served in the next interarrival interval depends only on the length of the interval and not on the time the present lead customer has been in service. The "snapshots" taken at arrival epochs then form an imbedded Markov chain.

Note that the imbedding process requires that either the arrival or service side of the system be Markovian. As noted by Kendall if both the input and service distributions are permitted to be perfectly general the only regeneration points for the process are those instants at which an arrival and departure occur simultaneously and those instants at which an arriving customer finds the service facility free. In both cases there are no simple means available for constructing a one-stage transition matrix to describe movement from one regeneration point to the next.

(M/D/1) QUEUES

The basic technique for forming an imbedded Markov chain can be easily demonstrated by first examining the constant service time case. We have already looked at one such model when we studied the fixed-cycle time model used to treat the constant service transient systems. What we will now do is to modify that approach slightly by looking at the system with a constant service time b only at departure epochs rather than at every b minutes whether or not a departure occurs. Furthermore we will also be concerned with infinite as well as finite capacity systems and their steady-state results.

Let f_i be the probability that i customers will arrive during a service interval of length b. Since the arrival process is assumed to be Poisson at a mean rate of λ customers per minute, the probability of i arrivals can be written as

$$f_i = \frac{(\lambda b)^i e^{-\lambda b}}{i!} \tag{9.3}$$

The single-stage transition matrix for a simple Markov chain can now be constructed in straightforward fashion. However, there is one important difference between the matrix in equation (9-4) and the one constructed for the constant cycle-time model of earlier chapters. In the present context the *time between transitions* is *not* a constant. The time between transitions is b minutes only when the system has customers waiting at a departure epoch. If the system is empty when a customer departs, the next transition cannot occur until a customer first arrives then is served. Our "snapshot camera" cannot operate until a departure takes place. The time between transitions, given that the system is empty at a departure epoch, consists of a random variable τ, representing the time until the next arrival event, plus the fixed service time b. It therefore follows that a move from state zero to state j or a move from state 1 to state j both require the arrival of j customers during a service interval. The result is that the first two rows of the transition matrix are identical. Moving from state i to j where $j \geq i - 1 > 0$ requires that $(j - i + 1)$ customers arrive during a service interval, the extra arrival being necessary to account for the customer known to be departing at the transition point. Transitions from i to j where $j < i - 1$ are clearly impossible as long as service occurs only one unit at a time.

$$\mathbf{T} = \begin{bmatrix} f_0 & f_1 & f_2 & f_3 & \cdots \\ f_0 & f_1 & f_2 & f_3 & \cdots \\ 0 & f_0 & f_1 & f_2 & \cdots \\ 0 & 0 & f_0 & f_1 & \cdots \\ \vdots & \vdots & \vdots & \vdots & \end{bmatrix} \tag{9.4}$$

If the capacity is unlimited, the transition matrix has an infinite number of rows and columns. If the system has some finite capacity k, the transition matrix will be $(k \times k)$ with the last column containing the complement of the distribution of arrivals as noted in equation (9-5). Note that the largest index in the matrix is $(k-1)$ since we are observing the system immediately after a departure.

$$
\mathbf{T} = \begin{array}{c}
 \\
0 \\
1 \\
2 \\
\\
k-1
\end{array}
\begin{array}{cccccc}
0 & 1 & 2 & \cdots & k-2 & k-1 \\
\left[\begin{array}{cccccc}
f_0 & f_1 & f_2 & \cdots & f_{k-2} & \sum_{i=k-1}^{\infty} f_i \\
f_0 & f_1 & f_2 & \cdots & f_{k-2} & \sum_{i=k-1}^{\infty} f_i \\
0 & f_0 & f_1 & \cdots & f_{k-3} & \sum_{i=k-2}^{\infty} f_i \\
\vdots & \vdots & \vdots & \vdots & & \\
& & & & f_0 & \sum_{i=1}^{\infty} f_i
\end{array}\right]
\end{array}
\qquad (9.5)
$$

Let us define the departure point distribution as

$$\pi_n = Pr \ (n \text{ customers present at a departure epoch})$$

We can then use our knowledge of elementary Markov chains to calculate the steady-state probability distribution. That is

$$\mathbf{\Pi} = \mathbf{\Pi}\,\mathbf{T}$$

where

$$\mathbf{\Pi} = (\pi_0, \pi_1, \pi_2 \cdots)$$

$$\qquad (9.6)$$

For the finite capacity system of equation (9-5) these equations become

$$
\pi_n = \begin{cases}
\pi_0 f_n + \sum_{i=1}^{n+1} \pi_i f_{n-i+1}, & n \leq k-2 \\[2em]
\pi_0 \widehat{F}_{k-2} + \sum_{i=1}^{k-1} \pi_i \widehat{F}_{k-i-1}, & n = k-1
\end{cases}
\qquad (9.7)
$$

where

$$\widehat{F}_i = \sum_{j=i+1}^{\infty} f_j = 1 - \sum_{j=0}^{i} f_i$$

These results, together with the usual normalization equation

$$\sum_{n=0}^{k-1} \pi_n = 1 \qquad (9.8)$$

can then be used to find the k unknown probabilities.

The theoretical structure of the infinite capacity system is identical to that of the finite capacity system. However, we are now led to an infinite set of difference equations which can be most easily solved by geometric transforms. Furthermore, since the (M/D/1) system is a special case of the (M/G/1) system, let us now turn our attention to the general service time problem which when solved will also yield the solution to the constant service time problem.

(M/G/1) QUEUES

The transition matrix for the (M/G/1) system is identical in form to equation (9.4). We are again concerned with changes in state between successive departure epochs. However, the time between every set of transitions is now a random variable whose distribution is the service time distribution $b(t)$.

Since the arrival process is Poisson, we can consider a conditional distribution for n arrivals given a service time t as

$$f(n \mid t) = \frac{(\lambda t)^n e^{-\lambda t}}{n!} \tag{9.9}$$

To obtain the marginal distribution for t arrivals in an arbitrary service interval t we must weight this function by $b(t)$ and integrate or sum over all t. That is for a continuously distributed service time

$$f_n = \int_t f(n \mid t)\, b(t)\, dt$$
$$= \int_0^\infty \frac{(\lambda t)^n}{n!}\, e^{-\lambda t}\, b(t)\, dt \tag{9.10a}$$

which for an integer valued service time becomes

$$f_n = \sum_{t=0}^\infty \frac{(\lambda t)^n}{n!}\, e^{-\lambda t}\, b(t) \tag{9.10b}$$

The matrix (9.4) can now be redefined with the elements from equation (9.10). The only conceptual difference between this Markov chain and our earlier one is that the transition matrix for the current chain is in a sense an expected value as opposed to the constant of the last model. The time between transitions is a random variable. The probability of moving from state i to state j has been obtained by integrating over all possible service times.

We can obtain the steady-state balance equations for the departure point distribution by performing the same sequence of calculations as before, namely

$$\Pi = \Pi\, T \tag{9.11}$$

which leads to

$$\pi_n = \pi_0 f_n + \sum_{i=0}^{n+1} \pi_i f_{n+1-i} \tag{9.12}$$

for all n.

Since we now have an infinite set of difference equations let us first reduce that set to a single equation in the geometric transform of $\{\pi_n\}$. Following our usual strategy of multiplying the nth equation by z^n and adding we have

$$\sum_{n=0}^{\infty} \pi_n z^n = \pi_0 \sum_{n=0}^{\infty} f_n z^n + \sum_{n=0}^{\infty} \left[\sum_{i=1}^{n+1} \pi_i f_{n+1-i} \right] z^n \qquad (9.13)$$

Note that equation (9.13) contains two discrete distributions. The distribution $\{f_n\}$ represents the number of customer arrivals during a service interval and $\{\pi_n\}$ represents the number of customers present at a departure point. The geometric transforms for these distributions are given by

$$\Pi\,(z) = \sum_{n=0}^{\infty} \pi_n z^n \qquad (9.14)$$

and

$$F(z) = \sum_{n=0}^{\infty} f_n z^n \qquad (9.15)$$

The object now is to rewrite equation (9.13) in a form which fits the transform definitions.

By rearranging summations equation (9.13) becomes

$$\Pi(z) = \pi_0 F(z) + z^{-1} \sum_{i=1}^{\infty} \pi_i z^i \sum_{n=i-1}^{\infty} f_{n+1-i} z^{n+1-i}$$

$$= \pi_0 F(z) + z^{-1} \left[\Pi_{(z)} - \pi_0 \right] F(z) \qquad (9.16)$$

Solving for the transform $\Pi(z)$ we have

$$\Pi(z) = \frac{\pi_0(z - 1)F(z)}{z - F(z)} \qquad (9.17)$$

When service time is a continuously distributed random variable, the geometric transform $F(z)$, representing the distribution for number of arrivals during a service interval, can be expressed as a simple function of the Laplace transform of the service time distribution by the following argument. From the definition of f_n in equation (9.10a)

$$F(z) = \sum_{n=0}^{\infty} f_n z^n = \sum_{n=0}^{\infty} \left[\int_0^{\infty} \frac{(\lambda t)^n}{n!} e^{-\lambda t}\, b(t)\, dt \right] z^n$$

$$= \int_0^{\infty} \left[\sum_{n=0}^{\infty} \frac{e^{-\lambda t}(\lambda t z)^n}{n!} \right] b(t)\, dt \qquad (9.18)$$

But the summation under the integral sign is simply the geometric transform of a Poisson distribution which can be written as

$$\sum_{n=0}^{\infty} \frac{e^{-\lambda t}(\lambda tz)^n}{n!} = e^{-\lambda t(1-z)} \tag{9.19}$$

Substituting in equation (9.18) we then have

$$F(z) = \int_0^{\infty} e^{-\lambda t(1-z)} b(t)\, dt \tag{9.20}$$

The integral above looks similar to the one which defines the Laplace transform of the service distribution. The only difference is that the integral above has the function $\lambda(1-z)$ in place of the Laplace transform variable s. This permits us to write the transform $F(z)$ as

$$F(z) = \int_0^{\infty} e^{-st} b(t)\, dt \,\bigg|_{s=\lambda(1-z)}$$

$$= \mathcal{L}(b(t)) \,\bigg|_{s=\lambda(1-z)} = B(\lambda(1-z)) \tag{9.21}$$

where $B(s)$ is the Laplace transform of the service time distribution. Note that the function $B(\lambda(1-z))$ is a geometric transform with a transform variable z. It is not a Laplace transform. We have only used the Laplace transform identity as a convenient way of obtaining a closed form solution for $F(z)$.

Returning to the line length distribution we have

$$\Pi(z) = \frac{\pi_0(z-1)B(\lambda(1-z))}{z - B(\lambda(1-z))} \tag{9.22}$$

The transform $\Pi(z)$ is now completely specified except for the initial condition π_0. The value of π_0 can be determined by recognizing that the geometric transform of a probability distribution evaluated at $z = 1$ should be equal to unity. That is

$$\Pi(z) \,\bigg|_{z=1} = \sum_{n=0}^{\infty} \pi_n 1 = 1 \tag{9.23}$$

Direct substitution of $z = 1$ in equation (9.22) leads to an indeterminant form. This can be circumvented by applying L'Hospital's rule, i.e., take the first derivative of numerator and denominator then substitute $z = 1$. This leads to

$$1 = \pi_0 \lim_{z \to 1} \frac{B(\lambda(1-z)) - \lambda(z-1)B'(\lambda(1-z))}{1 + \lambda B'(\lambda(1-z))}$$

$$= \frac{\pi_0 B(0)}{1 + \lambda B'(0)} \tag{9.24}$$

From the properties of Laplace transforms we know that

$$E(t^r) = (-1)^r \frac{d^r}{ds^r} \mathcal{L}(b(t)) \,\bigg|_{s=0}$$

It therefore follows that $B(0) = 1$ and $B'(0) = -E(t)$. This permits us to solve equation (9.24) for the unknown π_0 as

$$\pi_0 = 1 - \lambda E(t) \tag{9.25}$$

We are now in position to write the complete transform for the departure point distribution as

$$\Pi(z) = \frac{[1 - \lambda E(t)](z - 1)B(\lambda(1 - z))}{z - B(\lambda(1 - z))} \tag{9.26}$$

where

$$B(s) = \mathcal{L}(b(t))$$

and

$$E(t) = \int_0^\infty t b(t)\, dt$$

Although in theory we now have the complete solution for the (M/G/1) queue it is important to recognize that equation (9.26) has all the limitations of any transform. We must somehow invert it if we want to recover the distribution $\{\pi_n\}$. The ease of inversion will depend upon the particular service distribution $b(t)$ we happen to use. In some cases we may be able to find a closed-form solution for π_n. In other cases we may be forced to settle for a limited portion of the distribution by finding coefficients on z^n by Maclaurin series expansion or long division.[1]

This calculation is eased when part of the transform can be expressed as a ratio of two power series since it can be shown that the result is a single power series. (See Abramowitz and Stegun (1) p. 15 equation 3.6.22) Suppose that we can write

$$\Pi(z) = [1 - \lambda E(t)] \left[\frac{1 + \sum_{i=1}^{\infty} a_i z^i}{1 + \sum_{i=1}^{\infty} b_i z^i} \right] = [1 - \lambda E(t)] \sum_{i=0}^{\infty} c_i z^i \tag{9.27}$$

The coefficients c_i can be obtained in recursive fashion from

$$c_i = \begin{cases} a_i - \sum_{j=1}^{i} b_j c_{i-j} & (i = 1, 2, \ldots) \\ 1 & (i = 0) \end{cases} \tag{9.28}$$

One can then determine as much of the distribution as he desires by evaluating the low order coefficients on z^i.

---

[1]The Maclaurin series, which is a special case of the Taylor series, provides a method for expressing a continuous function of the variable z as a power series

$$f(z) = \sum_{i=0}^{\infty} \frac{z^i}{i!} \frac{d^i f(z)}{dz^i} \bigg|_{z=0}$$

The value of π_n is then the coefficient of the term z^n.

As an aside, it is important to note that for the unlimited (M/G/1) queue the steady-state departure point distribution π_n is equivalent to the distribution p_n defined at an arbitrary point in time. A simple proof of this relationship is given by Gross and Harris (42:235).

Example. A special tune-up station has been established at the end of an automotive assembly line to make adjustments on those vehicles which cannot meet federal exhaust gas emission standards. Vehicles failing to meet standards as assembled are routed to the station at an average rate of six per hour. Failures appear to be completely random and hence justify the Poisson arrival assumption. Each arrival is serviced by either a carburetor adjustment requiring twelve minutes or a timing adjustment requiring six minutes. These adjustments are done automatically and always require the same amount of time. Historical data indicate that 40 percent of the defective vehicles need carburetor adjustments and 60 percent need timing adjustments. In attempting to evaluate storage space requirements, management needs to know what portion of the time there will be more than two vehicles at the service point.

The service distribution in this case is discrete with

$$b(t) = \begin{cases} .6 & t = 6 \\ .4 & t = 12 \end{cases}$$

The arrival rate, expressed in vehicles per minute is

$$\lambda = 0.1$$

Combining this information we can write the geometric transform for the distribution of arrivals during a service interval as

$$F(z) = \sum_{n=0}^{\infty} \left[\sum_{t=0}^{\infty} \frac{(\lambda t)^n}{n!} e^{-\lambda t} b(t) \right] z^n$$

$$= \sum_{n=0}^{\infty} \left[\frac{.6(.6)^n}{n!} e^{-.6} + \frac{.4(1.2)^n}{n!} e^{-1.2} \right] z^n$$

$$= \sum_{n=0}^{\infty} d_n z^n$$

where the coefficient d_n is given by the terms in brackets. The probability of no vehicles in the system is given by

$$p_0 = 1 - \lambda E(t)$$
$$= 1 - 0.1\,(8.4) = 0.16$$

Putting this information together in equation (9.17) yields

$$P(z) = \frac{0.16(z - 1) \sum\limits_{n=0}^{\infty} d_n z^n}{z - \sum\limits_{n=0}^{\infty} d_n z^n}$$

Rewriting in standard format for a ratio of power series in z we have

$$P(z) = 0.16 \left[\frac{\displaystyle\sum_{n=0}^{\infty} d_n z^{n+1} - \sum_{n=0}^{\infty} d_n z^n}{-d_0 + (1 - d_1)\,z - \displaystyle\sum_{n=2}^{\infty} d_n z^n} \right]$$

$$= 0.16 \left[\frac{d_0 + \displaystyle\sum_{n=1}^{\infty} (d_n - d_{n-1})z^n}{d_0 + (d_1 - 1)z + \displaystyle\sum_{n=2}^{\infty} d_n z^n} \right]$$

$$= 0.16 \left[\frac{1 + \displaystyle\sum_{n=1}^{\infty} \left(\frac{d_n - d_{n-1}}{d_0}\right) z^n}{1 + \left(\dfrac{d_1 - 1}{d_0}\right) z + \displaystyle\sum_{n=2}^{\infty} \frac{d_n}{d_0} z^n} \right]$$

Using equation (9.28) we can now determine the necessary coefficients for powers of z as

$$c_0 = 1$$

$$c_1 = \left(\frac{d_1 - d_0}{d_0}\right) - \left(\frac{d_1 - 1}{d_0}\right) c_0$$

$$= \frac{1}{d_0} - 1$$

$$c_2 = \left(\frac{d_2 - d_1}{d_0}\right) - \left(\frac{d_1 - 1}{d_0}\right) c_1 - \left(\frac{d_2}{d_0}\right) c_0$$

$$= \left(\frac{1 - d_1}{d_0} - 1\right)\left(\frac{1}{d_0}\right)$$

Numerically

$$d_0 = 0.6e^{-0.6} + 0.4e^{-1.2} = 0.44976$$
$$d_1 = 0.36e^{-0.6} + 0.48e^{-1.2} = 0.34215$$

from which

$$c_1 = \frac{1}{.44976} - 1 = 1.2234$$

$$c_2 = \left(\frac{1 - .34215}{.44976} - 1\right)\left(\frac{1}{.44976}\right) = 1.0287$$

Finally we have

$$p_0 = 0.16$$
$$p_1 = 0.16\,c_1 = 0.16\,(1.2234) = 0.1957$$
$$p_2 = 0.16\,c_2 = 0.16\,(1.0287) = 0.1646$$

To answer the question originally posed we can now calculate the probability of more than two vehicles at the tune-up station as

$$\hat{P}(2) = 1 - \sum_{n=0}^{2} P_n = 0.4797$$

It should be obvious that even a simple two-point service distribution poses considerable computational burden when attempting to invert the transform $\Pi(z) = P(z)$.

PERFORMANCE MEASURES FOR (M/G/1)

The transform for the line-length distribution given by equation (9.26) may be awkward to invert but it can be used to develop as many moments of the distribution as the analyst may find useful. Recall from the properties of geometric transforms that one can develop the factorial moments of a distribution by taking derivatives with respect to z, then evaluating the result for $z = 1$. That is for $P(z)$, the transform of the arbitrary time distribution which is equivalent to $\Pi(z)$,

$$E(n^{(r)}) = \frac{d^r}{dz^r} P(z) \bigg|_{z=1} \tag{9.29}$$

where

$$n^{(r)} = \frac{n!}{(n-r)!} = n(n-1)(n-2)\dots(n-r+1) \tag{9.30}$$

Once the factorial moments are available it is a simple matter to convert them to the corresponding power moments by use of algebraic expansion or the Stirling number conversions.

The two moments most often used as measures of performance are the mean and variance of the distribution. Relating these moments to the geometric transform we can write

$$E(n) = \frac{d}{dz} P(z) \bigg|_{z=1} \tag{9.31}$$

and

$$\sigma_n^2 = \frac{d^2}{dz^2} P(z) \bigg|_{z=1} + E(n) - [E(n)]^2 \tag{9.32}$$

When evaluating higher order moments it is easier to solve the problem uniquely for each particular service distribution $b(t)$. However, in the case of the mean of the distribution, it is possible to develop a general equation in which the expected line length can be expressed as a function of moments of the service distribution.

Mean Value

Beginning with equation (9.26), in which $E(t)$ is replaced by $\bar{t}$, let us take the first derivative with respect to z. That is

$$P(z) = \frac{(1 - \lambda\bar{t})(z - 1)B(\lambda(1 - z))}{z - B(\lambda(1 - z))} \tag{9.33}$$

from which

$$\frac{dP(z)}{dz} = (1 - \lambda\bar{t})\{[B(\lambda(1 - z)) - \lambda B'(\lambda(1 - z))(z - 1)][z - B(\lambda(1 - z))]$$
$$- [1 + \lambda B'(\lambda(1 - z))][(z - 1)B(\lambda(1 - z))]\}/[z - B(\lambda(1 - z))]^2 \tag{9.34}$$

To evaluate this expression at $z = 1$ it is necessary to recall that from the properties of the Laplace transform

$$B(s)\Big|_{s=0} = 1 \tag{9.35}$$

and

$$\frac{d^r}{ds^r} B(s)\Big|_{s=0} = (-1)^r E(t^r) \tag{9.36}$$

Using that information in equation (9.34) we see that

$$\underset{z \to 1}{\text{Lim}} \frac{dP(z)}{dz} = \frac{0}{0} \tag{9.37}$$

which is an indeterminant form.

Consider the function $\dfrac{dP(z)}{dz}$ to be a new function, say

$$R(z) = \frac{C(z)}{D(z)} \tag{9.38}$$

L'Hospital's rule states that the limit as z goes to one of $R(z)$ is equal to the limit of the derivative of the numerator divided by the derivative of the denominator as z goes to one. That is

$$\underset{z \to 1}{\text{Lim}}\, R(z) = \underset{z \to 1}{\text{Lim}} \frac{C'(z)}{D'(z)} \tag{9.39}$$

Applying this rule to equation (9.34) we have

$$\frac{C'(z)}{D'(z)} =$$
$$\frac{(1 - \lambda\bar{t})\{-2\lambda\, B'(\lambda(1 - z))[z - B(\lambda(1 - z))] + z\lambda^2 B''(\lambda(z - 1))(z - 1)\}}{2[z - B(\lambda(1 - z))][1 + \lambda B'(1 - z))]} \tag{9.40}$$

The result is again an indeterminant form $\dfrac{0}{0}$ which requires one more application of L'Hospital's rule.
That is

$$\frac{C''(z)}{D''(z)} = (1 - \lambda\bar{t})\{2\lambda^2 B''(\lambda(1 - z))[z - B(\lambda(1 - z))]$$

$$+ [1 + \lambda B'(\lambda(1 - z))][-2\lambda B'(\lambda(1 - z))]$$
$$+ (2z - 1)\lambda^2 B''(\lambda(1 - z)) - \lambda^3 B'''(\lambda(1 - z))(z^2 - z)\}$$
$$\div \{2[1 + \lambda B'(\lambda(1 - z))][1 + \lambda B'(\lambda(1 - z))]$$
$$- 2\lambda^2 B''(\lambda(1 - z))[z - B(\lambda(1 - z))]\} \tag{9.41}$$

Again taking limits as $z \to 1$ and recalling that

$$B''(0) = \sigma^2 + \bar{t}^2 \tag{9.42}$$

we finally arrive at the general expression

$$E(n) = \frac{(1 - \lambda\bar{t})\{(1 - \lambda\bar{t})(2\lambda\bar{t}) + \lambda^2(\sigma^2 + \bar{t}^2)\}}{2(1 - \lambda\bar{t})^2}$$

or

$$L = \lambda\bar{t} + \frac{(\lambda\bar{t})^2 + \lambda^2\sigma^2}{2(1 - \lambda\bar{t})} \tag{9.43}$$

Equation (9.43), often referred to as the Pollaczek-Khintchine formula, is a very powerful result. It permits us to make estimates for expected line length in (M/G/1) systems from a knowledge of the arrival rate and the mean and variance of the service distribution. Since the first two moments of the service distribution are usually much easier to estimate than the complete probability function, this is a very handy equation for the practitioner to have available.

Waiting Times

The expected sojourn time, or time for a customer to pass through an (M/G/1) system can be easily determined from equation (9.43) and Little's formula. Recall that expected line length L and expected sojourn time W are related by

$$L = \lambda W \tag{9.44}$$

from which

$$W = \frac{L}{\lambda} \tag{9.45}$$

It therefore follows that for the (M/G/1) system

$$W = \bar{t} + \frac{\lambda(\bar{t}^2 + \sigma^2)}{2(1 - \lambda\bar{t})} \tag{9.46}$$

The expected delay time in queue, excluding service time, is easily calculated by noting that total sojourn time is made up of queue time plus service time. It therefore follows that

$$W_q = W - \bar{t}$$
$$= \frac{\lambda(\bar{t^2} + \sigma^2)}{2(1 - \lambda\bar{t})} \tag{9.47}$$

Once again we have an important measure of performance which can be evaluated with very limited knowledge, i.e., the first two moments, of the service distribution.

There is also a simple relationship between the transform of the system state distribution $P(z)$ [or $\Pi(z)$] and the transform of the total delay distribution, $W(s)$. Again taking the viewpoint of a departing customer one can argue that the probability of n customers in the system at the departure point is simply the probability of n arrivals during the departing customers sojourn weighted by the length of its sojourn time and integrated over all values of t. That is

$$\pi_n = p_n = \int_0^\infty pr(n \text{ arrivals} \,|\, \text{sojourn } t) \, pr(\text{sojourn } t) \, dt$$

$$= \int_0^\infty \frac{(\lambda t)^n}{n!} e^{-\lambda t} w(t) \, dt \tag{9.48}$$

We can now use this expression to develop an alternate form for the geometric transform $P(z)$.

By definition

$$P(z) = \sum_{n=0}^\infty p_n z^n$$

$$= \sum_{n=0}^\infty \left[\int_0^\infty \frac{(\lambda t)^n}{n!} e^{-\lambda t} w(t) \, dt \right] z^n$$

$$= \int_0^\infty \left[\sum_{n=0}^\infty \frac{(\lambda t)^n}{n!} z^n e^{-\lambda t} \right] w(t) \, dt \tag{9.49}$$

By using the same arguments previously employed to find a compact expression for $F(z)$ (see equation 9.21) we can now express the transform $P(z)$ as a function of the transform of the delay time distribution. Since

$$\sum_{n=0}^\infty \frac{(\lambda t z)^n}{n!} e^{-\lambda t} = e^{-\lambda t(1-z)} \tag{9.50}$$

from the properties of the transform of the Poisson distribution it follows that $P(z)$ can be written as

$$P(z) = \int_0^\infty e^{-\lambda t(1-z)} \, w(t) \, dt$$

$$= \int_0^\infty e^{-st} w(t) \, dt \, \bigg|_{s=\lambda(1-z)} \tag{9.51}$$

But this expression is nothing more than the Laplace transform of the distribution $w(t)$ with the transform variable s replaced by $\lambda(1 - z)$. That is

$$P(z) = W(\lambda(1 - z)) \tag{9.52}$$

where

$$W(s) = \mathcal{L}(w(t))$$

$$= \int_0^\infty e^{-st} w(t)\, dt$$

We are now in position to combine the results of equations (9.26) and (9.52) to obtain an explicit expression for the Laplace transform of the sojourn time distribution. That is

$$W(\lambda(1 - z)) = \frac{(1 - \lambda\bar{t})(z - 1)B(\lambda(1 - z))}{z - B(\lambda(1 - z))} \tag{9.53}$$

By changing variables such that

$$s = \lambda(1 - z) \tag{9.54}$$

we can rewrite this relationship as

$$W(s) = \frac{(1 - \lambda\bar{t})\left(-\dfrac{s}{\lambda}\right)B(s)}{1 - \dfrac{s}{\lambda} - B(s)}$$

$$= \frac{(1 - \lambda\bar{t})sB(s)}{s - \lambda[1 - B(s)]} \tag{9.55}$$

The sojourn time distribution can now be obtained by inverting the Laplace transform $W(s)$ which is itself a function of the transform of the service time distribution.

The distribution for the time in queue can be developed by using the convolution property of transforms. Let the sojourn time, queue time, and service time be given by t_w, t_q and t_b respectively. then

$$t_w = t_q + t_b \tag{9.56}$$

represents a sum of independent random variables. The distribution for t_w which we previously developed in equation (9.55) can also be obtained from the convolution of the distribution of t_q with that of t_b. In transform domain the convolution operation is replaced by a product of transforms. Therefore

$$W(s) = W_q(s)B(s) \tag{9.57}$$

which yields

$$W_q(s) = \frac{W(s)}{B(s)}$$

$$= \frac{(1 - \lambda\bar{t})s}{s - \lambda[1 - B(s)]} \tag{9.58}$$

As was true of the line length distribution it may be difficult to invert these Laplace transforms for some particular service distributions $b(t)$. However, we can obtain the moments of the waiting time distributions by taking an appropriate number of derivatives of the transforms evaluated at $s = 0$.

(M/M/1) Waiting Times

One special case of interest occurs when service is exponentially distributed in which case the (M/G/1) results reduce to (M/M/1). For that case

$$b(t) = \mu e^{-\mu t}$$

from which

$$B(s) = \int_0^\infty e^{-st}\mu e^{-\mu t}\, dt$$

$$= \frac{\mu}{s + \mu} \tag{9.59}$$

Substituting into equation (9.55) we have

$$W(s) = \frac{(1 - \lambda\bar{t})s\mu/(s + \mu)}{s - \lambda\left[1 - \dfrac{\mu}{s + \mu}\right]}$$

$$= \frac{\mu - \lambda}{s + \mu - \lambda} \tag{9.60}$$

But this transform has an easily recognized inverse. It is of the same form as equation (9.59). It therefore follows that the throughput or sojourn time distribution for an (M/M/1) system is exponential with a mean of $1/(\mu - \lambda)$. That is

$$w(t) = (\mu - \lambda)e^{-(\mu-\lambda)t} \tag{9.61}$$

The distribution for time in queue for the (M/M/1) system is slightly more complex. Direct substitution of $B(s) = (\mu/s + \mu)$ into equation (9.58) and inversion yields

$$w_q(t) = \frac{\lambda}{\mu}(\mu - \lambda)e^{-(\mu-\lambda)t} + \left(1 - \frac{\lambda}{\mu}\right)\delta(t) \tag{9.62}$$

where $\delta(t)$ is the Dirac delta function at the origin. This says that the density is a mixture of a discrete portion at $t = 0$ and a continuous portion for $t > 0$. The cumulative distribution looks more reasonable since integrating equation (9.26) will give a function with a step at the origin then a smoothly increasing portion as t becomes large. That is

$$W_q(t) = \int_0^t w_q(\tau)\, d\tau$$

$$= \begin{cases} \left(1 - \dfrac{\lambda}{\mu}\right) & , \quad t = 0 \\[2ex] 1 - \left(\dfrac{\lambda}{\mu}\right)e^{-(\mu-\lambda)t}, & \quad t > 0 \end{cases} \tag{9.63}$$

Example. An aircraft assembly plant has one work station devoted to fitting axle stubs to landing gear mounting pads. This operation requires hand filing of the pad until the axle can be properly seated. Due to a high degree of variability in material roughness, caused by warpage during heat treatment, the time required to hand-fit these stubs can only be estimated in a probability sense. Past data indicate that fitting time averages thirty minutes with a variance of 100. The station produces an average of twelve pads per day. Since not all gears require this special operation, management feels that it is reasonable to suppose that the arrival process at the station is Poisson. Every gear which must undergo this fitting process causes production delays in other departments. The cost of such delays is estimated at eighty dollars per hour per gear. A machine tool company has offered to develop a grinder which they estimate can perform this task in four minutes with a variance of one. The same operator could run the new equipment. The question to be answered is what can the company afford to pay for the machine if they demand a payback period of one year?

From equation (9.46) we can estimate the average delay times for each proposal as

$$W_{current} = 30 + \frac{.025((30)^2 + 100)}{2(1 - .025(30))}$$

$$= 80 \text{ minutes}$$

$$W_{new} = 4 + \frac{.025(4^2 + 1)}{2(1 - .025(4))}$$

$$= 4.47 \text{ minutes}$$

At an average demand of twelve pads per day and an assumed 250 working days per year the difference in delay costs for the two systems is

$$D = (W_c - W_n)(12)(250)(\$80/60)$$
$$= (80 - 4.47)(4000) = \$302,120.$$

This implies that in terms of expected value of savings the company could pay as much as \$300,000 to have the new equipment installed.

(GI/M/1) QUEUES

Let us now turn briefly to another class of queues which are easily modeled by imbedded Markov chain procedures. This time we will preserve the Markov property by assuming that we have a single-channel system with exponential service while permitting any general independent arrival time distribution. As discussed early in the chapter, the regeneration points for this system are the instants just prior to the arrival of a customer.

Let k_m be the probability that m customers are served between successive arrival events. If i customers are present at some particular arrival epoch, j will be present at the next epoch if and only if $i + 1 - j$ are served during the interarrival period. The probability of this transition can be expressed as

$$p_{ij} = k_{i+1-j} \tag{9.64}$$

Since arrivals occur one at a time it is impossible for the system to move from state i to j where $j > i + 1$, hence

$$p_{ij} = 0 \text{ for } j > i + 1 \tag{9.65}$$

Moving from state i to state zero requires that $(i + 1)$ services take place. The probability of this event can be determined from the probability of $(i + 1)$ or more potential service opportunities during the interarrival period. This does not mean that more than $(i + 1)$ actual services took place but rather that they could have taken place if customers would have been present to be served. Equivalently we could write this as the complement

$$p_{i0} = 1 - \sum_{m=0}^{i} k_m \tag{9.66}$$

Combining this information in a transition matrix we can write the one-stage transition probabilities as

$$
\mathbf{T} =
\begin{array}{c|ccccc}
 & 0 & 1 & 2 & 3 & \cdots \\
\hline
0 & 1 - k_0 & k_0 & 0 & 0 & \cdots \\
1 & 1 - \displaystyle\sum_{m=0}^{1} k_m & k_1 & k_0 & 0 & \cdots \\
2 & 1 - \displaystyle\sum_{m=0}^{2} k_m & k_2 & k_1 & k_0 & \cdots \\
\vdots & \vdots & \vdots & \vdots & \vdots &
\end{array}
\tag{9.67}
$$

Let $\mathbf{q} = \{q_n\}$ be the probability that n customers are present at the time of an arrival event, then we can calculate the steady-state distribution by invoking the usual Markov chain argument

$$\mathbf{q} = \mathbf{qT} \tag{9.68}$$

If the system is truncated at some finite capacity it is possible to replace the k_m with their numerical values, perform the matrix multiplication and solve the resulting system of linear equations for the unknown $\{q_n\}$. If the queue is unlimited we can again resort to transforms to find a solution. The system of equations to be solved is

$$
q_n =
\begin{cases}
\displaystyle\sum_{i=0}^{\infty} \left[q_i \left(1 - \sum_{m=0}^{i} k_m \right) \right], & n = 0 \\[3ex]
\displaystyle\sum_{i=0}^{\infty} q_{n+i-1} k_i, & n > 0
\end{cases}
\tag{9.69}
$$

Note that

$$k_n = \int_0^\infty \frac{e^{-\mu t}(\mu t)^n}{n!}\, a(t)dt \tag{9.70}$$

where $a(t)$ is the interarrival time distribution. This follows from the relationship of the exponential distribution for service to the Poisson distribution for number of services one can complete in an elapsed time t.

The solution to this system of equations is not so straightforward as the one obtained for the (M/G/1) system. Because of its complexity we will not pursue the details of that argument here but will only relate the final results (For alternative ways to do the calculation see Gross and Harris (42:276) or Cox and Smith (29:16). The end result is that the queue size distribution is geometric of the form

$$q_n = (1 - r_0)r_0{}^n \tag{9.71}$$

for $\lambda/\mu < 1$.

The parameter r_0 is the unique real root in the range $0 < r_0 < 1$ of the transcendental equation

$$z = A(\mu(1 - z)) \tag{9.72}$$

where

$$A(s) = \mathcal{L}(a(t)) = \int_0^\infty e^{-st}\, a(t)\, dt \tag{9.73}$$

There is at most one such root $z = r_0$ in the interval $(0, 1)$ due to the convexity of the transform

$$B(z) = A(\mu(1 - z)) \tag{9.74}$$

$B(z)$ represents the transform for the distribution of services during an interarrival period. There will be no roots between zero and one if $\lambda/\mu > 1$. In that instance the system is unstable and there is no steady-state solution.

Finding the required root of the transcendental equation (9.72) may require the use of numerical techniques such as the Newton-Raphson approximation method. However we are assured that the approximation will always converge because of the convexity of $B(z)$.

It is important to note that equation (9.71), which is of exactly the same form as the (M/M/1) solution, is valid only at arrival epochs. It is no longer true that $q_n = p_n$, the arbitrary time state distribution, as was the case with the (M/G/1) system. The distributions q_n and p_n are equivalent if, and only if, the arrivals are Poisson. The special timing of the distribution q_n is of no concern, however, when calculating the waiting time distributions. We really want to take the view of an arriving customer and it is from that point of view that q_n was developed.

We can use the similarity of equation (9.71) to the standard (M/M/1) results to quickly establish the waiting time distributions. All that is necessary is to replace the argument (λ/μ) with the new parameter r_0 in the results of equations (9.61) and (9.63). The sojourn or throughput time density for the $(GI/M/1)$ system is then exponential

$$w(t) = \mu(1 - r_0)\, e^{-\mu(1-r_0)t} \tag{9.75}$$

while the cumulative queue time distribution is given by

$$W_q(t) = \begin{cases} 1 - r_0 & , & t = 0 \\ 1 - r_0 e^{-\mu(1-r_0)t}, & t > 0 \end{cases} \tag{9.76}$$

Example. A poultry farmer is considering liquidating his stock by slaughtering his entire flock of 10,000 chickens, dressing, cutting, and quick freezing them for sale in local retail stores over the next several months. He is prepared to run a two-man operation twenty-four hours a day until the task is complete. The first man will slaughter and pluck the chickens. The second man will cut them up and place them in the freezer. Slaughter time, which includes catching the chicken, is estimated to be exponentially distributed with a mean of five minutes. Plucking time is exponentially distributed with a mean of ten minutes. Cutting is also exponential with a mean ten minutes. The problem is to determine the time a chicken carcass must wait between plucking and freezing in order to guarantee that it will not spoil. Any which exceed sixty minutes must be thrown away. Before undertaking this task the farmer wants to know how many chickens he can expect to lose due to delays in cutting.

For purposes of model building let us consider the 10,000 chickens to be an infinite population for which a long range solution is appropriate. The interarrival time at the cutting station is really the sum of two independent random variables, slaughter time and plucking time. That is

$$t = t_1 + t_2$$

Using the convolution property of transforms it follows that the Laplace transform of the interarrival distribution is given by

$$A(s) = F_1(s)F_2(s)$$
$$= \left(\frac{0.2}{s + 0.2}\right)\left(\frac{0.1}{s + 0.1}\right)$$

Note that the interarrival distribution is not exponential but rather a convolution of exponential distributions which puts the model in the (GI/M/1) class.

Applying equation (9.72) we must first find a root r_0 between 0 and 1 from

$$z = A(\mu(1 - z)) = A(0.1(1 - z))$$

or

$$z = \left(\frac{0.2}{.1(1 - z) + 0.2}\right)\left(\frac{0.1}{0.1(1 - z) + 0.1}\right)$$

Combining terms yields the cubic equation

$$0.01z^3 - 0.05z^2 + .06z - .02 = 0$$

The positive root of interest is approximately

$$z = r_0 \approx 0.586$$

The throughput time distribution of equation (9.75) can be written in complementary cumulative form as

$$\widehat{W}(t) = \int_t^\infty w(\tau)\, d\tau$$
$$= e^{-\mu(1-r_0)t}$$
$$= e^{-0.1(1-0.586)t}$$
$$= e^{-.0414t}$$

The probability that an arriving carcass will wait more than sixty minutes is then

$$\widehat{W}(t = 60) = e^{-.0414(60)} \approx .0834$$

By this calculation it then appears that 834 chickens from the flock of 10,000 can be expected to be thrown away. The farmer's production plan may need reworking.

SYNOPSIS

We have used imbedded Markov chain arguments to study the (M/G/1) and (GI/M/1) systems in some detail. The reader should note that these are general and very powerful results. Many of the specialized models such as those involving constant service (D), Erlang service (E_k), constant interarrival times, or Erlang arrivals can be solved by these equations. All that is necessary is to properly define the required transforms and moments to achieve complete solutions.

We have not mentioned multi-channel systems for good reason. There are only limited results available for (M/G/C) configurations. More is known about (GI/M/C) systems but such configurations are not commonly encountered in practice. The reader who is interested in such multi-channel extensions should consult one of the more theoretical texts listed in the bibliography.

PROBLEMS

9.1 Passengers for check-in at an airline ticket counter arrive by automobile. The autos appear in Poisson fashion at a rate of six per hour. However, each auto contains a random number of passengers x distributed by

$$g(x) = \begin{cases} .3 & x = 1 \\ .1 & x = 2 \\ .2 & x = 3 \\ .4 & x = 4 \end{cases}$$

Customer service time measured in minutes is uniformly distributed in the interval of one to five minutes.

a) Develop a closed-form expression for the probability generating function of number of customers arriving in time t.

b) Express the probability generating function for the number of arrivals during the service of a customer as a function of the Laplace Transform of the service distribution.

    c) Calculate an approximation for the expected line length. Hint: Treat the number of cars as the measure of performance.

    d) Show how to develop the generating function for the waiting time distribution.

9.2 Poke-n-Prod is a training clinic for medical doctors. The normal complement of trainees is four interns. An average of six patients per hour seek services at the clinic. When a patient arrives he must be examined by one or more of the interns. When only one patient is present one intern will see that patient with probability 0.3, two with probability 0.6, and three with probability 0.1. They examine in series and each follows a service time distribution given by $3e^{-3t}$. When two patients are present the first one in the examining room will see one doctor with probability 0.6 and two with probability 0.4. When three or more patients are present all doctors know that they will have a shot at someone if they rotate in the examining room. Therefore the lead patient under those conditions will see only one doctor. Since there is only one examining room no new patients can be admitted to the examining room until all doctors who are scheduled to see the current patient complete their examination. Waiting room space is limited to a total of six patients not including the one in the examination room.

    a) Develop an appropriate model to describe the probability distribution for the number of patients at the clinic. State all assumptions.

    b) Explain in detail how you would obtain a solution for your model and accompanying performance measures.

9.3 Consider a single channel service system in which service is uniformly distributed in the interval [0, 5] minutes. Arrivals occur in Poisson fashion at the rate of 0.08 per minute.

    a) Find the mean number of customers in the system.

    b) Find the expected sojourn time per customer.

    c) Find the probability that there will be more than one customer present.

    d) Find the variance for the number of customers in the system.

9.4 Consider a system in which the time between arrivals is constant at t minutes. Single channel service occurs in exponential fashion at a rate of μ customers per minute.

    a) Develop a general expression for number of customers present at the regeneration point. State all assumptions.

    b) Suppose that $t = 5$, $\mu = 15$. Find the probability that an arriving customer will find less than two customers already present.

9.5 A methods analyst has rigged a camera to shoot one frame of film at the instant an automobile paint shop discharges a car. The paint shop serves a large body shop and as a result may be required to do everything from minor touch-up to a full body paint job. When the paint shop can run continuously they can handle an average of 1.5 jobs per hour in exponential fashion. However, each time the queue becomes empty the spray heads must be cleaned. When the next job arrives it then takes a fixed period of twenty minutes to set up after which the usual production time occurs.

Space is limited so that no more than four cars can be present at once. Cars arriving when the system is full are sent to a contract shop for painting. Arrivals occur in Poisson fashion at the rate of one per hour.

a) Find the distribution for number of cars recorded per frame of film.

b) Find the expected number of cars which will be sent to an outside contractor each day.

9.6 Consider a fixed cycle service system, such as a conveyor system, in which service opportunities occur every c minutes. If any customers are present at a service opportunity time, one will be instantaneously served. If none are present at that instant, no one can be served for another c-minutes even if one arrives before that time. The state of the system is recorded immediately following a service event.

a) Find the transform for the state distribution.

b) Compare this solution with one obtained from the (M/D/1) model in the text. Explain any differences and offer guidelines for use of each solution.

9.7 Figure 8-3 in the transient analysis chapter suggests that there is a reasonably tight performance envelope between transient systems with exponential and deterministic service times. Compare that observation with the steady-state results for the (M/G/1) system depicted by equation 9.43. Explain any discrepancies in the conclusions drawn.

9.8 Consider the expected waiting time for an (M/E$_k$/1) system in which $\mu = 1$. Plot W_q for a family of Erlang service distributions with the shape parameter running from 1 to 20 for $\lambda = 0.1$, 0.5, and 0.9. What can you say about using approximate versus exact service distributions when modeling?

9.9 In analyzing certain classes of (M/G/1) queues, namely (M/E$_r$/1), it may be convenient to mathematically simulate service-system behavior by using the method of stages. In this procedure the service facility is viewed as a "black box" which contains a number of pseudo-servers operating in series. Each pseudo-server operates in exponential fashion. If the overall service rate for the black box is μ and there are r pseudo-servers, the service rate of each pseudo-server will be $r\mu$. The net effect is that the mean time required by a customer to pass through all r channels will be the same as that observed for the service facility viewed as a black box. When a customer enters service no others may follow him until he has cleared all of the pseudo-servers, i.e., only one customer at a time is permitted on the black box.

a) Find the service distribution which can be replaced by r pseudo-servers in series each operating in exponential fashion with rate $r\mu$. (This distribution is called the Erlangian distribution E_r.)

b) Define a new state variable j which represents j stages of service to be completed. If the system contains n customers and the lead customer is in the ith stage or channel, the state would be $j = (n - 1)r + (r - i + 1)$. Use this definition to write a set of steady-state balance equations. How would you convert to our more usual state definition for (M/G/1) systems?

9.10 Consider an (M/G/1) system in which the arrival rate is ten per hour and the service distribution is given by

$$b(t) = \frac{45(45t)^2 e^{-45t}}{2!}, \qquad t \geq 0$$

a) Describe the service distribution by the method of stages discussed in problem 9.9.

b) Calculate L and W_q for this system.

9.11 The hyperexponential is a service distribution sometimes fit to empirical data. The hyperexponential distribution H_r is given by

$$b(t) = \sum_{i=1}^{r} a_i \mu_i e^{-\mu_i t} \qquad t \geq 0$$

where

$$\sum_{i=1}^{r} a_i = 1.$$

a) Show how one might model a hyperexponential "black box" server as a set of parallel pseudo-servers each operating in exponential fashion. (A customer is served by only one pseudo-server.)

b) Develop a set of balance equations to describe the probability distribution for n_i when service is H_r, there are n customers present and the lead customer is in the ith parallel pseudo-channel. How would you convert this solution to the more usual (M/G/1) results?

9.12 Prove that all members of the Erlang family of distributions have coefficients of variation

$$C = \sigma / \bar{t}$$

less than or equal to one. What does this suggest about using E_r as an approximation for an empirical interarrival or service distribution?

9.13 Prove that all members of the hyperexponential family of distributions have coefficients of variation

$$C = \sigma / \bar{t}$$

greater than or equal to one. What does this suggest about using H_r as an approximation for an empirical interarrival or service distribution?

9.14 In analyzing certain classes of (GI/M/1), namely (E$_r$/M/1), queues, it may be convenient to mathematically simulate the arrival process by using the method of stages. Arriving customers are treated as if they passed a series of timing channels before reaching the queue. Each channel is exponential with rate $r\lambda$. The mean arrival rate for the entire process is then λ. When a customer exits the last timing channel he is treated as an arrival to the queue. At that time another customer immediately enters the first timing channel. No more than one may occupy this arrival "black box" mechanism at one time.

a) Find the interarrival distribution which can be replaced by r pseudo-timing channels in series each operating in exponential fashion at rate $r\lambda$.

b) Define a new state variable j which represents j stages of arrival in the system. If the system contains n customers and the next arrival is in the ith timing channel the state would be $j = rn + i - 1$. Use this definition to write a set of steady-state balance equations. How would you convert to our more usual definition for (GI/M/1) systems?

9.15 A shipping clerk is required to inspect all bushings sent to him prior to packing them for shipment. Bushings arrive in Poisson fashion at the rate of fifteen per hour. Past experience indicates that 20 percent of the arriving bushings must be deburred before packing. The rest can be packed as is. The clerk performs both deburring and packing duties on one bushing at a time. The average deburring time and the average packing time is three minutes. Both operations may be assumed to be exponential.

a) Find the service time distribution for the shipping clerk's activities viewed as a single channel operation.

b) Find the probability distribution for the number of bushings in the shipping area.

c) Find the mean and variance for the time an arriving bushing will be delayed prior to shipment assuming that all are processed in order of arrival.

9.16 A federal flu shot program is to be conducted in the lobby of the local high school. Arrivals to the lobby are expected to occur at such a rate that the lobby will always be full. Two people are to be involved in administering the treatment. The first person will help the patients fill out forms, hang up his coat and prep his arm for injection. It is estimated that each of these operations will be exponential with mean times of 1.0, 0.4, and 1.6 minutes, respectively. After these operations the patients enter a queue in front of the man with the gun. The shot time and associated hand-holding until the patient demonstrates that he can walk on without fainting is estimated to average 1.2 minutes. The program director is willing to assume that this time will also be exponentially distributed.

a) Find the interarrival time distribution at the shot station.

b) Find the distribution for number of patients waiting for a shot.

c) Find the probability that a patient must queue before the gun for more than three minutes.

10

BULK QUEUES

In all of the models so far discussed, we have assumed that customers will arrive singly and be served singly. Such assumptions may not be reasonable for some special classes of queueing systems. For example, consider the shipping department of a small machine shop. As castings are machined they are stacked on pallets for movement to the packaging area. When a pallet is full and a fork truck is available, an entire pallet is moved. Castings then arrive at the shipping area in bulk where they are invoiced and packaged one at a time. Similar characteristics can be noted at an airline terminal. Each arriving flight delivers a batch of passengers who will require service at the baggage pick-up area. In both cases one must account for bulk arrivals coupled with single-channel service in order to develop a reasonable queueing model.

Bulk or batch operations are also common on the service side of some systems. Consider, for example, passenger pickups at a city bus stop. Passengers normally arrive singly and form a queue. However, when service occurs an entire group of the waiting passengers, equal to the maximum of the number present or the available seats on the bus, is served. A single-service model would be a poor choice for analyzing such a bulk-service problem.

What we will now explore are some basic methods for analyzing systems with bulk arrivals or bulk service. We will treat both the simple Markov systems and the more complicated models requiring use of imbedded Markov chains.

POISSON BATCH ARRIVALS

For our first bulk queueing problem let us reconsider the multiple-arrival Poisson process introduced in Chapter 4 (equation 4.42). Recall from that discussion that arrival events are assumed to occur in Poisson fashion at rate λ but that each event represents the arrival of a batch of b customers. When $b = 1$ this arrival process reduces to a standard

Poisson stream. Service is assumed to be single-channel exponential with customers served one at a time from the queue.

This system can be analyzed by the same techniques employed for the birth-death process. The only difference is that we must now admit jumps to states other than immediately adjacent states. The transition rate diagram in Figure 10-1 shows that the system moves into state j from

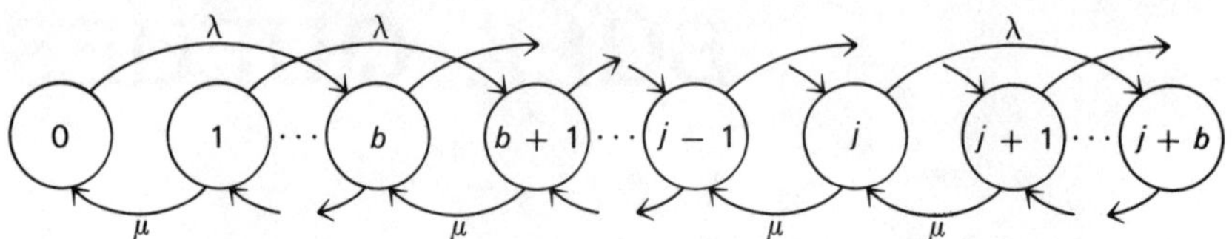

FIGURE 10-1 Flow Rate Diagram for Bulk Arrivals.

state $(j - b)$ by virtue of an arrival event or from state $(j + 1)$ by virtue of a service event. It moves out of state j into $(j + b)$ or $(j - 1)$ when an arrival or service takes place.

Using the usual rate of change arguments we can write the appropriate set of differential difference equations as

$$\frac{dp_0(t)}{dt} = \mu p_1(t) - \lambda p_0(t)$$

$$\frac{dp_n(t)}{dt} = \mu p_{n+1}(t) - \lambda p_n(t) - \mu p_n(t), \qquad 0 < n < b$$

$$\frac{dp_n(t)}{dt} = \mu p_{n+1}(t) - \lambda p_n(t) - \mu p_n(t) + \lambda p_{n-b}(t), \qquad n \geq b \quad (10.1)$$

Due to the inherent complexity of the transient solution, let us once more proclaim interest only in the equilibrium condition which exists as long as $b\lambda < \mu$. Rewriting for the steady-state by setting the derivatives equal to zero we have

$$\begin{aligned} \lambda p_0 &= \mu p_1 \\ (\lambda + \mu) p_n &= \mu p_{n+1} &, \qquad 0 < n < b \\ (\lambda + \mu) p_n &= \mu p_{n+1} + \lambda p_{n-b}, & n \geq b \end{aligned} \qquad (10.2)$$

The resulting set of difference equations is most easily solved by using geometric transforms. Multiplying the nth equation in the set of equations (10.2) by z^n and adding yields the single equation

$$\sum_{n=0}^{\infty} \lambda p_n z^n + \sum_{n=1}^{\infty} \mu p_n z^n = \sum_{n=0}^{\infty} \mu p_{n+1} z^n + \sum_{n=b}^{\infty} \lambda p_{n-b} z^n \qquad (10.3)$$

Let us now change subscripts and rearrange terms to write this expression as a function of the transform

$$P(z) = \sum_{n=0}^{\infty} p_n z^n \qquad (10.4)$$

That is

$$\lambda P(z) + \mu[P(z) - p_0] = \frac{\mu}{z}[P(z) - p_0] + \lambda z^b P(z) \tag{10.5}$$

from which

$$P(z) = \frac{\mu p_0(1 - z)}{\mu + \lambda z^{b+1} - (\lambda + \mu)z} \tag{10.6}$$

The constant p_0 can be obtained in the usual fashion by recognizing that $P(1) = 1$ from the definition of the geometric transform. Since direct substitution leads to an indeterminant form, we must again use L'Hospital's rule to find

$$p_0 = 1 - \frac{b\lambda}{\mu} \tag{10.7}$$

The final form of the geometric transform for the number of customers in the system at some arbitrary time t is then

$$P(z) = \frac{(\mu - b\lambda)(1 - z)}{\mu(1 - z) - \lambda z(1 - z^b)} \tag{10.8}$$

Note that the traffic intensity, which is also equal to the offered load for this system can be written as

$$\rho = \frac{\text{arrivals during a service interval}}{\text{number of channels}}$$

$$= \frac{\lambda b}{\mu} \tag{10.9}$$

Further note that one can cancel the term $(1 - z)$ from numerator and denominator. With these changes it is sometimes convenient to revise equation (10.8) as follows

$$P(z) = \frac{\mu(1 - \rho)}{\mu - \lambda \sum_{i=1}^{b} z^i} \tag{10.10}$$

One can now obtain the moments of the distribution by taking derivatives of the transform. In addition, since this transform is expressed as a ratio of polynomials in z one can in theory factor the denominator and obtain the inverse by partial fraction expansion. Suppose that the b roots of the denominator, i.e., poles of the transform, are designated as $z_1, z_2, \ldots, z_b$. We can then write the denominator in factored form as

$$d(z) = (z_1 - z)(z_2 - z) \cdots (z_b - z) \tag{10.11}$$

By employing the theory of partial fractions we can then obtain the partial fraction expansion for $P(z)$ as

$$P(z) = (\mu - b\lambda) \sum_{i=1}^{b} \frac{C_i}{(z_i - z)} \tag{10.12}$$

The coefficient C_i are obtained from the relationship

$$C_i = \lim_{z \to z_i} (z_i - z)\, P(z)/(\mu - b\lambda) \qquad (10.13)$$

Each term in the partial fraction expansion (10.12) involves a transform which can be written in the form.

$$F(z) = \frac{d}{1 - dz} \qquad (10.14)$$

where $d = 1/z_i$. The inverse of $F(z)$ is

$$f_n = d^{n+1} \qquad (10.15)$$

It therefore follows that the probability distribution for our Poisson bulk arrival system is given by

$$p_n = (\mu - b\lambda) \sum_{i=1}^{b} C_i z_i^{-(n+1)} \qquad (10.16)$$

An alternative to this approach is of course to revert to Maclaurin series expansion or to long division and pick off the coefficients of z^n in the resulting series. However, those techniques do not often indicate a general functional form.

Example. A shoe repair shop receives worn and damaged shoes, one pair at a time. Repairs are made as needed, one shoe at a time. Past records indicate that shoes are received at the rate of forty pairs per eight-hour work day. The average time to repair a single shoe is estimated to be three minutes. The problem is to estimate the probability distribution for number of shoes awaiting repair.

Let us assume that arrival events can be described by a Poisson process with a mean rate of $\lambda = 5$ per hour and that service is exponential at the rate $\mu = 20$ per hour. The arrival batch size is $b = 2$.

The transform of the state distribution is

$$P(z) = \frac{(\mu - b\lambda)}{\mu - \lambda \sum_{i=1}^{b} z^i}$$

$$= \frac{(20 - 10)}{20 - 5(z + z^2)}$$

The denominator can be written in factored form as

$$d(z) = \left[z - \left(\frac{-1 + \sqrt{17}}{2} \right) \right]\left[z - \left(\frac{-1 - \sqrt{17}}{2} \right) \right] (-5)$$

It therefore follows that $P(z)$ reduces to

$$P(z) = \frac{-2}{\left(\dfrac{-1 + \sqrt{17}}{2} - z \right)\left(\dfrac{-1 - \sqrt{17}}{2} - z \right)}$$

from which

$$p_n = \frac{2}{\sqrt{17}} \left\{ \left(\frac{-1 + \sqrt{17}}{2} \right)^{-(n+1)} - \left(\frac{-1 - \sqrt{17}}{2} \right)^{-(n+1)} \right\}$$

RANDOM BULK ARRIVAL SIZE

Let us now generalize the bulk-arrival model by permitting the size of each arriving group to be a random variable. Here the batch size b is distributed by the probability mass function

$$Pr(\text{batch} = b) = g_b \tag{10.17}$$

The arrival stream over time is a compound Poisson process of the type described in equation (4.61).

We could view such an arrival process as a combination of separate Poisson streams with the arrival rate for the ith stream being λg_i. This means that transitions into state j due to arrivals can occur from any state below it since an arriving batch may be of any size including zero. A zero batch size implies that the system remains in its current state. Transition into state j due to service occurs only from state $(j + 1)$ as we still assume a single-channel system. Transition out of j due to service goes only to $(j - 1)$. Transition out of j due to arrivals may lead to any higher order state. However, the total rate of departure flow from state j is still λ due to

$$\sum_{i=0}^{\infty} \lambda g_i = \lambda \sum_{i=0}^{\infty} g_i = \lambda \tag{10.18}$$

Putting this information into the usual rate of change equations leads to

$$\frac{dp_0(t)}{dt} = \mu p_1(t) - \lambda p_0(t) + \lambda g_0 p_0(t)$$

$$\frac{dp_n(t)}{dt} = \mu p_{n+1}(t) - \lambda p_n(t) - \mu p_n(t) + \sum_{i=0}^{n} \lambda g_{n-i} p_i(t), \qquad n \geq 1$$

$$\tag{10.19}$$

By setting the derivatives equal to zero we have the equilibrium equations

$$\lambda p_0 = \mu p_1 + \lambda g_0 p_0$$

$$(\lambda + \mu)p_n = \mu p_{n+1} + \sum_{i=0}^{n} \lambda g_{n-i} p_i, \qquad n \geq 1 \tag{10.20}$$

The solution strategy is again to reduce the infinite set to a single equation in the transform $P(z) = \sum_{n=0}^{\infty} p_n z^n$ by multiplying the nth equation by z^n and adding. This yields

204

$$\lambda \sum_{n=0}^{\infty} p_n z^n + \mu \sum_{n=1}^{\infty} p_n z^n = \mu \sum_{n=0}^{\infty} p_{n+1} z^n + \sum_{n=0}^{\infty} \sum_{i=0}^{n} \lambda g_{n-i} p_i z^n \qquad (10.21)$$

Changing the order of summation and rearranging terms we have

$$\lambda P(z) + \mu[P(z) - p_0] = \frac{\mu}{z}[P(z) - p_0] + \lambda \sum_{i=0}^{\infty} p_i z^i \sum_{n=i}^{\infty} g_{n-i} z^{n-i} \qquad (10.22)$$

But the term

$$\sum_{n=i}^{\infty} g_{n-i} z^{n-i} = \sum_{k=0}^{\infty} g_k z^k = G(z) \qquad (10.23)$$

is simply the transform of the bulk-size distribution. With that substitution we can now express the transform of the state distribution for the random bulk-size Poisson arrival system as

$$P(z) = \frac{\mu p_0 (1 - z)}{\mu + \lambda z G(z) - (\lambda + \mu)z} \qquad (10.24)$$

Resorting to $P(1) = 1$ and L'Hospital's rule leads to evaluation of the constant as

$$p_0 = 1 - \frac{\lambda G'(1)}{\mu} \qquad (10.25)$$

But

$$G'(1) = E(b) \qquad (10.26)$$

Therefore we can write the final form of the transform as

$$P(z) = \frac{[\mu - \lambda E(b)](1 - z)}{\mu(1 - z) - \lambda z[1 - G(z)]} \qquad (10.27)$$

Note that the constant batch size model of equation 10.8 is a special case of the current model. For a fixed batch size b

$$G(z) = z^b$$

and

$$E(b) = b$$

which when substituted into (10.27) yields equation (10.8).

The transform can now be used directly to obtain the moments of the state distribution or inverted to obtain any desired probabilities. As is typical of transforms, the ease of inversion is highly dependent upon the choice of the distribution $g(b)$.

PERFORMANCE MEASURES

We can obtain a general expression for the expected number of customers in the system, L, by taking the first derivative of $P(z)$ evaluated at

$z = 1$. Since this leads to an indeterminant form it is necessary to apply L'Hospital's rule twice to obtain the following expression

$$P'(z = 1) = \frac{[\mu - \lambda E(b)][2\lambda G'(1) + \lambda G''(1)]}{2[\mu - \lambda G'(1)]^2} \tag{10.28}$$

But we know from the transform of the arriving bulk size that

$$G'(1) = E(b) \tag{10.29}$$

$$G''(1) = \sigma_b{}^2 - E(b) + [E(b)]^2 \tag{10.30}$$

After making these substitutions and canceling terms we find that the expected number of customers in the system is

$$L = \frac{\lambda[\sigma_b{}^2 + E(b) + [E(b)]^2]}{2[\mu - \lambda E(b)]} \tag{10.31}$$

If we define the utilization factor as

$$\rho = \frac{\lambda E(b)}{\mu} \tag{10.32}$$

this function can also be expressed as

$$L = \frac{\rho[1 + E(b) + \sigma_b{}^2/E(b)]}{2(1 - \rho)} \tag{10.33}$$

The expected number in the queue can be calculated by substracting the expected number in service, which is equivalent to the utilization factor, from the number in the system. Therefore

$$L_q = L - \frac{\lambda E(b)}{\mu} \tag{10.34}$$

To find the waiting times for this system we can again invoke Little's formula to deduce that the total sojourn time, given an effective arrival rate of $\lambda_e = \lambda E(b)$, is

$$W = L/\lambda_e$$
$$= \frac{1 + E(b) + \sigma_b{}^2/E(b)}{2\mu(1 - \rho)} \tag{10.35}$$

The expected queue time for an arriving customer is simply his total expected wait minus his expected service time.

$$W_q = W - \frac{1}{\mu} \tag{10.36}$$

We now have several easily calculated measures of performance for the Poisson bulk-arrival system in which bulk size is a random variable. Note that it is not necessary to know the entire distribution for the arriving bulk size if one is only interested in expected value measures of performance. All that is required are the first two moments of that distribution.

Example. A private pheasant hunting preserve plans to offer hunters the service of cleaning all pheasants they shoot during their hunt. They expect hunters to arrive at the cleaning station in Poisson fashion at a rate of ten per hour. The number of birds shot by a hunter is expected to have a mean and variance of three and four respectively. The pheasants are to be cleaned one at a time with the average cleaning time expected to be 1.5 minutes. Because of competitive pressures the management of the preserve wants to estimate the average delay experienced by a hunter while he waits for his birds to be cleaned.

This is a bulk-arrival Poisson queue in which

$$\lambda = 1/6 \text{ per minute}$$
$$\mu = 1/1.5 \text{ per minute}$$
$$\sigma_b^2 = 4, \quad E(b) = 3$$

The utilization factor is

$$\rho = \frac{\lambda E(b)}{\mu} = \frac{(1/6)(3)}{1/1.5} = 0.75$$

The expected time that a bird will be in the cleaning system is

$$W = \frac{1 + E(b) + \sigma_b^2/E(b)}{2\mu(1 - \rho)}$$
$$= \frac{1 + 3 + 4/3}{2\left(\dfrac{1}{1.5}\right)(1 - 0.75)} = 16 \text{ minutes}$$

EXPONENTIAL BULK SERVICE

Let us now consider systems in which customers arrive singly in Poisson fashion but are served in bulk. When service takes place an entire group of c may be served. The duration of service for the group is assumed to be exponentially distributed. Two variations in such service mechanisms may be encountered. In the first version the server will not act until a group of c customers has accumulated. He will then serve the entire group. In the second version service takes place on groups which cannot be larger than c. If less than c customers are present, all are served. If more than c are present a group of c will be served.

Constant Size Service Batch

The flow rate diagram for the constant bulk size case appears in Figure 10-2. Flow into state n occurs with an arrival from state $(n - 1)$ or bulk

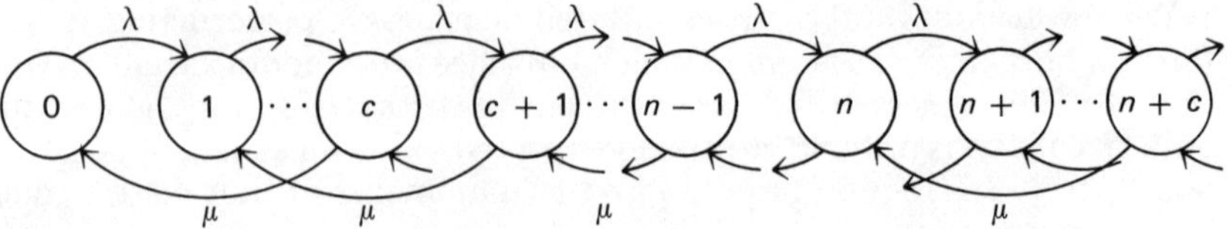

FIGURE 10-2 Flow Rate Diagram for Bulk Service.

service from state $(n + c)$. Flow out of state n goes to $(n + 1)$ with an arrival or to $(n - c)$ with a bulk service whenever $n \geq c$. The proper differential-difference equations can then be formulated as

$$\frac{dp_0(t)}{dt} = \mu p_c(t) - \lambda p_0(t)$$

$$\frac{dp_n(t)}{dt} = \mu p_{n+c}(t) + \lambda p_{n-1}(t) - \lambda p_n(t), \; 1 \leq n < c$$

$$\frac{dp_n(t)}{dt} = \mu p_{n+c}(t) + \lambda p_{n-1}(t) - \mu p_n(t) - \lambda p_n(t), \; n \geq c \quad (10.37)$$

By setting the derivatives equal to zero we obtain the usual steady-state balance equations.

$$\lambda p_0 = \mu p_c$$
$$\lambda p_n = \mu p_{n+c} + \lambda p_{n-1}, \qquad 1 \leq n < c$$
$$(\lambda + \mu) p_n = \mu p_{n+c} + \lambda p_{n-1}, \qquad n \geq c \quad (10.38)$$

To solve these equations we will again multiply the nth equation by z^n and add in an attempt to achieve a single equation in the transform $P(z)$. That is

$$\lambda \sum_{n=0}^{\infty} p_n z^n + \mu \sum_{n=c}^{\infty} p_n z^n = \mu \sum_{n=0}^{\infty} p_{n+c} z^n + \lambda \sum_{n=1}^{\infty} p_{n-1} z^n$$

which can be rewritten in the transform as

$$\lambda P(z) + \mu \left[P(z) - \sum_{i=0}^{c-1} p_i z^i \right] = \frac{\mu}{z^c} \left[P(z) - \sum_{i=0}^{c-1} p_i z^i \right] + \lambda z P(z)$$

$$(10.39)$$

Solving for $P(z)$ we have

$$P(z) = \frac{\mu(1 - z^c) \displaystyle\sum_{i=0}^{c-1} p_i z^i}{\lambda z^{c+1} - (\lambda + \mu) z^c + \mu} \quad (10.40)$$

For computational convenience let us define $\rho = \lambda/\mu$, divide by μ and rewrite the transform as

$$P(z) = \frac{(1 - z^c) \displaystyle\sum_{i=0}^{c-1} p_i z^i}{\rho z^{c+1} - (\rho + 1) z^c + 1} \quad (10.41)$$

This transform contains c unknowns, i.e. the low order probabilities $(p_0, p_1 \cdots p_{c-1})$. The problem now is to be able to specify values for those quantities.

In order for $P(z)$ to converge on and inside the unit circle, which is required from its definition, the zeros of the numerator must coincide with the zeros of the denominator which fall within the unit circle. The denominator in equation (10.41) has $c + 1$ roots, one of which is unity. Using Rouche's theorem[1] it can be shown that exactly $(c - 1)$ of the remaining roots lie in the range $|z| < 1$ and one, say z_0 lies in $|z| > 1$.

The numerator contains a factor $(1 - z^c)$ the roots of which all have absolute value of unity. Therefore the zeros in the numerator which match the $(c - 1)$ zeros with $|z| < 1$ in the denominator must all be contained in the summation term. After removing the factors containing unity and z_0 from the denominator the result should differ by at most a multiplicative constant from the sum in the numerator. That is

$$K \left[\frac{\rho z^{c+1} - (\rho + 1)z^c + 1}{(z - 1)(z - z_0)} \right] = \sum_{i=0}^{c-1} p_i z^i \tag{10.42}$$

By making this substitution in (10.41) and cancelling common terms we are led to the reduced form

$$P(z) = \frac{K(1 - z^c)}{(z - 1)(z - z_0)}$$

$$= \frac{K}{z_0 - z} \sum_{i=0}^{c-1} z^i \tag{10.43}$$

The constant K can be evaluated from the relationship $P(1) = 1$ which leads to

$$K = \frac{z_0 - 1}{c} \tag{10.44}$$

Finally we have the transform

$$P(z) = \frac{(z_0 - 1) \sum_{i=0}^{c-1} z^i}{(z_0 - z)c}$$

$$= \frac{(1 - 1/z_0) \sum_{i=0}^{c-1} z^i}{(1 - z/z_0)c} \tag{10.45}$$

This function is easily inverted by using the shifting property of geometric transforms. That is

if $$G(f(n)) = G(z)$$

then $$G(f(n - r)U(n - r)) = z^r G(z) \tag{10.46}$$

where $$U(n - r) = \begin{cases} 0 & n < r \\ 1 & n \geq r \end{cases}$$

---

[1]Rouche's theorem states: If $f(z)$ and $g(z)$ are analytic inside and on a closed contour C and if $|g(z)| < |f(z)|$ on C, then $f(z)$ and $f(z) + g(z)$ have the same number of zeros inside C. For proof of this theorem see any complex variable book such as Churchill, (17) or Spiegel, (85).

What we have is a basic transform

$$\frac{1}{1 - z/z_0} = \frac{1}{1 - dz} \tag{10.47}$$

multiplied by a constant and several terms involving powers of z. The inverse of $(1 - z/z_0)^{-1}$ is

$$f(n) = \left(\frac{1}{z_0}\right)^n \tag{10.48}$$

It therefore follows that

$$\begin{aligned}
p_n = {}& \left(\frac{1 - 1/z_0}{c}\right)\left[\left(\frac{1}{z_0}\right)^n \right.\\
& \left. + \left(\frac{1}{z_0}\right)^{n-1} U(n - 1) + \cdots + \left(\frac{1}{z_0}\right)^{n-c+1} U(n - c + 1)\right]
\end{aligned} \tag{10.49}$$

which may be written in more familiar notation as

$$p_n = \begin{cases} \dfrac{1 - (1/z_0)^{n+1}}{c} & 0 \leq n < c \\[2em] \dfrac{\left[1 - \left(\dfrac{1}{z_0}\right)^c\right]\left(\dfrac{1}{z_0}\right)^{n-c+1}}{c} & n \geq c \end{cases} \tag{10.50}$$

To summarize, the state distribution for the fixed bulk size (c) exponential service, Poisson arrival system is found by first locating the root of the equation

$$\rho z^{c+1} - (\rho + 1)z^c + 1 = 0 \tag{10.51}$$

which lies outside the unit circle, say z_0. This value is then used in equation (10.50) to find the necessary probabilities.

The moments of the distribution are most easily obtained by taking derivatives of the transform as expressed in equation 10.45. For example, the mean number of customers present at a random point in time is given by

$$P'(z = 1) = \frac{\left(1 - \dfrac{1}{z_0}\right)}{c}\left[\frac{\displaystyle\sum_{i=1}^{c-1} iz^{i-1}\left(1 - \dfrac{z}{z_0}\right) + \dfrac{1}{z_0}\displaystyle\sum_{i=0}^{c-1} z^i}{\left(1 - \dfrac{z}{z_0}\right)^2}\right]\Bigg|_{z=1} \tag{10.52}$$

Therefore

$$L = \frac{c - 1}{2} + \frac{1}{z_0 - 1} \tag{10.53}$$

The expected number in the queue can be obtained by subtracting the expected number in service from L. Service is always in bulk with a batch size c. By using $\displaystyle\sum_{i=0}^{c-1} p_i = \frac{c - \rho}{c}$ from equation (10.42) evaluated at $z = 1$, it therefore follows that the expected number in service is

$$E(s) = c\left(1 - \sum_{i=0}^{c-1} p_i\right) = \rho \tag{10.54}$$

It then follows that

$$L_q = L - \rho \tag{10.55}$$

Corresponding waiting times can be found by again applying Little's formula. That is

$$W = \frac{1}{\lambda}\left[\frac{c-1}{2} + \frac{1}{z_0 - 1}\right] \tag{10.56}$$

and

$$W_q = W - \frac{1}{\mu} \tag{10.57}$$

Example. A ferry operator controls the only passage across a remote river. His policy is never to leave the riverbank without a full load. His new boat can handle two cars. Cars arrive at the crossing at the rate of four per hour. The trip across the river averages six minutes. The problem is to estimate the delay time experienced by travelers who use the ferry.

Let us suppose that the usual Poisson and exponential assumption apply. The parameters are then $c = 2$, $\lambda = 4$, $\mu = 10$. For the current fixed bulk-service size policy we must first find the roots of the polynomial

$$0.4z^3 - 1.4z^2 + 1 = 0$$

We know that $z = 1$ is one of the roots so we can factor that out leaving

$$(z - 1)(0.4z^2 - z - 1) = 0$$

The positive root of the remaining quadratic term is

$$z_0 = \frac{1 + \sqrt{1 + 1.6}}{0.8} = 3.2656$$

From equation (10.56) it then follows that an arriving traveler can expect a river crossing delay of

$$W = \frac{1}{4}\left[\frac{2 - 1}{2} + \frac{1}{3.2656 - 1}\right] = 0.36 \text{ hours}$$

The boat operator used to have a single-car ferry which could cross the river in exactly the same time as the new boat. The problem now is to see whether or not the new boat has improved performance. With single-car service the expected delay time is

$$W = \frac{1}{\mu - \lambda} = \frac{1}{10 - 4} = 0.167 \text{ hours}$$

Customer service has obviously deteriorated.

Variable Service Batch Size

Let us now consider the second major operating policy for bulk service. In this system whenever the server is free it will accept c customers for bulk service if they are available. If less than c are present when the server becomes free it will serve in bulk all who are present. We will assume Poisson arrivals and an exponentially distributed time to serve a batch, regardless of its size.

Following the usual flow rate arguments one can quickly construct the proper set of differential-difference equations as

$$\frac{dp_0(t)}{dt} = \mu \sum_{i=1}^{c} p_i(t) - \lambda p_0(t)$$

$$\frac{dp_n(t)}{dt} = \mu p_{n+c}(t) + \lambda p_{n-1}(t) - \mu p_n(t) - \lambda p_n(t), \qquad n \geq 1 \quad (10.58)$$

The equilibrium equations, obtained by setting $\dfrac{dp_n(t)}{dt} = 0$ are then

$$\lambda p_0 = \mu \sum_{i=1}^{c} p_i$$

$$(\lambda + \mu)p_n = \mu p_{n+c} + \lambda p_{n-1} , \qquad n \geq 1 \qquad (10.59)$$

We will again reduce the infinite set of equations to a single equation in the transform by multiplying by z^n and adding. That is

$$\lambda \sum_{n=0}^{\infty} p_n z^n + \mu \sum_{n=1}^{\infty} p_n z^n = \mu \left[\sum_{i=1}^{c} p_i + \sum_{n=1}^{\infty} p_{n+c} z^n \right] + \lambda \sum_{n=1}^{\infty} p_{n-1} z^n \quad (10.60)$$

Rewriting as a function of $P(z) = \sum_{i=0}^{\infty} p_i z^i$ we have

$$\lambda P(z) + \mu \left[P(z) - p_0 \right] = \frac{\mu}{z^c} \left[P(z) - \sum_{i=0}^{c} p_i z^i \right] + \lambda z P(z) + \mu \sum_{i=1}^{c} p_i$$

$$(10.61)$$

from which

$$P(z) = \frac{\mu z^c \displaystyle\sum_{i=0}^{c} p_i - \mu \displaystyle\sum_{i=0}^{c} p_i z^i}{(\lambda + \mu)z^c - \mu - \lambda z^{c+1}}$$

$$= \frac{\displaystyle\sum_{i=0}^{c-1} p_i (z^i - z^c)}{\rho z^{c+1} - (1 + \rho)z^c + 1} \qquad (10.62)$$

where

$$\rho = \lambda/\mu$$

We are again faced with the prospect of determining the c unknowns $(p_0, p_1 \cdots p_{c-1})$ in order to completely specify the transform $P(z)$. This time our task is somewhat simplified by noting that the denominator of equation (10.62) is identical to the denominator of (10.41). Using the same arguments as before we can factor the terms $(z - 1)$ and $(z - z_0)$ from the denominator where $|z_0| > 1$. Note that $z = 1$ is also a zero of the numerator. Matching the remaining $(c - 1)$ zeros of the denominator against the portion of the numerator which remains after cancelling $(z - 1)$ we have a proportionality of the form

$$K\left[\frac{\rho z^{c+1} - (1 + \rho)z^c + 1}{(z - 1)(z - z_0)}\right] = \frac{\displaystyle\sum_{i=0}^{c-1} p_i(z^i - z^c)}{(z - 1)} \tag{10.63}$$

By substituting in equation (10.62) and cancelling common terms we are led to

$$P(z) = \frac{K}{z - z_0} \tag{10.64}$$

The constant K is evaluated from the condition that $P(1) = 1$. Therefore

$$K = 1 - z_0 \tag{10.65}$$

and

$$P(z) = \frac{1 - z_0}{z - z_0}$$
$$= \frac{1 - 1/z_0}{1 - z/z_0} \tag{10.66}$$

The inverse of this transform is especially simple. The distribution for the number of customers in this bulk-service system is

$$p_n = \left(1 - \frac{1}{z_0}\right)\left(\frac{1}{z_0}\right)^n \tag{10.67}$$

where z_0 is the root of the equation

$$\rho z^{c+1} - (1 + \rho)z^c + 1 = 0 \tag{10.68}$$

which lies outside the unit circle.

Note that this distribution is of exactly the same form (geometric) as the distribution obtained for the $(M/M/1)$ system. The only difference is that the parameter ρ has been replaced by $(1/z_0)$. This permits us to jump immediately to expressions for the expected number of customers present and the expected delay time. That is

$$L = \frac{1/z_0}{1 - 1/z_0} = \frac{1}{z_0 - 1} \tag{10.69}$$

and

$$W = \frac{1}{\lambda(z_0 - 1)} \tag{10.70}$$

Example. Let us now return to our ferry operator cited in the last example. Suppose that he decides to alter his policy for the new boat and agrees to take one or two cars depending upon the number present when he reaches the riverbank. The parameters which define equation (10.68) are unchanged from that example. Therefore $z_0 = 3.2656$ and $\lambda = 4$ from which delay time under the new policy is expected to be

$$W = \frac{1}{4(3.2656 - 1)} = 0.11 \text{ hours}$$

By following this procedure we can see that customer service is improved over both his old policy and his former one-car ferry operation. The probability that he will now be unable to serve all who are in the queue when he starts across the river is only

$$P(\text{customers left}) = 1 - p_0 - p_1 - p_2 = 0.0287$$

GENERAL DISTRIBUTIONS FOR BULK SERVICE

The next task is to generalize the bulk-service problem by permitting service time to be continuously distributed in some arbitrary fashion, say $b(t)$. We will retain the Poisson arrival and maximum group size assumptions of the previous section. The techniques employed are those of the imbedded Markov chain introduced in Chapter 9 and matching zeros in transforms as used extensively in this chapter. Our discussion closely parallels that of Bailey and Downton who developed the original models (5, 31).

An imbedded Markov chain is created by examining regeneration points which occur immediately prior to service epochs. If c or less customers are present all waiting customers will be served in bulk. If more than c are present the bulk size is limited to c. We will further assume that the system operates in conveyor fashion similar to the way we earlier modeled the constant service-time transient systems. This means that even though an arriving customer may find the system empty he must wait until the beginning of the next service epoch before he can be served. The server continues to cycle whether or not customers are there to be served.

Let k_j be the probability that j customers will arrive during a service interval. Since arrivals occur singly in Poisson fashion, we can express this probability

$$k_j = \int_0^\infty \frac{e^{-\lambda t}(\lambda t)^j}{j!}\, b(t)\, dt \tag{10.71}$$

Taking geometric transforms leads to the compact expression

$$K(z) = \sum_{j=0}^{\infty} k_i z^i$$

$$= \left. \mathcal{L}(b(t)) \right|_{s=\lambda(1-z)}$$

$$= B(\lambda(1 - z)) \tag{10.72}$$

where $B(s)$ is the Laplace transform of the service time distribution. The mean of this distribution is

$$m = \lambda \bar{t} \tag{10.73}$$

where $\bar{t}$ is the average length of a service interval.

The single-stage transition matrix for state changes between successive occasions of service is given by

$$
\mathbf{T} = \quad
\begin{array}{c}
\\
\\
0 \\
1 \\
2 \\
\vdots \\
c \\
c+1 \\
c+2 \\
\vdots
\end{array}
\begin{array}{ccccc}
0 & 1 & 2 & \cdots \\
\hline
k_0 & k_1 & k_2 & \cdots \\
k_0 & k_1 & k_2 & \cdots \\
k_0 & k_1 & k_2 & \cdots \\
\\
k_0 & k_1 & k_2 & \cdots \\
0 & k_0 & k_1 & \cdots \\
0 & 0 & k_0 & \cdots \\
\end{array}
\tag{10.74}
$$

Note that the first $(c + 1)$ rows of this matrix are identical. Any number n customers present, as long as $n < c$, will be eliminated by the upcoming service event. Moving to state j then implies that j arrivals must occur before the next occasion of service. If more than c are present only c will be served. Moving to state $j \geq c$ then requires $j + c - n$ arrivals.

The process is ergodic as long as $m < c$. We can therefore perform the usual steady-state calculation

$$\mathbf{\Pi} = \mathbf{\Pi T} \tag{10.75}$$

to obtain the infinite set of difference equations

$$\pi_0 = k_0 \sum_{i=0}^{c} \pi_i$$

$$\pi_n = k_n \sum_{i=0}^{c} \pi_i + \sum_{i=1}^{n} \pi_{c+i}\, k_{n-i} \,, \qquad n \geq 1 \tag{10.76}$$

Multiplying by z^n and adding we have

$$\sum_{n=0}^{\infty} \pi_n z^n = k_0 \sum_{i=0}^{c} \pi_i + \sum_{i=0}^{c} \pi_i \left[\sum_{n=1}^{\infty} k_n z^n \right] + \sum_{n=1}^{\infty} \left[\sum_{i=1}^{n} \pi_{c+i} k_{n-i} \right] z^n \tag{10.77}$$

The last term above can be recognized as the transform of the convolution of two sequences $\{\pi_{c+i}\}$ and $\{k_i\}$. Furthermore the transform of the sequence $[\pi_{c+i}\}$ can be related to the transform of the desired sequence $\{\pi_n\}$ from the shifting theorem as

$$G(\pi_{c+i}) = z^{-c} \left[G(\pi_n) - \sum_{i=0}^{c-1} \pi_i z^i \right] \tag{10.78}$$

Recognizing that we have two transforms in the expression,

$$K(z) = \sum_{i=0}^{\infty} k_i z^i$$

and

$$P(z) = \sum_{i=0}^{\infty} \pi_i z^i$$

we can now rewrite equation (10.77) as

$$P(z) = k_0 \sum_{i=0}^{c} \pi_i + \sum_{i=0}^{c} \pi_i \left[K(z) - k_0 \right]$$

$$+ K(z) \left\{ z^{-c} \left[P(z) - \sum_{i=0}^{c-1} \pi_i z^i \right] \right\} - K(z)\pi_c \tag{10.79}$$

Solving for $P(z)$ we have

$$P(z) = \frac{\displaystyle\sum_{i=0}^{c-1} \pi_i(z^c - z^i)}{z^c/K(z) - 1} \tag{10.80}$$

Here we are again faced with finding a way to eliminate the c-unknown terms $(\pi_0, \pi_1 \cdots \pi_{c-1})$ from the transform. The strategy is exactly the same as before, namely match those zeros of the denominator which fall within the unit circle with corresponding zeros in the numerator. The resulting equations, expressed as functions of those zeros falling outside the unit circle can then be used to eliminate the unknown values $(\pi_0, \pi_1 \cdots \pi_{c-1})$. The problem is more complicated than those of earlier sections, however, since the number of zeros falling outside the unit circle is strongly influenced by the form of the service time distribution. (Recall that for the exponential service case in the last section we had a single zero, z_0, which fell outside the unit circle).

Bailey has developed results for the special case of a k-Erlang service distribution, i.e., chi-square with $2k$ degrees of freedom

$$b(t) = \frac{\mu^k}{\Gamma(k)} t^{k-1} e^{-\mu t}, \qquad 0 \le t < \infty \tag{10.81}$$

This is a very useful service distribution since it can be made to assume a variety of shapes all the way from exponential to constant densities depending upon the choice of k. Selecting $k = 1$ for example given the usual exponential density. For the Erlang distribution we have

216

$$\mathcal{L}(b(t)) = B(s) = \int_0^\infty \frac{\mu^k}{\Gamma(k)} t^{k-1} e^{-\mu t} e^{-st} dt$$

$$= \left(\frac{\mu}{s+\mu}\right)^k \tag{10.82}$$

from which

$$K(z) = B(\lambda(1-z)) = \left[\frac{\mu}{\lambda(1-z)+\mu}\right]^k$$

$$= [1 + m(1-z)/k]^{-k} \tag{10.83}$$

where

$$m = \lambda \bar{t} = \lambda k/\mu$$

The state distribution at service epochs for bulk Erlang service then reduces to

$$P(z) = \frac{\displaystyle\sum_{i=0}^{c-1} \pi_i(z^c - z^i)}{z^c[1 + m(1-z)/k]^k - 1} \tag{10.84}$$

Applying Rouches' theorem to the denominator we find that k out of $(c+k)$ zeros in the denominator fall outside the unit circle. After factoring, cancelling common terms in numerator and denominator, and evaluating the resulting proportionality constant one can write

$$P(z) = \prod_{i=0}^{k-1} \frac{z_i - 1}{z_i - z} \tag{10.85}$$

where $(z_0, z_1, \cdots z_{k-1})$ are the zeros of the denominator in equation (10.84) which fall outside the unit circle. The transform can now be inverted by partial fraction expansion and table look-up or the moments can be obtained by taking derivatives.

From the transform it can be shown that the mean and variance, for number of customers present at a service event for k-Erlang bulk service with batch size c, are

$$E(n) = \sum_{i=0}^{k-1} (z_i - 1)^{-1} \tag{10.86}$$

and

$$\sigma_n^2 = \sum_{i=0}^{k-1} z_i(z_i - 1)^{-2} \tag{10.87}$$

Note that $\{\pi_n\}$ is the distribution for number present at specific instants in time and is not necessarily equal to $\{p_n\}$ the distribution for number present at an arbitrary time.

Developing a waiting time distribution for the general bulk-service problem is considerably more complicated. Readers who are interested should consult Downton (31). For the special case of k-Erlang service and arrival rate $\lambda = 1$, Downton shows that the mean and variance for waiting time will be

$$E(W') = \sum_{i=0}^{k-1} (z_i - 1)^{-1} - \frac{k-1}{2\mu} \tag{10.88}$$

and

$$\sigma_{w'}^2 = \sum_{i=0}^{k-1} (z_i - 1)^2 + \frac{(k-1)(k-5)}{12\mu^2} \tag{10.89}$$

Scaling these values for an arbitrary arrival rate λ we have

$$W = E(W')/\lambda \tag{10.90}$$

$$\sigma_w^2 = \sigma_{w'}^2/\lambda^2 \tag{10.91}$$

Example. One passenger waiting for crossing at the ferry operation has learned of the modeling attempts by the operator which assumed exponential service times. After observing the operation briefly the passenger-analyst concludes that service time has the mean of six minutes but is more nearly distributed by a 2-Erlang distribution. The question now is how much error did the operator introduce by assuming exponential service.

The parameters for this problem are $\lambda = 4$, $\mu = 20$, $k = 2$, $c = 2$. The polynomial for which we must find roots such that $|z_i| > 1$ is

$$z^c \left[1 + \frac{\lambda(1 - z)}{\mu} \right]^k - 1 = z^2 [1 + 0.2(1 - z)]^2 - 1 = 0$$

The roots of this equation are $z \approx -.74,\ 1.0,\ 5.0,\ 6.74$. Using only the roots outside the unit circle we can now calculate the expected delay time as

$$W = \frac{1}{4} \left[(5 - 1)^{-1} + (6.74 - 1)^{-1} - \frac{2 - 1}{2(20)} \right]$$

$$\approx .0998 \text{ hours}$$

This number compares with $W = 0.11$ obtained under the exponential assumption for an error of about 10 percent.

TRANSIENT SOLUTIONS
FOR BULK QUEUES

Let us conclude our discussion of bulk queues with a brief indication of the feasibility of attaining transient solutions.

First, the sets of differential-difference equations (10.1) and (10.19) posed for bulk-arrival problems would indeed be formidable should we seek the full-time varying probability distribution $P_n(t)$. Furthermore they do not lend themselves well to the truncation arguments we employed in Chapter 8 for achieving approximate numerical solutions for time varying queues. This is true since truncating at some point m would imply that entire batch would be turned away if the number of customers present fell in the interval $(m - b) < n \leq m$. The truncated system would then be a poor approximation for the unlimited queue case. Let us hasten to add however, that we could achieve meaningful transient solutions to such a truncated system if in fact it behaved that way.

Transient solutions for the bulk-service problem are somewhat easier to come by. Consider first the constant batch size exponential service model. If we truncate the queue at m and permit arrival and service rates to vary, the constant size batch equations (10.37) are modified to

$$\frac{dp_0(t)}{dt} = \mu(t)p_c(t) - \lambda(t)\,p_0(t)$$

$$\frac{dp_n(t)}{dt} = \mu(t)p_{n+c}(t) + \lambda(t)p_{n-1}(t) - \lambda(t)p_n(t), \quad 1 \leq n < c$$

$$\frac{dp_n(t)}{dt} = \mu(t)p_{n+c} + \lambda(t)p_{n-1}(t) - \mu(t)p_n(t) - \lambda(t)p_n(t), \quad c \leq n \leq m - c$$

$$\frac{dp_n(t)}{dt} = \lambda(t)p_{n-1}(t) - \mu(t)p_n(t) - \lambda(t)p_n(t), \quad m - c < n < m$$

$$\frac{dp_m(t)}{dt} = \lambda(t)p_{m-1}(t) - \mu(t)p_m(t) \tag{10.92}$$

where we assume that $m \geq 2c$.

By segmenting the time-varying arrival and service rates into discrete steps we have a finite set of linear differential equations with constant coefficients. These equations can then be solved by Runge-Kutta methods within each discrete interval to develop the entire transient solution. The technique is precisely the one used on equations (8.1) in the transient analysis chapter.

We can treat the variable batch size exponential service equations (10.58) in similar fashion. When the system is truncated at $n = m$, these equations become

$$\frac{dp_0(t)}{dt} = \mu(t) \sum_{i=1}^{c} p_i(t) - \lambda(t)p_0(t)$$

$$\frac{dp_n(t)}{dt} = \mu(t)p_{n+c}(t) + \lambda(t)p_{n-1}(t) - \mu(t)p_n(t) - \lambda(t)p_n(t), \quad 1 \leq n \leq m - c$$

$$\frac{dp_n(t)}{dt} = \lambda(t)p_{n-1}(t) - \mu(t)p_n(t) - \lambda(t)p_n(t), \quad m - c < n < m$$

$$\frac{dp_m(t)}{dt} = \lambda(t)p_{m-1}(t) - \mu(t)p_m(t) \tag{10.93}$$

where $m \geq c + 1$.

Here again we can use the Runge-Kutta method to obtain the complete time varying solution in the manner demonstrated in Chapter 8.

The general service distribution bulk-service model requires slightly different tactics to achieve a transient solution. The strategy here is to use the basic Markov chain iteration

$$\mathbf{\Pi}(t + \bar{t}) = \mathbf{\Pi}(t)\mathbf{T}(t) \tag{10.94}$$

where the transition matrix of equation (10.74) is modified to reflect a truncation of $m > c$. The entries in the modified matrix must be calculated directly from equation (10.71). To have a meaningful time scale we will limit our attention to the constant service time case.

The special case of a constant service time of $\bar{t}$ can be solved by the general constant time model contained in the Appendix. The transition matrix now is

$$\mathbf{T}(t) = \begin{array}{c c} & \begin{array}{c c c c c} 0 & 1 & 2 & \cdots & m \end{array} \\ \begin{array}{c} 0 \\[30pt] 1 \\[30pt] \vdots \\[20pt] c \\[30pt] c+1 \\[30pt] \vdots \\[20pt] m \end{array} & \left[\begin{array}{c c c c c} k_0(t) & k_1(t) & k_2(t) & \cdots & 1 - \sum_{i=0}^{m-1} k_i(t) \\[20pt] k_0(t) & k_1(t) & k_2(t) & & 1 - \sum_{i=0}^{m-1} k_i(t) \\[10pt] \vdots & \vdots & & & \\[10pt] k_0(t) & k_1(t) & k_2(t) & & 1 - \sum_{i=0}^{m-1} k_i(t) \\[20pt] 0 & k_0(t) & k_1(t) & & 1 - \sum_{i=0}^{m-2} k_i(t) \\[10pt] \vdots & \vdots & & & \\[10pt] 0 & 0 & 0 & \cdots & 1 - \sum_{i=0}^{m-c-1} k_i(t) \end{array} \right] \end{array}$$

$$\tag{10.95}$$

where

$$k_i(t) = \frac{(\lambda(t)\bar{t})^i e^{-\lambda(t)\bar{t}}}{i!}$$

The problem is then solved iteratively with the state probabilities calculated only at multiples of the service time.

Development of proper programs and further analysis of transient solutions for the bulk service queueing problem are left as exercises for the reader.

PROBLEMS

10.1 Consider the problem of trimming the flash from plastic baskets made on injection molding machines. A single trimmer serves a large group of machines. Each shot from a single machine produces three baskets which are attached to a common sprue. This group of three is then placed on a conveyor for transport to the trimmer. The rate at which these rough products reach the trimmer is thirty sprues per hour. The trimmer trims baskets one at a time in an average time of thirty seconds.

a) Construct a model to describe the distribution for number of baskets waiting to be trimmed. State all assumptions.

b) Calculate the probability that there will be no unbroken sprues at the trim station.

c) Calculate the average number of baskets at the trim station.

d) Estimate the average time a basket will remain at the trim station.

10.2 A marriage counseling service receives couples for counseling at the rate of two couples per hour. The couples are interviewed as couples by a receptionist. This takes an average of twenty minutes. After this interview they are counseled individually by a social worker for an average of twelve minutes each. State all assumptions made in answering the following questions.

a) What is the probability distribution for the number of couples waiting to be interviewed by the receptionist?

b) What is the probability distribution for the number of persons waiting to see the counselor?

c) What is the probability distribution for total number of persons at the counseling service?

10.3 A surveillance camera has been set up at the check approval machine in a supermarket in an attempt to identify known forgers. Customers arrive at the machine at the rate of sixty per hour. Approval requires an average of forty seconds. The hidden camera takes a snapshot of the queue at random intervals at the rate of five shots per hour. These shots are immediately available to a security agent who examines the faces in the picture one at a time. He spends an average of five minutes per face.

a) Find the probability distribution for the number of customers per shot.

b) Find the distribution for n faces to be examined.

c) What is the expected elapsed time from taking the picture until a customer has been identified by the agent?

10.4 Consider a materials handling system in which baskets of completed subassemblies are loaded onto pallets. The pallets are in turn transported by fork lift to an inspection station where the subassemblies are inspected one at a time. Inspection time averages forty-five seconds per subassembly. Sample data have been collected which show that the mean and variance for the number of subassemblies per basket are six and nine, respectively. The number of baskets on a pallet is equally likely to fall anywhere between one and five. The rate at which fork trucks deliver pallets to the inspection station averages four pallets per hour. State all assumptions before answering the following:

a) Find the average number of subassemblies at the inspection station.

b) Find the expected time that an arriving unit will be delayed in queue.

10.5 Consider the constant bulk service model of equation (10.37).

a) Modify that model to account for limited waiting space, say k customers where $k > c$.

b) Suggest a technique for solving the resulting equations for a system in which arrival and service rates may vary over time.

c) Find the steady-state solution for the truncated system when $c = 2$ and $k = 4$.

10.6 Consider the "penny-a-pound" ride being sponsored by a university flying club. Their aircraft will accommodate three passengers. The announced policy is to withhold departure until all seats are full. Because of uncertain passenger response to rough air the length of the ride is a random variable which averages fifteen minutes. Customers arrive at the rate of ten per hour. The mean and variance for customer weight are 170 pounds and 625 pounds, respectively. State all assumptions in answering the following:

a) Find the probability distribution for the number of customers in the queue.

b) Find the expected time a customer must wait before he can get in the airplane.

c) If the airplane has a legal gross weight limit for passengers of 600 pounds, what portion of the flights will be over gross?

d) If the operation lasts eight hours, what are the mean and variance for gross income?

10.7 Work problem 10.6 under an operating policy in which any number of passengers up to and including three will be served whenever the aircraft is ready for boarding and passengers are present.

10.8 Consider the variable bulk service model of equation (10.58).

a) Modify that model to account for limited waiting space, say k customers where $k > c$.

b) Suggest a technique for solving the resulting equations for a system in which arrival and service rates may vary over time.

c) Find the steady-state solution for the truncated system when $c = 2$ and $k = 5$.

10.9 Consider a system in which Poisson batch arrivals occur at rate λ with fixed batch size b. Suppose that bulk service takes place in exponential fashion at rate μ with a constant batch size c where $c < b$.

a) Construct a flow rate diagram for this system.

b) Develop a proper set of differential-difference equations to describe the state distribution.

c) Find the transform for the steady-state distribution.

10.10 Consider a system with Poisson batch arrivals with fixed batch size b and exponential bulk service with variable batch size c where $c < b$.

  a) Construct a flow rate diagram for this system.

  b) Develop a proper set of differential-difference equations to describe the state distribution.

  c) Find the transform for the steady-state distribution.

10.11 Consider the bulk arrival problem described by equation (10.8).

  a) Find the transform of the distribution for the time spent by an arriving customer from the time of his arrival to the start of service for his group.

  b) For a randomly selected customer C, find the transform for the distribution of time spent from the start of service of his group to C's own service completion.

  c) Find the transform for total time spent in the system.

  d) Use the transform in part (*c*) to find the expected time a customer will spend in the system.

10.12 Consider a system with bulk k-Erlang service and Poisson arrivals in which $\lambda = 10$, $\mu = 10$, $k = 2$, $c = 2$.

  a) Find the mean and variance for the number of customers present.

  b) Treat a similar system in which service is exponential ($k = 1$). Compare the results of this calculation with part (*a*) and with similar results using equation (10.62). Explain any differences observed.

10.13 Treat the problem posed in 10.12 as if service time was constant at $\mu = 10$. Compare these results with the calculations performed in 10.12. What conclusions can you draw about model precision? (Hint: use the numerical program in Appendix D.)

10.14 An educational movie is shown every ten minutes. When each showing ends, all students are discharged from the makeshift theater and a new batch is permitted to enter. Because of limited facilities the present batch size is only three students, and waiting facilities outside of the theater are limited to five students. The cumulative distribution for number of arrivals in the ten minute show interval is given by

$$
G(x) = \begin{array}{ll}
.10 , & x = 0 \\
.20 , & x = 1 \\
.40 , & x = 2 \\
.60 , & x = 3 \\
.80 , & x = 4 \\
.90 , & x = 5 \\
.95 , & x = 6 \\
.98 , & x = 7 \\
.99 , & x = 8 \\
1.00 , & x = 9
\end{array}
$$

The distribution in any interval is independent of all other intervals.

a) Assuming that the theater comes under surveillance immediately after the 9:00 PM showing begins and that one student is in the queue, what will be the probability distribution for the queue length immediately prior to the 9:20 PM showing?

b) Develop a steady-state probability distribution for line length immediately following the admission of each batch.

10.15 Consider a bus-stop which is served by twenty seat buses. The time between buses averages fifteen minutes. The Poisson arrival rate for customers varies throughout the day in the following fashion:

Time	Customers/Hour
0000–0600	0
0600–0700	60
0700–0800	100
0800–0900	80
0900–1500	40
1500–1700	80
1700–1800	120
1800–2200	20
2200–0000	0

State all assumptions before answering the following:

a) Plot the expected line length as a function of time.

b) Calculate the average load factor [E(seats filled)/Total Seats] for the hours of 0600 to 2200.

c) What changes would you suggest in system design? Why?

11

NETWORKS OF QUEUES

A common type of queueing system is one in which an arriving unit must be served by a variety of service centers before being discharged. Each service center may become an input device for subsequent centers. This phenomenon occurs at registration time in many universities when students must queue to see a counselor for course planning, then queue for registration, for fee payment, for purchase of athletic privilege cards, *ad nauseam*. A similar process occurs in job-shop manufacturing. A casting may go to the grinding area, from there to a turret lathe, then to a drill press, on to inspection, then perhaps be fed back for more lathe work, and so forth. At every station a queue of work pieces is formed. The problem is to describe the overall system load for a network of work stations and to estimate the total throughput time for a unit about to enter the process.

GENERAL NETWORKS

The general network flow problem is a very challenging one and a topic of current research interest by theoreticians. Simple configurations such as two servers in series in which one or both have finite capacity lead to some very awkward equations to solve. When more realistic networks involving many stations with general service time distributions, feedback, nonrandom dispatching rules, and special queue disciplines are encountered, simulation may be the most viable analysis technique. What we hope to do in this chapter is explore some simple approximations which can be used for initial analysis.

JACKSON NETWORKS

Fortunately there are results which yield tractable equations for some special network configurations. The networks we will initially be concerned with have the following characteristics:

1. The networks contain more than one service center.
2. Each service center is a multi-channel queue with each channel at that center having an identical exponential service time distribution (standard model).
3. Arrivals at any given center may come from outside the system or from other centers in the network.
4. Arrivals from outside the network occur in Poisson fashion.
5. When a unit completes service at a particular center it may leave the system or be routed to another center, its path being controlled by a fixed probability distribution associated with the center it is leaving.
6. There is unlimited waiting space at every service center.
7. Total arrival rate at every center is less than its potential service rate.

Such networks are referred to as Jackson networks after the man who formulated the basic model (50).

Jackson's main result is that if one properly defines the mean arrival rate at the various service centers then the steady-state distributions at those centers look exactly like the standard multi-channel systems with which we are familiar. This very powerful result suggests that we can decompose a complex network into a number of much simpler subsystems. There is an interesting theoretical problem involving Jackson networks in which feedback occurs. Feedback destroys the Poisson characteristic of the input stream, yet those centers continue to behave as though the input was Poisson. We will leave the unraveling of that mystery to the theoreticians and revert to the pragmatic posture that says "if it seems to work, use it."

Variations of Jackson networks have been studied by a variety of authors. For example, Gordon and Newell considered closed systems in which a fixed and finite number of customers are trapped in the system with no new entries or removals permitted (41). Keonigsberg introduced the idea of cyclic queues in which service stages are arranged in tandem with the last stage serving as input to the first (62). James R. Jackson himself extended his early work to include state dependent arrival rates and service rates (49). R.R.P. Jackson treated tandem queues in great detail (51).

The common thread among such efforts is the maintenance of the Markovian assumption on arrivals and service. The problems become much more complex when something is done to alter the departure process at any given center. Authors such as Burke (16), Reich (78) and Finch (38) have studied the departure process to establish when independence and Poisson assumptions are warranted. For a more complete survey of such pioneering efforts the reader should consult Saaty (82).

Decomposition

Node-by-node decomposition in a network can be justified on the basis of three important properties of Poisson/exponential streams.

1. Combining a finite number of independent Poisson streams yields a Poisson stream whose mean is equal to the sum of the means of the component streams.
2. Splitting a Poisson stream having rate λ in random fashion with a portion r_i of the flow going to stream i yields a set of new streams which are Poisson with rates $r_i\lambda$.
3. Passing a Poisson arrival stream through an exponential service facility yields a departure stream which is Poisson with the same parameter as the input stream. This property also applies to (M/M/C) systems (Burkes theorem, 16).

Combining Streams

Proof of the combination of Poisson streams is most easily shown by transforms. Let

$$A = a_1 + a_2 + \cdots + a_n \tag{11.1}$$

represent the total number of arrivals per unit time where a_i is Poisson distributed with mean λ_i. If we assume independence it follows that the distribution of A can be obtained from the convolution of the a_i distributions. In transform domain this is accomplished by multiplication. That is

$$G_A(z) = G_{a_1}(z)\, G_{a_2}(z) \cdots G_{a_n}(z)$$
$$= e^{-\lambda_1(1-z)}e^{-\lambda_2(1-z)} \cdots e^{-\lambda_n(1-z)} \tag{11.2}$$

Adding exponents it follows that

$$G_A(z) = e^{-(\lambda_1+\lambda_2\cdots+\lambda_n)(1-z)} \tag{11.3}$$

which is the geometric transform of a Poisson distribution with a mean of

$$\lambda = \lambda_1 + \lambda_2 + \cdots + \lambda_n \tag{11.4}$$

Splitting Streams

To analyze a Poisson stream which is split at random into k paths first consider the conditional state vector

$$\mathbf{N}(t) = (n_1, n_2 \cdots n_k \,|\, n) \tag{11.5}$$

where n_i represents the number of units going into stream i through time t and $n = \sum_{i=1}^{k} n_i$ represents the total number of (Poisson) arrivals through time t. If the split occurs at random with r_i portion going to stream i, the probability of achieving this particular state vector is given by the multinomial distribution

$$p(n_1, n_2, \cdots, n_k \,|\, n) = \frac{n! \displaystyle\prod_{i=1}^{k} r_i^{n_i}}{\displaystyle\prod_{i=1}^{k} n_i!} \tag{11.6}$$

228

To obtain the joint probability $p(n_1, n_2, \cdots, n_k, n)$ we must multiply the conditional statement by the marginal probability

$$p(n) = \frac{(\lambda t)^n e^{-\lambda t}}{n!} \tag{11.7}$$

This yields

$$p(n_1, n_2, \cdots, n_k, n) = \frac{n! \prod_{i=1}^{k} r_i^{n_i}}{\prod_{i=1}^{k} n_i!} \left[\frac{(\lambda t)^n e^{-\lambda t}}{n!} \right]$$

$$= \prod_{i=1}^{k} \frac{(r_i \lambda t)^{n_i}}{n_i!} e^{-r_i \lambda t} \tag{11.8}$$

where

$$\sum_{i=1}^{k} n_i = n$$

Since the joint distribution factors into a product of Poisson distributions, the output streams must be Poisson with rates $r_i \lambda$.

The Exponential Window

An exponential server operating on a Poisson arrival stream has the remarkable property of acting as a clear window. The departure process looks exactly like the arrival process. For purposes of downstream modeling it is as if the server did not exist. We will prove this pass-through property for single-server systems, although Burke's theorem is even more powerful in that he shows that the steady-state output for an (M/M/C) system is Poisson (16).

Consider a departing customer from the first channel of a two tandem-channel system with exponential servers. Suppose that he leaves a positive number n of fellow customers behind as he departs the first channel. The time until the next customer, following our test customer, reaches the second channel will then be exponentially distributed due to his service time in the first channel. We can write the transform of the arrival time distribution for the second channel given a nonzero queue at the first channel as

$$\mathcal{L}(a_2(t) \mid n_1 > 0) = \frac{\mu_1}{\mu_1 + s} \tag{11.9}$$

If, on the other hand, our test customer leaves an empty queue behind as he departs the first channel, the next departure (and consequently arrival at the second channel) will not occur until an arrival takes place at the first channel and that arrival is subsequently served. The time to the next departure is then

$$t = t_a + t_s \tag{11.10}$$

which leads to the transform

$$\mathcal{L}(a_2(t)\,|\,n_1 = 0) = \mathcal{L}(a_1(t))\,\mathcal{L}(b(t))$$

$$= \left(\frac{\lambda}{\lambda + s}\right)\left(\frac{\mu_1}{\mu_1 + s}\right) \tag{11.11}$$

From the properties of the (M/M/1) queue existing at the first service center we know that

$$pr(n_1 > 0) = \frac{\lambda}{\mu_1} \tag{11.12}$$

and

$$pr(n_1 = 0) = 1 - \frac{\lambda}{\mu_1} \tag{11.13}$$

Combining this information with equations (11.9) and (11.11) we can then calculate the transform for the marginal interarrival time distribution at the second channel as

$$\mathcal{L}(a_2(t)) = \mathcal{L}(a_2(t)\,|\,n > 0)\,pr(n > 0) + \mathcal{L}(a_2(t)\,|\,n = 0)\,pr(n = 0)$$

$$= \left(\frac{\mu_1}{\mu_1 + s}\right)\left(\frac{\lambda}{\mu_1}\right) + \left(\frac{\lambda}{\lambda + s}\right)\left(\frac{\mu_1}{\mu_1 + s}\right)\left(1 - \frac{\lambda}{\mu_1}\right)$$

$$= \frac{\lambda}{\lambda + s} \tag{11.14}$$

Inverting the transform we see that

$$a_2(t) = \lambda e^{-\lambda t} \tag{11.15}$$

which is the same interarrival time distribution seen at the first channel. Exponential service has not altered the Poisson stream.

Modeling Jackson Networks

The use of the above three properties of Poisson streams assures us that we can connect many multi-channel service centers together in feedforward fashion and still preserve the simple (M/M/C) character at each node considered separately. These proofs unfortunately do not support the apparent decomposition also available in Jackson networks. Feedback in those networks destroys the Poisson properties of departure streams. However the individual service centers continue to behave as if they were being fed by Poisson streams and that is the important fact for the pragmatic analyst to know.

The state variable for network problems is a vector the elements of which represent the number of customers present at each node in the network. For a network consisting of N service centers or nodes, let

$$p(k_1, k_2 \cdots k_N;\, t) = \text{Probability of } k_i \text{ customers at node } i \text{ at time } t$$
$$\lambda_i = \text{Arrival rate at node } i \text{ due to externally generated demand}$$

$$e_i = \text{Effective arrival rate at node } i \text{ from all sources}$$
$$\mu_i = \text{Service rate per channel at node } i$$
$$c_i = \text{Number of channels at node } i$$
$$r_{ij} = \text{Probability that upon completing service at node}$$
$$i \text{ customers will go to node } j$$
$$N = \text{Number of nodes in the network}$$
$$\left(1 - \sum_{j=1}^{N} r_{ij}\right) = \text{Probability that a customer will leave the system after completing service at node } i.$$

It is obvious that in the steady state the total flow rate into node i can be written as

$$e_i = \lambda_i + \sum_{k=1}^{N} r_{ki} e_k \tag{11.16}$$

Using these variables and the usual birth and death arguments it is possible to develop a set of differential equations for $\dfrac{d}{dt} p(k_1, k_2, \ldots, k_N; t)$.

The steady-state solution is then obtained by setting the derivative equal to zero. The notation for the general solution becomes somewhat awkward so we will not pursue it here. The interested reader is urged to consult Jackson (50). The important result is that the final equation can be factored into a product to marginal probabilities

$$p(k_1, k_2, \ldots, k_N) = p_1(k_1) p_2(k_2) \ldots p_N(k_N) \tag{11.17}$$

where

$$p_i(k) = \begin{cases} \dfrac{p_i(0)(e_i/\mu_i)^k}{k!}, & k = 0, 1, \ldots, c_i \\[3mm] \dfrac{p_i(0)(e_i/\mu_i)^k}{c_i!(c_i)^{k-c}}, & k \geq c_i \end{cases} \tag{11.18}$$

$$p_i(0) = \left[\sum_{k=0}^{c_i-1} \frac{(e_i/\mu_i)^k}{k!} + \frac{(e_i/\mu_i)^{c_i}}{c_i!(1 - e_i/c_i\mu_i)}\right]^{-1} \tag{11.19}$$

This is of course the standard multi-channel result one would expect for a c_i channel system subjected to a Poisson input at rate e_i. This then permits us to decompose the network into a series of simple multi-channel subsystems to complete the analysis.

The mean number in the queue, mean number in the system, expected queue time, and expected sojourn time at each service center can be calculated in the usual fashion from the standard multi-channel results and Little's formula. That is

$$L_{q_i} = \frac{p_i(0)(e_i/\mu_i)^{c_i}(e_i/c_i\mu_i)}{c_i!(1 - e_i/c_i\mu_i)^2} \tag{11.20}$$

$$L_i = L_{q_i} + e_i/\mu_i \tag{11.21}$$

$$W_{q_i} = L_{q_i}/e_i \tag{11.22}$$

$$W_i = W_{q_i} + \frac{1}{\mu_i} \tag{11.23}$$

The probability distribution for the total waiting time required by a customer to pass completely through the network can be obtained from the convolution of the waiting time distributions at each node. This problem has been treated by Nelson (70). His results suggest that the cumulative probability distribution for the total waiting time at series of N centers can be written as

$$P\left(\sum_{j=1}^{N} x_j \le t\right) = 1 - \sum_{j=1}^{N} A_{j,n} exp\,(B_j t) \tag{11.24}$$

The $A_{j,n}$ are constants depending upon the processing centers j and number of nodes N. That is

$$A_{j,n} = K_j \left[\prod_{\substack{k=1 \\ k \ne j}}^{N} \left(\frac{1 - K_k B_j}{B_j - B_k}\right)\right], \qquad j = 1, 2, \ldots, N \tag{11.25}$$

The K_j and B_j parameters are standard multi-channel results

$$B_j = -c_j \mu_j (1 - e_i/c_j \mu_j) \tag{11.26}$$

$$K_j = \frac{(e_j/\mu_j)^{c_j}}{c_j!(1 - e_j/c_j\mu_j)} \bigg/ \left[\sum_{k=0}^{c_j-1} \frac{(e_j/\mu_j)^k}{k!} + \frac{(e_j/\mu_i)^{c_j}}{c_j!(1 - e_j/c_j\mu_j)}\right]$$

$$\tag{11.27}$$

The K_j parameter is simply the probability that there are c_j or more at node j and hence that an arrival must wait. The probability that a customer must wait longer than t at node j is given by

$$P(t_j > t) = K_j exp\,(B_j t) \tag{11.28}$$

Example.   Consider a radio repair shop which specializes in aircraft transponder certification. (A transponder is a piece of radio gear carried by an aircraft which responds to radar interrogations from ground stations and aids in positive identification of the aircraft). Every transponder must be checked biennially to insure that it is operating within proper frequency tolerance. The shop serves a large population of customers, receiving an average of sixteen transponders per day for certification. Past practice has been for each of the six employees in the transponder repair section to perform every operation required to certify a transponder. Under this procedure the average time required from receipt of a transponder until it is ready for customer pick-up has been three days. Management now wants to reorganize the shop so that each man will specialize in only one operation. Two men will now inspect the incoming units, two will perform any necessary overhauls, one will do the calibration, and one will do the certification paperwork. The estimated flow rates and single-channel service rates for the new process are noted in Figure 11-1. The question is whether or not the proposed reorganization will improve customer service.

Station #1, inspection, acts as a two-channel system with $e_1 = 16$, $\mu_1 = 9$, $c_1 = 2$. Station #2, calibration, is a single-channel system which receives input directly from inspection and from the overhaul station

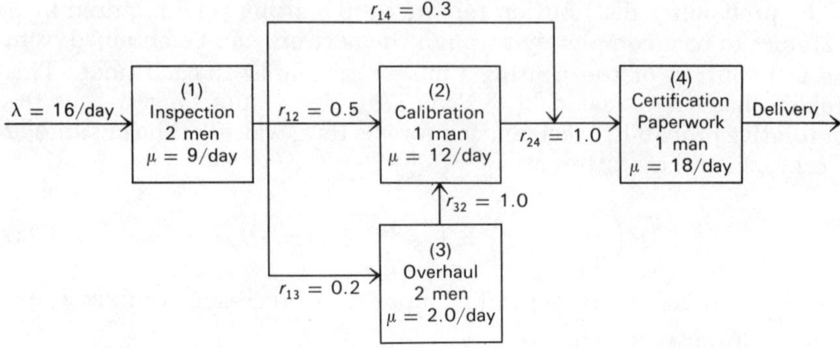

FIGURE 11-1   Flow-process Diagram for Transponder Certification.

after units have received necessary repairs. The parameters for station #2 are then $e_2 = 11.2$, $\mu_2 = 12$, $c_2 = 1$. Station #3, overhaul, is a two-channel system which receives 20 percent of all incoming units. The appropriate parameters are then $e_3 = 3.2$, $\mu_3 = 2$, $c_3 = 2$. Since all units must ultimately be certified, station #4 receives the entire flow. Its parameters are $e_4 = 16$, $\mu_4 = 18$, $c_4 = 1$. By assuming Poisson arrivals and exponential service, we are now in position to use the results for Jackson networks represented by equations (11.18) through (11.23). The outcome of applying those equations is summarized in Table 11-1.

TABLE 11-1
OPERATING CHARACTERISTICS FOR TRANSPONDER REPAIR SHOP

	(1) Inspection	(2) Calibration	(3) Overhaul	(4) Certification
$\lambda_i = e_i$	16	11.2	3.2	16
μ_i	9	12	2	18
c_i	2	1	2	1
$p_i(0)$	.059	.067	.11	.11
L_{q_i}	6.69	13.06	2.84	7.11
W_{q_i}	0.42	1.17	0.89	0.44
L_i	8.47	14	4.44	8
W_i	0.53	1.25	1.39	0.50

From the flow process diagram of Figure 11-1 we know that 30 percent of the incoming jobs will pass through stations 1 and 4; 50 percent will pass through stations 1, 2, and 4; and 20 percent will pass through every station. It therefore follows that the total expected processing time for a randomly arriving transponder with unknown defects will be

$$W_{system} = 0.3(W_1 + W_4) + 0.5(W_1 + W_2 + W_4) + 0.2(W_1 + W_2 + W_3 + W_4)$$
$$= W_1 + W_4 + 0.7W_2 + 0.2W_3$$
$$= 0.53 + 0.50 + 0.7(1.25) + 0.2(1.39) = 2.18 \text{ days}$$

The proposed system appears to cut the expected delay time per transponder by about 0.82 days for a 27 percent improvement in customer service.

TRANSIENTS IN NETWORKS

A logical next question is how can one account for time-varying demand rates and transient response in networks? In some special cases this can be accomplished by the same numerical techniques discussed in Chapter 8. What is required is to write the necessary differential difference equations for a system with a finite number of states, define the arrival rate function as a series of steps, and solve the resulting finite system of linear differential equations with constant coefficients by Runge-Kutta methods. Runge-Kutta methods are discussed in Chapter 8.

The computational difficulty here is one of dimensionality. For example, the state descriptor in a four-node network with each node capacity 5 is a four-dimensional vector. If one were to map this vector description back into a single-dimensional array as required for solving the differential equations, a total of 625 equations would be in the solution set. In addition there are many alternatives for handling the complex blocking phenomena which would occur between interconnected nodes in a finite network. Large-scale problems soon reach unmanageable dimensions for such techniques. However, we can use this approach for some specialized networks. The following example illustrates one such application.

Radar Control Communications

Radio communications are required between pilots and air traffic controllers in a terminal radar environment.[1] There is some evidence that it is the communications backlog which is a major contributor to air traffic delays at large terminals. Dunlay has collected data and structured a model which asserts that stochastic variation in total communication workload is due more to fluctuations in the number of required communication transmissions rather than fluctuations in the transmission lengths themselves (33). Using this as a clue, William W. Bundy, working under the direction of the author, has structured air traffic control communications as a time-varying queue with feedback (15).

Aircraft entering the terminal area are treated as customers who may demand multiple services. Each identifiable conversation between pilot and controller is treated as one service. Radar vectors or other instructions lead to subsequent conversations before ultimate hand-off from the

---

[1]Adapted with permission of the author from William W. Bundy, "Queues with Delayed Probabilistic Feedback as a Model of Air Traffic Control Communications," M.S. Thesis, The Ohio State University, 1976.

radar controller to the local controller. This process is treated as a feedback system which returns the customer to the communications queue, after some random delay necessary to execute the instructions given. While in a feedback loop each aircraft is treated as though it was the only occupant of a single-channel system. Since there is an upper bound to the number of aircraft a controller can handle at one time (currently in the neighborhood of ten) the system is truncated with some specified capacity M. That capacity includes aircraft in the communications queue plus those in the feedback loops. After any given conversation with the controller the aircraft may require further service hence, feedback, with probability PF or depart the system with probability $(1 - PF)$. When the system reaches saturation arriving aircraft must be diverted to another facility. This process is summarized in Figure 11-2.

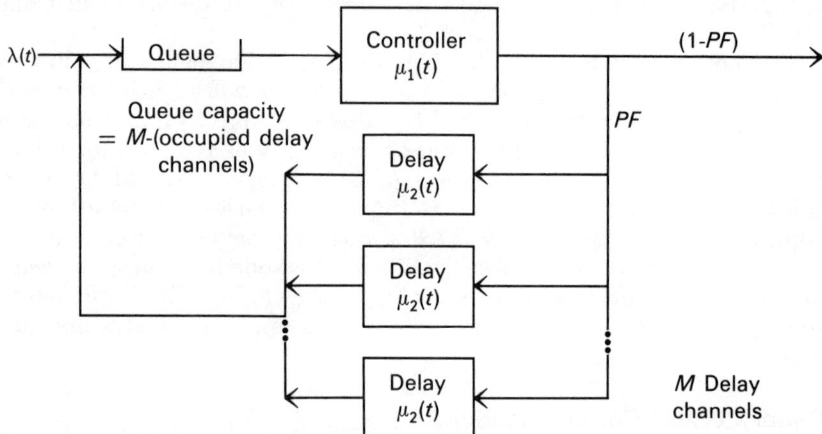

FIGURE 11-2 Delayed Probabilistic Feedback in a Communications Queue.

Because the system has a fixed number of single-channel subsystems with capacity one each, it is possible to describe it by a three-dimensional vector, the third dimension being time. Let

$p(i, j; t)$ = Probability of i in communications queue plus service and j in
feedback loops at time t
$\lambda(t)$ = Arrival rate of new customers
PF = Probability of feedback
$\mu_1(t)$ = Controller service rate (single conversation)
$\mu_2(t)$ = Pseudo service rate in delay loops
M = Capacity of system

By taking advantage of our knowledge of Poisson/exponential processes let us write a rate matrix for the special case of capacity $M = 2$. There are six states in which the system can exist. The elements of an appropriate rate matrix can be easily determined by constructing a flow rate diagram. For example, moving from state $(0, 1)$, which says that one aircraft is in the delay loop, to state $(1, 0)$ which says that one is in the communication queue and none in the loop, implies that one pseudo service must take place. The rate at which this happens is $\mu_2(t)$. The reader is urged to verify the other flows. The resulting rate matrix is

$$\mathbf{R}(t) =$$

	0, 0	0, 1	0, 2	1, 0	1, 1	2, 0
0, 0	$-\lambda(t)$	0	0	$\lambda(t)$	0	0
0, 1	0	$-[\lambda(t)+\mu_2(t)]$	0	$\mu_2(t)$	$\lambda(t)$	0
0, 2	0	0	$-2\mu_2(t)$	0	$2\mu_2(t)$	0
1, 0	$\mu_1(t)(1-PF)$	$\mu_1(t)PF$	0	$-[\lambda(t)+\mu_1(t)]$	0	$\lambda(t)$
1, 1	0	$\mu_1(t)(1-PF)$	$\mu_1(t)PF$	0	$-[\mu_1(t)+\mu_2(t)]$	$\mu_2(t)$
2, 0	0	0	0	$\mu_1(t)(1-PF)$	$\mu_1(t)PF$	$-\mu_i(t)$

$$(11.29)$$

One can now write the corresponding set of differential-difference equations as

$$\frac{d\mathbf{P}(i,j;\, t)}{dt} = \mathbf{P}(i,j;\, t)\mathbf{R}(t) \qquad (11.30)$$

where

$$\mathbf{P}(i,j;\, t) = [p(0, 0;\, t), p(0, 1;\, t), \ldots, p(2, 0;\, t)]$$

This is a set of six linear differential equations. By defining $\lambda(t)$ as a step function in our usual fashion it is possible to use Runge-Kutta methods to determine the time varying solution.

Bundy has generalized this technique for capacity M. The number of equations necessary to describe a system of capacity M is

$$N = \frac{M^2}{2} + \frac{3}{2}M + 1 \qquad (11.31)$$

Let $PD = (1 - PF)$ then the equations can be written as

$$p'(0, 0;\, t) = -\lambda(t)p(0, 0;\, t) + PD\mu_1(t)p(1, 0;\, t)$$

$$p'(0, M;\, t) = -M\mu_2(t)\, p(0, M;\, t) + PF\mu_1(t)\, p(1, M - 1;\, t)$$

$$p'(M, 0;\, t) = \lambda(t)p(M - 1, 0;\, t) + \mu_2(t)p(M - 1, 1;\, t) - \mu_1(t)p(M, 0;\, t)$$

$$\begin{aligned}
p'(0, k;\, t) = &-\lambda(t)p(0, k;\, t) - k\mu_2(t)p(0, k;\, t) \\
&+ PF\mu_1(t)p(1, k - 1;\, t) \\
&+ PD\mu_1(t)p(1, k;\, t), \qquad 0 < k < M
\end{aligned}$$

$$\begin{aligned}
p'(I, 0;\, t) = &\lambda(t)p(I - 1, 0;\, t) + \mu_2(t)p(I - 1, 1;\, t) - \lambda(t)p(I, 0;\, t) \\
&- \mu_1(t)p(I, 0;\, t) + PD\mu_1(t)p(I + 1, 0;\, t), \qquad 0 < I < M
\end{aligned}$$

$$\begin{aligned}
p'(a, b;\, t) = &\lambda(t)p(a - 1, b;\, t) + (b + 1)\mu_2(t)p(a - 1, b + 1;\, t) \\
&- \lambda(t)p(a, b;\, t) - \mu_1(t)p(a, b;\, t) - b\mu_2(t)p(a, b;\, t) \\
&+ PF\mu_1(t)p(a + 1, b - 1;\, t) + PD\mu_1(t)p(a + 1, b;\, t), \\
&0 < a < M, \qquad 0 < b < M, \qquad 0 < a + b < M
\end{aligned}$$

$$\begin{aligned}
p'(c, d;\, t) = &\lambda(t)p(c - 1, d;\, t) + (d + 1)\mu_2(t)p(c - 1, d + 1;\, t) \\
&- \mu_1(t)p(c, d;\, t) - d\mu_2(t)p(c, d;\, t) \\
&+ PF\mu_1(t)p(c + 1, d - 1;\, t), \\
&0 < c < M, \qquad 0 < d < M, \qquad c + d = M
\end{aligned}$$

The program used to solve these equations includes a transformation which maps the $p(i,j;t)$ values into a single dimension array. Otherwise it is quite similar to the time-varying system model contained in Appendix C.

Bundy has checked this model against simulation results reported by Hunter and Hsu for a New York Air Route Traffic Control Center sector at three different traffic intensities (47). In matching figures for queue length, number in the system, and channel utilization for average arrival rates of 16.5, 33, and 41 aircraft per hour he found differences averaging about 10 percent variation from the simulation.

The simulation used empirical data while the model used exponential approximations results. The model does seem to be a reasonable representation and has been used to study the effects of a number of system parameter changes.

PROBLEMS

11.1 Consider a sweeper repair service which receives an average of twenty sweepers per eight hour day for repair. The first operation is an inspection which averages fifteen minutes per sweeper. Two inspectors are available. When inspected 50 percent are found to need new brushes, 30 percent need new belts, and 20 percent need rebuilt motors. Single servers exist for each of these functions with service rates of 2, 3, and 0.8 per hour, respectively. All sweepers receiving rebuilt motors will as a matter of policy also be routed to the brush and belt installation stations for those components. After necessary mechanical repairs are complete, each sweeper is routed to a clean-and-polish man who installs new bags and polishes the machine for good customer relations. This task requires an average of twenty minutes with a variance of nine. All other tasks are exponential.

  a) Draw a schematic for flows in this network.

  b) Find the expected number of units at each station in the repair facility.

  c) Calculate the expected elapsed time before a randomly arriving customer can return to pick up his sweeper.

11.2 Prove that the departure process from an M/M/c queue is Poisson (Burkes Theorem (16)).

11.3 Consider an unlimited two-node system with each node containing single channel exponential servers operating at rates μ_1 and μ_2. Poisson arrivals at rate λ arrive at the first node, queue as necessary, are served, proceed to the second node, queue as necessary, are served and finally leave the system. The state of the system is represented by $p_{k_1 k_2}(t)$, i.e., the probability that there are k_1 customers in the first subsystem and k_2 in the second at time t.

  a) Develop a set of differential-difference equations from which one could derive $p_{k_1 k_2}(t)$.

b) Find the steady-state solution for this set of equations. Verify your answer by comparing results with equation 11.18.

11.4 Consider two single-channel servers in series with limited waiting room. Arrivals are Poisson at rate λ and service rates are μ_1 and μ_2. Suppose that each server can accommodate only a single customer at once. If service is completed at the first channel but the second one is still occupied, blocking occurs. This means that the customer at the first channel cannot move on even though his service there is complete. Furthermore no new arrivals can enter until he does move on. This requires an added state definition; say $p_{b_1,1}(t)$ is the probability that the first channel is blocked and the second is serving a customer.

a) Develop a set of differential-difference equations for this system.

b) Find the steady-state distribution.

c) Find the expected number of customers in the system for the special case of $\mu_1 = \mu_2$.

11.5 Consider the problem faced by a county auditor's office at tax collection time. Property tax returns arrive at the rate of 3200 per eight-hour day. A group of eight clerks open the envelopes, log the returns and route them to either business or personal auditing. Each clerk can process an average of sixty returns per hour. Approximately 80 percent of the returns are personal and 20 percent are business. There are twenty-two business auditors who require an average of fifteen minutes to give a return a quick audit. One percent of the returns are then selected for an indepth audit which requires a senior analyst one hour to perform. There are eighty auditors for personal returns each of which requires an average of twelve minutes to process a return. Thirty percent of the personal returns and 50 percent of the business returns are correct as submitted and require no further action after audit. Of the remainder 80 percent are routed to the billing department for preparation of tax due bills and 20 percent to the treasurer for preparation of refund checks. Both the billing and treasurer's operations can be approximated as single channel servers with processing rates of 200 and 50 per hour, respectively.

a) Construct a schematic diagram for flows in this system correctly identifying rates on each branch.

b) Find the probability distribution for the number of returns in the treasurer's office.

c) Explain how one could estimate the average time to process a business return.

11.6 Consider an admission's office at a small university which receives hand-carried applications for both undergraduate and graduate school. Applications arrive at the rate of twelve per hour. They are logged in and given a document check by a clerk who takes an average of four minutes to process one application. Thirty percent of the applications are routed to the graduate evaluator and the remainder to the undergraduate evaluator. These persons require fifteen and six minutes, respectively, to evaluate an application. After evaluation 20 percent of the graduate and 40 percent of the undergraduate applications are rejected by the admission's office who then notifies the student. The remainder are routed to the registrar who prepares entrance information before notifying the student of his admission. This operation requires six minutes per application.

a) Draw a schematic of this operation.

b) Find the probability distribution for n applications (total) in the system.

c) Find the expected time for a graduate application to be processed.

d) Find the average number of students waiting for a decision on their application.

11.7 Prepare a plausible queue network diagram for vehicles in the service department of a large automobile dealership. Explain what models you might use to analyze this system. Discuss how you would collect data to validate your model.

11.8 Prepare a plausible queue network diagram for a computer installation. Consider major components such as the CPU, a number of remote terminals, tape drives, card readers, disc packs, punches, and printers. Construct a numerical example to illustrate the application of network modeling techniques to the analysis of this system. Discuss shortcomings of the model and any difficulties you would anticipate in validating model results.

11.9 Consider a small machine shop which performs work on demand for a large population of customers. The shop has two turret lathes, one milling machine, one shaper, two drill presses, and one layout table. Data have been collected on the job flow and capabilities of each center as follows:

From \ To	Jobs per Day						Average Processing Time per job
	Layout	Lathe	Mill	Shaper	Drill	Customer	
Layout		3		2	3		45 min.
Lathe			3	3	3	10	40 "
Mill		4		2		5	40 "
Shaper					10	5	30 "
Drill		6	4			11	15 "
Customer	8	6	4	8	5		

a) Find the probability distribution for the number of jobs at each work center.

b) Find the expected delay time for a randomly arriving lathe job brought in by a customer.

c) What changes could you suggest to better balance work load in the shop? Quantify the effects of your suggested changes.

11.10 Consider the air traffic control communication problem. Arriving aircraft approach the terminal area for service by a radar controller. Control is exercised by voice command. A particular conversation between pilot and controller, the length of which is a random variable, may terminate the need for additional service with probability 0.2 or may require that the pilot feedback after he performs some maneuver, e.g., vectoring to the final approach course, with probability 0.8. The time to perform the maneuver is a random variable. The controller can handle at most eight aircraft in his airspace, either awaiting communication time (in queue) or executing previous instructions.

a) Develop a schematic representation of at least two ways in which such a system might be modeled. Explain all assumptions.

b) Develop systems of differential-difference equations in both two and one dimension which could be used to find a time-varying solution to this problem.

c) Explain how you would go about developing the data necessary to validate your model.

12

SPECIAL TOPICS

The purpose of this chapter is to introduce to the reader a few of the more specialized topics which inhabit the queueing literature. In keeping with our theme of robustness and ease of application the models which follow are the simplest of their respective families. We will examine in cursory fashion such diverse topics as queue disciplines, priority systems, optimization, and simulation. The hope is that such a montage will indicate to the student some of the extensions which are available from the fundamental Markov-based models of earlier chapters to more complex systems.

QUEUE DISCIPLINES AND PRIORITIES

When one speaks of queue discipline he is really considering priorities. In every model so far discussed in this textbook, the implied priority scheme has been first-come-first-served (FCFS). The customer with most seniority in the system is the top priority customer. Other commonly encountered queue disciplines which are not normally considered to be priority schemes, but which in fact do imply priorities, are last-come-first-served (LCFS) and random selection for service (RSS). Last-come-first-served implies that the most recent arrival is the top priority customer. Random selection for service implies that all customers have the same priority when it comes time to select the next one for service.

The three basic queue disciplines FCFS, LCFS, and RSS all exhibit the property we will refer to as *non-preemptive priority*. Non-preemptive priority means that even though a customer may appear who has higher priority than the one currently in service, he cannot interrupt that service. He must wait until the service in progress at the time of his arrival has been completed before claiming the service facility. Contrariwise, one might consider *preemptive priority*. In this scheme an arriving customer who possesses a higher priority rating (a lower order number) than the customer currently in service may bump that customer from the service facility and immediately begin his own processing. The displaced customer then returns to the queue to await further service. When he again can claim the service facility he may be required to complete his service from the point of interruption, in which case we say that the discipline is preemptive-resume. Contrariwise he may be required to begin service anew in which case the system is preemptive-repeat. Even the repeat service may require special modeling considerations. The customer may repeat with the same service time, a random number, that he had on his last trip to the facility or he may be required to select a new service time, another random number, for this trip. It is also possible to create preemptive priority schemes between the extremes of resume and repeat by requiring that some portion of the work completed at the time of interruption be repeated.

In most queueing literature priority schemes are conceived as segmenting the population into a number of customer classes. A class-1 customer has priority over all other classes and a class-r customer has priority over all classes with a higher index number. Within each subclass customers are normally treated in FCFS fashion. The relationship among classes may be either preemptive or non-preemptive.

We will develop a few of the more easily used performance measures for a limited number of priority systems. Readers interested in pursuing this subject in depth are urged to consult Jaiswal (52), Conway, Maxwell and Miller (22) or Cox and Smith (29).

Selection For Service

Let us now consider the effect of the three standard disciplines (FCFS, LCFS, and RSS) on waiting times in an (M/G/1) system. The first observation to make is that the transform for the distribution of number of customers in the system equation (9.26) was developed without reference to queue discipline. Since the state distribution is identical for all three disciplines it follows that the expected number of customers in the system is the same. Furthermore one can invoke Little's formula from which the expected sojourn time and expected queue time, given by equations (9.46) and (9.47), can be determined. The conclusion is that W and W_q are independent of whether the discipline is FCFS, LCFS, or RSS.

The probability distributions for waiting time, and consequently higher order moments, do, however, differ as functions of the queue discipline. Recall from Chapter 9 that we developed the Laplace transform for the queue time distribution based upon a FCFS argument. Reproducing that equation for convenience we have

$$W_q(s) = \frac{(1 - \lambda \bar{t})s}{s - \lambda[1 - B(s)]} \tag{12.1}$$

where $B(s)$ is the Laplace transform of the service time distribution and $\bar{t}$ is its mean.

By using "busy period"[1] arguments Conway, Maxwell and Miller

---

[1] Busy period refers to an unbroken period of time during which there is one or more customers in the system. Successive busy periods are separated by idle periods. The study of the statistical properties of busy periods forms an important part of many theoretical developments in priority queueing systems.

(22:158) show the corresponding transform for the LCFS discipline as

$$W_{q_L}(s) = 1 - \lambda \bar{t} + \frac{\lambda[1 - \eta(s)]}{s + \lambda[1 - \eta(s)]} \tag{12.2}$$

The transform $\eta(s)$ is the Laplace transform of the busy period distribution. That transform is related to the transform of the service time distribution $B(s)$ by

$$\eta(s) = B(s + \lambda - \lambda \eta(s)) \tag{12.3}$$

Since $\eta(s)$ appears both on the left and as a part of the functional argument on the right, it may be extremely difficult to obtain an explicit expression for it. However, one can obtain the moments of the busy period distribution by differentiation.

There are queue time distribution results available for RSS assuming particular service distributions, e.g., constant and exponential. However, there does not appear to be a general expression for the RSS discipline in $(M/G/1)$ systems.

Without inverting the transforms one can develop expressions for the moments of the queue waiting time distributions from

$$E(t_w) = (-1)^r \frac{d^r W_q(s)}{ds^r} \bigg|_{s=0} \tag{12.4}$$

We already know that the first moments for all three disciplines are identical. That is

$$W_q = E(t_w) = \frac{\lambda(\bar{t}^2 + \sigma^2)}{2(1 - \lambda \bar{t})}$$

$$= \frac{\lambda E(t^2)}{2(1 - \rho)} \tag{12.5}$$

where $E(t^2)$ is the second moment of the service distribution. By taking second derivatives of equations (12.1) and (12.2) one can establish that the second moment for queue time under FCFS rules is

$$E(t_{wF}^2) = \frac{\lambda E(t^3).}{3(1 - \rho)} + \frac{[\lambda E(t^2)]^2}{2(1 - \rho)^2} \tag{12.6}$$

and under LCFS rules it is

$$E(t_{wL}^2) = \frac{\lambda E(t^3)}{3(1 - \rho)^2} + \frac{[\lambda E(t^2)]^2}{2(1 - \rho)^3} \tag{12.7}$$

The important thing to note here is that the second moments are related in a simple fashion. That is

$$E(t_{wL}^2) = \frac{E(t_{wF}^2)}{(1 - \rho)} \tag{12.8}$$

Since $\rho = \lambda \bar{t}$ must be less than one for a steady-state solution to exist, it follows that the second moment and hence variance for waiting time under LCFS rules will always be larger than that for FCFS. Furthermore the magnitude of the difference between the two systems increases dramatically under high load conditions, i.e., when ρ approaches one.

Riordan (80) has developed moments which can be used to show that, for the special case of exponentially distributed service times and random selection for service, the second moment for queue time is related to the like moment under FCFS rules by

$$E(t_{wR}^2) = \frac{E(t_{wF}^2)}{1 - \rho/2} \tag{12.9}$$

Conway, Maxwell and Miller note that if one conjectures that a similar relationship must hold for more general service distributions, we then have an indication that queue time variance under random selection procedures will fall somewhere between the minimum variance case of FCFS and the maximum variance case of LCFS (22). However, as we have indicated the mean is the same for all three selection policies.

Example. A manufacturer of custom-made award trophies uses an epoxy cement to attach the main structure to a plexiglass base. The epoxy resin is precoated on the main structure with the hardener added at the time of assembly. For best results the joint should be made less than twenty minutes after precoating with resin. A single cementer serves a large number of independent structure builders, which lends credence to assuming a Poisson arrival process. The rate at which structures arrive at the cementing station is estimated to be thirty per hour. Due to the great variety of trophies manufactured, with many requiring very simple joints but a few requiring extensive work, an exponential service time can be assumed. The cementing rate is estimated to be forty per hour. The easiest way for the cementer to select the next work piece is by the LCFS rule due to the way new arrivals are stacked in the holding bin. With some added effort on his part a RSS policy could be instituted. To follow FCFS rules would require a redesign of the holding bin. The problem is to estimate the probability of meeting the good bonding time interval of up to twenty minutes in order to evaluate the merits of redesigning the cementing workplace.

From the properties of the exponential distribution we know that

$$E(t^2) = \bar{t}^2 + \sigma^2 = 2\bar{t}^2 = .00125$$

By using equation (12.5) it follows that the average delay time per trophy under any of the selection rules will be

$$W_q = \frac{\lambda E(t^2)}{2(1 - \rho)} = \frac{30(.00125)}{2(1 - 0.75)}$$
$$= .075 \text{ hrs.} = 4.5 \text{ minutes}$$

The third moment of the service time distribution is

$$E(t^3) = (-1)^3 \frac{d^3 B(s)}{ds^3}\bigg|_{s=0}$$

$$= (-1)^3 \frac{d^3}{ds^3}\left(\frac{40}{s+40}\right)\bigg|_{s=0}$$

$$= \frac{240}{(40)^4} = 0.9375(10^{-4})$$

Applying equation (12.6) we can calculate the second moment of the delay time distribution under the FCFS policy as

$$E(t^2_{wF}) = \frac{\lambda E(t^3)}{3(1-\rho)} + \frac{[\lambda E(t^2)]^2}{2(1-\rho)^1}$$

$$= \frac{30(.9375)(10^{-4})}{3(1-0.75)} + \frac{[30(.00125)]^2}{2(1-0.75)^2}$$

$$= 0.015$$

From equations (12.8) and (12.9) the corresponding moments for LCFS and RSS policy are

$$E(t^2_{wL}) = \frac{E(t^2_{wF})}{1-\rho} = \frac{.015}{1-.75} = 0.06$$

and

$$E(t^2_{wR}) = \frac{E(t^2_{wF})}{1-\rho/2} = \frac{.015}{1-0.375} = 0.024$$

Using these moments and the computational formula for variance

$$\sigma^2 = E(w^2) - [E(w)]^2$$

we have

$$\sigma_F^2 = .015 - (.075)^2 = .009375 \text{ hr.}^2 = 33.75 \text{ min.}^2$$

$$\sigma_L^2 = .06 - (.075)^2 = .054375 \text{ hr.}^2 = 195.75 \text{ min.}^2$$

$$\sigma_R^2 = .024 - (.075)^2 = .018375 \text{ hr.}^2 = 66.15 \text{ min.}^2$$

As a rough measure of performance let us use Tchebycheff's inequality which places a bound on probability statements as follows:

$$P(|x - \mu| \geq K) \leq \frac{\sigma^2}{K^2} \qquad (12.10)$$

In the present context the resin is likely to be no good whenever $t_w \geq 20$ or $t_w - E(t_w) \geq 15.5$. Under the three different selection policies we can then calculate

$$P_F(|t_w - 4.5| \geq 15.5) \leq \frac{33.75}{(15.5)^2} = 0.14$$

$$P_L(|t_w - 4.5| \geq 15.5) \leq \frac{195.75}{(15.5)^2} = 0.81$$

$$P_R(|t_w - 4.5| \geq 15.5) \leq \frac{66.15}{(15.5)^2} = 0.28.$$

Even with these admittedly weak bounds, the odds of exceeding the twenty-minute time limit seems rather remote if a FCFS discipline is followed. However, the chances of exceeding the limit for the present LCFS policy is significantly higher. Since random selection is a no-cost alternative it appears to offer a reasonable compromise between the expense of new facilities under FCFS and the risk of bad bonding with attendant reprocessing costs under LCFS.

NON-PREEMPTIVE PRIORITY

Let us now turn our attention to those systems which have a finite number of identifiable classes of customers, say $(1, 2, \ldots, r)$ where class 1 represents the highest priority. We will initially assume that class-k customers arrive in Poisson fashion at rate λ_k and that we have a single-channel service facility. Class-k customers are served according to the service distribution $b_k(t)$. A class-k customer will be 1) served prior to any class-$(k + r)$ customers, 2) served in FCFS fashion within his class and, 3) required to wait until service is complete on any customer who may be in service at the time of his arrival before exercising his priority rights.

Similar schemes are often used in facilities such as computer installations. In one maintaining business records, for example, class-1 runs might be those required to execute current production activities such as preparing routing slips for new jobs. Class-2 runs might be those required to generate periodic reports such as a weekly inventory status report. Class-3 runs might be those used to analyze long-term planning problems such as an econometric model designed to forecast market share over the next twelve months. Another commonly used scheme is to assign priorities to computer runs based upon estimated run times. Here class-1 runs might be those estimating less than five minutes CPU time and class-2 runs those estimating five to ten minutes, etc.

From the viewpoint of the server this system behaves exactly like an ordinary (M/G/1) queue in which $\lambda = \sum_{i=1}^{r} \lambda_i$ and $b(t) = \sum_{k=1}^{r} \frac{\lambda_k}{\lambda} b_k(t)$. The non-preemptive priority scheme only influences the waiting time experienced by randomly arriving customers of different classes. The development of the complete waiting time distributions requires a great amount of mathematical argument which is admirably detailed by Conway, Maxwell and Miller (22:160). We will limit our attention to simpler expected value arguments for equilibrium conditions.

Consider the wait expected by a particular class-k customer, say $W_k^{(0)}$. His waiting time is made up of a sum of phases:

1. The time required to complete service for the customer in service at the point of arrival of the test customer
2. The time required to serve any customers in classes 1 through k who were present when the test customer arrived

3. The time required to serve any new arrivals in classes 1 through $(k-1)$ who appear before the test customer can seize the service facility

The time required to complete service on a customer currently occupying the service channel can be related to a concept from renewal theory known as "residual lifetime" or "forward recurrence time." It is well known that the probability density function for residual lifetime given a random sampling point generated by a Poisson process is

$$f(t_r) = \frac{\displaystyle\int_{t_r}^{\infty} b(t)\, dt}{E(t)} \tag{12.11}$$

where $b(t)$ is the original lifetime (or service) density function (see Cox 26:62). The Laplace transform for this function is

$$\mathcal{L}(f(t_r)) = \frac{1 - \mathcal{L}(b(t))}{sE(t)} \tag{12.12}$$

Assuming that a class-j customer is in service the mean residual lifetime can then be obtained from

$$E_j(t_r) = -\frac{d}{ds}\left[\frac{1 - B_j(s)}{sE_j(t)}\right]\Bigg|_{s=0} \tag{12.13}$$

This calculation leads to an indeterminant form requiring the use of L'Hospital's rule twice to obtain

$$E_j(t_r) = \frac{B_j''(0)}{2E_j(t)} = \frac{E_j(t^2)}{2E_j(t)} \tag{12.14}$$

Since the probability of encountering a class-j customer in service is

$$\rho_j = \lambda_j/\mu_j = \lambda_j E_j(t) \tag{12.15}$$

it follows that the total expected delay time experienced by our test customer for this phase is

$$E(t^{(1)}) = \sum_{j=1}^{r} \rho_j E_j(t_r)$$

$$= \sum_{j=1}^{r} \frac{\lambda_j E_j(t^2)}{2} \tag{12.16}$$

where $E_j(t^2)$ is the second moment of the class-j service time distribution.

To calculate the second phase delay we need to know the number present for each higher priority group and the service time for each. Since these variables are independent we can find the expected second phase delay by multiplying the expected service time for class j by the expected number of customers in class j then summing over all j. The expected service time is $E_j(t)$. The expected number in queue at the arrival instant can be related to their own waiting times by Little's formula.

$$L_{q_j} = \lambda_j W_j$$

It therefore follows that the total expected second phase delay is

$$E(t^{(2)}) = \sum_{j=1}^{k} (\lambda_j W_j)E_j(t) \tag{12.17}$$

For the third phase of our test customer's delay we need to know the expected number of higher priority arrivals during his delay time and the expected service times for those arrivals. The expected number of class-j arrivals during the waiting time of our class-k customer is

$$N_j = \lambda_j W_k \tag{12.18}$$

each of whom has an expected service time of $E_j(t)$. It therefore follows that the total expected delay in the third phase is

$$E(t^{(3)}) = \sum_{j=1}^{k-1} \lambda_j W_k E_j(t) \tag{12.19}$$

We can now write the expression for total expected delay of a class-k customer as a function of delays in other classes as

$$W_k = E(t^{(1)}) + E(t^{(2)}) + E(t^{(3)})$$

$$= \sum_{j=1}^{r} \frac{\lambda_j E_j(t^2)}{2} + \sum_{j=1}^{k} (\lambda_j W_j)E_j(t)$$

$$+ \sum_{j=1}^{k-1} \lambda_j W_k E_j(t) \tag{12.20}$$

The result is a set of r simultaneous linear equations which can now be solved to find the expected waiting times $\{W_k : k = 1, 2, \ldots, r\}$. Solving this set recursively and using a proof by induction for the general term we can show that the expected delay for a class-k customer can be written as

$$W_k = \frac{\displaystyle\sum_{j=1}^{r} \lambda_j E_j(t^2)}{2\left(1 - \displaystyle\sum_{j=1}^{k-1} \rho_j\right)\left(1 - \displaystyle\sum_{j=1}^{k} \rho_j\right)} \tag{12.21}$$

If one desires to establish an optimal priority classification it is possible to show by interchange arguments that total expected delay time across all classes is minimized by assigning priorities in reverse order of class mean service times, i.e., do the jobs with the shortest expected service time first. In more general terms if there is a delay cost C_j associated with each unit of time that a class-j job must wait, then the optimal priority assignment is to order the classes such that

$$\frac{E_1(t)}{C_1} \le \frac{E_2(t)}{C_2} \cdots \le \frac{E_r(t)}{C_r} \tag{12.22}$$

Under a non-preemptive discipline one can go a step further and create an infinite number of priority class to show that for systems in which the processing time for an arriving customer (or job) is known prior to its entry into service, the optimal policy, in terms of mean sojourn time, is to select the job with shortest processing time as the next to be served. Proof of this and other facts concerning optimal policies is given by Conway, Maxwell and Miller (22).

Example. An academic department at a university has established a non-preemptive priority system for typing faculty material. Classroom work has top priority, professional correspondence second, and manuscript development third. Classroom jobs arrive at the rate of sixteen per eight-hour workday, correspondence twelve per day, and manuscripts one per week. The estimated mean typing time for these jobs are 15, 6, and 600 minutes, respectively. The department chairman believes that all the world is Markovian and hence is willing to assume that arrivals are Poisson and services exponential. The problem is to estimate the mean delay times under this scheme and to evaluate possible alternatives.

The necessary data for this analysis are as follows:

	λ_i	μ_i	$E_i(t^2)$	ρ_i
1. Classroom	2/hr.	4/hr.	0.125	0.5
2. Professional	1.5/hr.	10/hr.	0.02	0.15
3. Manuscripts	0.025/hr.	0.1/hr.	200	0.25

From $\Sigma \rho_i$ we can immediately determine that the secretary will be busy 90 percent of the time regardless of the priority assignment used. By applying equation (12.21) we can estimate the average delay times as

$$W_1 = \frac{2(.125) + 1.5(.02) + 0.025(200)}{2(1 - 0.5)} = 5.28 \text{ hrs./class job}$$

$$W_2 = \frac{5.28}{2(1 - 0.5)(1 - 0.65)} = 15.09 \text{ hrs./correspondence}$$

$$W_3 = \frac{5.28}{2(1 - 0.65)(1 - 0.9)} = 75.43 \text{ hrs./manuscript}$$

In order to optimize the system a junior faculty member suggests that the jobs be reordered by interchanging the current class-1 and class-2 priorities. Under this proposal the expected delays would be

$$W_1 = \frac{5.28}{2(1 - 0.15)} = 3.09 \text{ hrs./correspondence}$$

$$W_2 = \frac{5.28}{2(1 - 0.15)(1 - 0.65)} = 8.88 \text{ hrs./class job}$$

$$W_3 = \frac{5.28}{2(1 - 0.65)(1 - 0.9)} = 75.43 \text{ hrs./manuscript}$$

Correspondence response has increased dramatically at the expense of about 3.5 hours delay in classroom work. Manuscripts still take two weeks to complete.

The dean has heard of this scheme and would like to improve faculty manuscript output. The question he has is what would be the effect of giving manuscripts top priority followed by classroom work and correspondence. The expected delays now become

$$W_1 = \frac{5.28}{2(1 - 0.25)} = 3.52 \text{ hrs./manuscript}$$

$$W_2 = \frac{5.28}{2(1 - 0.25)(1 - 0.75)} = 14.06 \text{ hrs./class job}$$

$$W_3 = \frac{5.28}{2(1 - 0.75)(1 - 0.9)} = 105.58 \text{ hrs./correspondence}$$

The big loser in this scheme is correspondence which must wait over thirteen days before receiving attention. The dean is ready to try it but has been cautioned by the assistant dean that he should attempt to estimate the variance before implementing such a plan. At that point we must consult more advanced work to obtain the necessary information.

PREEMPTIVE PRIORITY

Let us now consider preemptive priority disciplines in which an arriving customer with higher priority than the customer in service may displace that customer. We will limit our attention to single-channel systems in which arrivals for each class of customers are Poisson and services are exponential. When preempted a customer is assumed to return to the head of the queue in his priority class to await further servicing. Note that by assuming exponential service our results will be valid for both preemptive-resume and preemptive-repeat policies. The exponential service distribution has the convenient property of forgetfulness which means that the p.d.f. for remaining service on an interrupted customer is the same as his original p.d.f.

The first thing to note is that results for class-1 customers can be obtained by ignoring all lower priority classes. Since they can preempt all others, from their viewpoint the system behaves as a simple (M/M/1) queue in which $\lambda = \lambda_1$ and $\mu = \mu_1$. This means that we can immediately apply equation (6.14) to estimate the mean delay time as

$$W_1 = \frac{\rho_1}{\mu_1(1 - \rho_1)} = \frac{\lambda_1}{\mu_1(\mu_1 - \lambda_1)} \tag{12.23}$$

When there are only two customer classes it can be shown that the expected waiting time for the second class of customers is

$$W_2 = \frac{1}{\mu_2(1 - \rho_1 - \rho_2)} \left[\rho_1 + \rho_2 + \frac{\mu_2 \rho_1}{\mu_1(1 - \rho_1)} \right] \tag{12.24}$$

(see Cox and Smith, 29:89). Corresponding statements for larger numbers of customer classes, general service distributions, and multiple channels are much more complicated. The interested reader is urged to consult

Conway, Maxwell and Miller (22) or Jaiswal (52) for an indication of the modeling effort involved.

There is one important exception to the modeling complication, however. That occurs when we have a number of different priority classes each of which has a common mean service rate μ. For that case, after putting over a common denominator then adding and subtracting unity in the numerator, equation (12.24) reduces to

$$W_2 = \frac{1}{\mu\left(1 - \sum\limits_{i=1}^{2} \rho_i\right)}\left[\sum_{i=1}^{2} \rho_i + \frac{\rho_1}{1 - \rho_1}\right]$$

$$= \frac{1}{\mu\left(1 - \sum\limits_{i=1}^{2} \rho_i\right)\left(1 - \sum\limits_{i=1}^{1} \rho_i\right)} - \frac{1}{\mu} \tag{12.25}$$

where $\rho_i = \lambda_i/\mu$. When written in this form the first term represents the total expected sojourn time for a class-2 customer and the second term $1/\mu$ represents his expected service time. The difference between the two is, of course, his waiting time.

Now suppose that we add a third preemptive class. Under equilibrium conditions $\left(\sum\limits_{i=1}^{r} \rho_i < 1\right)$, the first two classes continue to behave as though the third class does not exist, hence equations (12.23) and (12.25) still hold. Furthermore from the viewpoint of the third class customer classes 1 and 2 can be lumped into a single pseudo class-1 group. Now by renumbering so that

$$\rho_1 = \sum_{i=1}^{2} \lambda_i/\mu = \sum_{i=1}^{2} \rho_i \tag{12.26}$$

and

$$\rho_2 = \rho_3 \tag{12.27}$$

equation (12.25) can be also used to find the expected waiting time for a third class customer as

$$W_3 = \frac{1}{\mu\left(1 - \sum\limits_{i=1}^{3} \rho_i\right)\left(1 - \sum\limits_{i=1}^{2} \rho_i\right)} - \frac{1}{\mu} \tag{12.28}$$

The generalization should now be obvious. If there are multiple preemptive classes in a single-channel system in which all classes have a common exponential service time, the expected waiting time for a class-k customer is

$$W_k = \frac{1}{\mu\left(1 - \sum\limits_{i=1}^{k} \rho_i\right)\left(1 - \sum\limits_{i=1}^{k-1} \rho_i\right)} - \frac{1}{\mu} \tag{12.29}$$

As before, we can determine the expected number in line and the expected number in the system for each class by using Little's formula $L = \lambda W$.

MULTI-CHANNEL PRIORITIES

Results for general multi-channel priority systems are more difficult to achieve. One exception is a non-preemptive rule in which all customers have identical exponential service distributions. For that special case it can be shown that the expected waiting time for a class-k customer is

$$W_k = \frac{A}{\left(1 - \sum_{i=1}^{k-1} \rho_i\right)\left(1 - \sum_{i=1}^{k} \rho_i\right)} \tag{12.30}$$

where

$$A = \frac{(c\rho)^c/c\mu}{c!(1 - \rho) \sum_{j=0}^{c-1} (c\rho)^j/j! + (c\rho)^c} \tag{12.31}$$

$$\lambda = \sum_{i=1}^{r} \lambda_i$$

$$\rho = \lambda/c\mu$$

(see for example Taha, 86:544, or Saaty, 82:234).

Example. One of the students in the operations research class has watched the typing queue in the academic department of the last example long enough to realize that the non-preemptive nature of the current priority scheme is causing excessive delays for routine typing tasks. Manuscripts simply take too long to allow them to dominate the workload. His proposal is to create a new preemptive priority class 1 consisting of classroom and professional correspondence and putting manuscripts in class 2. Since the original data was itself very sketchy he is willing to assume that the new combined class will have an exponential service distribution with rate $\mu = 5.4$. All other parameters remain unchanged. The data for his analysis are then

	λ_i	μ_i	$E_i(t^2)$	ρ_i
1. Classroom & Professional	3.5/hrs.	5.4/hrs.	0.0686	0.648
2. Manuscripts	0.025/hrs.	0.1/hrs.	200	0.25

If one continues to follow a non-preemptive policy with the new classification scheme the mean delay times will be

$$W_1 = \frac{3.5(.0686) + .025(200)}{2(1 - 0.648)} = 7.44 \text{ hrs.}$$

$$W_2 = \frac{3.5(.086) + .025(200)}{2(1 - 0.648)(1 - .898)} = 72.97 \text{ hrs.}$$

By combining classroom work and correspondence the average delay is now approximately one day as compared to five and fifteen hours, respectively, under the three class system. Manuscript delay is essentially unchanged. However, under a preemptive policy we must use equations (12.23) and (12.24). The corresponding times then become

$$W_1 = \frac{3.5}{5.4(5.4 - 3.5)} = 0.34 \text{ hrs.}$$

$$W_2 = \frac{1}{0.1(1 - 0.898)} 0.898 + \frac{0.1(0.648)}{5.4(1 - 0.648)} = 91.38 \text{ hrs.}$$

Under the preemptive policy, expected delay time for routine classroom work and correspondence has been drastically reduced from nearly a full day to approximately twenty minutes. The penalty, of course, is that manuscript delay has increased by nearly twenty hours. The department chairman now has extensive information with which to evaluate typing priorities.

OPTIMIZATION AND CONTROL

Most of the material we have studied has involved *descriptive* models. Such models seek to describe system behavior for any given specification of parameters. Another important segment of the mathematical modeling world involves *prescriptive* models. The goal of prescriptive models is to prescribe a certain behavior on the part of the decision maker. Usually this means select some subset of operating parameters such that the system is "optimized" in the sense of producing maximum profit or minimum cost.

Some authors decry the fact that there is not an extensive literature on cost or profit functions for queueing systems. However, it is the view of this author that the shortage is understandable since such models are highly dependent upon the particular physical environment in which they are developed and are generally a rather straightforward superposition of a cost structure on the classical static models. The goal is usually to determine the number of servers at a facility, server efficiency, or the number of service facilities needed to minimize system costs. The real challenge is in developing proper estimates for the costs, not in superimposing the cost structure on existing descriptive models. Hillier and Lieberman have an excellent discussion of basic economic queueing models (45). The reader will be asked to use the basic queueing models in an economic decision-making context in problems at the end of this chapter.

One exception to the seeming lack of general solutions for optimum queue configurations is that of the machine assignment problem. Palm has developed a set of tables for determining the number of machines to assign a worker in an (M/M/1) system (73). Ashcroft did a similar task for the (M/D/1) system (3). Both authors base their results on a cost ratio

$$ r = \frac{\text{cost of station idle}}{\text{cost of server}} $$

and a specified system utilization λ/μ. Their results can be read directly without intermediate calculation beyond that necessary to develop the cost ratio.

Other areas for prescriptive modeling involve optimizing queue discipline and selecting scheduling rules. The best source of information in that area is Conway, Maxwell and Miller (22). Some of that material was summarized in our discussion of priority queues.

Another area which receives some attention is rate-control models. Here the goal is to control either service rates or arrival rates as functions of system state. The policy of charging high landing fees at major air terminals during peak traffic hours in an effort to discourage additional traffic is an arrival rate control policy. The policy of opening additional service windows at the local bank when the crowd in the lobby becomes large is a service rate control policy. A brief survey of available rate-control models is given by Gross and Harris (42).

Example. A manufacturer of recreational vehicles is concerned with the utilization of his fork truck fleet. A single fork truck is assigned to a given production area. Calls for the truck are generated by individual work stations within the area in random fashion. An average call takes six minutes to service. Calls are served on a FCFS basis. When a call for service is made and the truck is unable to provide that service, production is delayed. The estimated cost of delay is thirty dollars per hour. The total cost of operating a fork truck is estimated to be ninety dollars per day. Management has data indicating an average 50 percent utilization of their fork trucks and feels that this is too low. They would like to redefine service areas in such a way as to generate sufficient calls to provide an "optimum" utilization. The problem is to determine an appropriate utilization level.

Let us assume that an immediate answer is needed and there is no time to collect data beyond that specified above. Since the work stations appear to operate independently from each other and in random fashion we will assume a Poisson arrival process. We have no data on service variability but must account for a reasonable range of alternatives. Management tends to think of service as constant. The engineering staff thinks of it as highly variable with the exponential distribution being their choice. The obvious solution then is to provide two curves, one for exponential service and one for constant service, in the hopes that they bound the real situation.

The first task is to force arrival rates which will produce a range of specified utilization factors. At 50 percent utilization and a service rate of ten per hour, for example, the implied calling rate is

$$\lambda = \rho\mu$$
$$= 0.5(10) = 5/\text{hr.}$$

We will assume that the fork truck will be provided with whatever numbers of customers are necessary to obtain the desired utilization. Although it may be more exact to account for a finite population, we will use an infinite population approximation.

Optimum utilization can be defined as the minimum total expected cost per service call. Assuming an eight-hour day this cost can be obtained by summing the delay cost and the allocated fork truck cost per call. That is approximately

$$E(\text{cost}) = W_q(\$30) + \frac{\$90}{\lambda(8)}$$

The delay time for any given utilization factor is obtained directly from the (M/M/1) or (M/D/1) model as appropriate. That is

$$W_{q_M} = \frac{\rho}{\mu(1 - \rho)}$$

and

$$W_{q_D} = \frac{\rho}{2\mu(1 - \rho)}$$

The total expected cost for fork truck utilization ranging from 30 to 80 percent under each assumed service distribution is given in Table 12-1. Those results are also plotted in Figure 12-1.

TABLE 12-1
COST PER SERVICE CALL

	Exponential Service		Constant Service	
ρ	W_q	$E(\text{cost})$	W_q	$E(\text{cost})$
.3	0.0428 hrs.	$5.03	0.0214	$4.39
.4	0.0667	4.81	0.0333	3.81
.5	0.1	5.25	0.05	3.75
.6	0.15	6.38	0.075	4.12
.7	0.2333	8.61	0.1167	5.11
.9	0.4	13.40	0.2	7.41

From the table and the graph it appears that if there is very little variation in service time, 50 percent utilization is in fact near optimum. If, on the other hand, the service call process exhibits random characteristics approaching the exponential distribution, an even lower utilization in the neighborhood of 40 percent will reduce overall costs. In any event it does not appear that increased utilization is warranted. In fact both curves rise rather sharply for high utilization factors. The obvious caveat here is that if the system exhibits even modest variation in service times one is ill advised to demand high utilization of the service equipment.

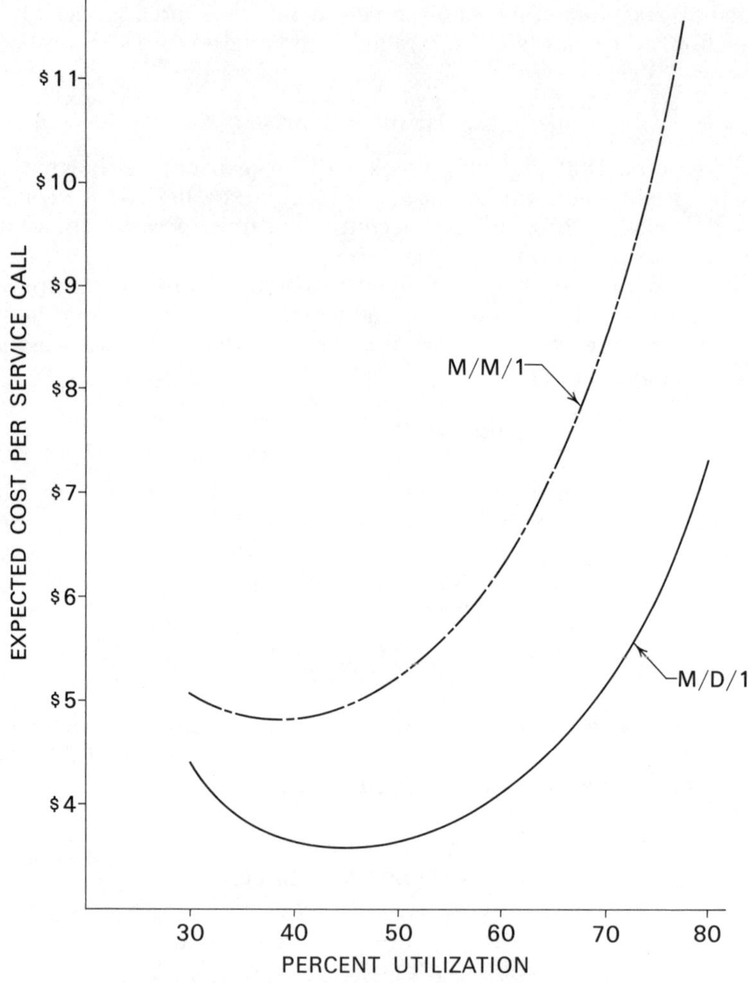

FIGURE 12-1 Service Cost Envelope

SIMULATION

Simulation refers to the dynamic operation of a model of a system. When used in the context of queueing systems the model is a collection of mathematical statements relating various system performance measures in the sense of balance equations. For example, the model for number in a single channel queue at the end of period t might be

$$n_t = n_0 + \sum_{i=1}^{t} a_i - \sum_{i=1}^{t} b_i \qquad (12.32)$$

where a_i represents arrivals in period i and b_i represents the number of services. "Operating the model" then consists of selecting numbers a_1 and

b_1, usually by some random number generating process, modifying services to account for zero queues, plugging those values into the balance equation to find n_1, selecting new numbers a_2 and b_2 to find n_2, and so forth. By plotting n_t one then has a time history of the state of the system. By using the n_t values to create a histogram one has a measure of the relative frequency of time the system occupied each state. By making many iterations of this process one hopefully can develop meaningful statistics on long-range system behavior.

There are many specialized programming languages available to help the analyst formulate and manipulate such simulation models on the digital computer. Computer simulation is a very powerful tool. When properly utilized it can be used to develop insights into very complex queueing systems which seem to defy more formal means of analysis. For example, in the network problems considered in this text we were limited to Poisson arrivals, exponential service, and random scheduling. With simulation we are not nearly so constrained. We can easily simulate a large network of queues in which every node has a unique, perhaps even empirical, service distribution, blocking may occur, and traffic control among nodes may vary over time.

The advantage which simulation has over more formal analysis is that it often has a higher degree of isomorphism with the real-world phenomena it represents. The basic equations in the model may be simple summations like equation (12.32). These can be linked in a bewildering array of interconnections which emulate the ebb and flow of activities in a queueing system. Furthermore it is easy to build a rudimentary simulation model. One can quickly create an impressive number of computer program statements which will generate a staggering amount of printed output purporting to model the system under study. Simulation by modern computer techniques is indeed a powerful tool.

However, simulation is not without defects. It is essentially an experimental device. One day's worth of simulated activity in a queue is a single data point. One needs to invoke all the techniques of experimental design and statistical analysis to make meaningful statements about simulation results. It is easy to generate numbers by simulation. It is much more difficult to draw meaningful inferences from those numbers. Furthermore simulation is expensive. Although the basic model may be created quickly and cheaply, iterating it a sufficient number of times then performing the requisite statistical analyses can be costly in both time and money.

The fact that it is relatively easy to generate numbers by simulation is both a strength and a weakness of the technique. It is a strength in the sense that it permits one to obtain numerical answers to seemingly intractable analytic problems. It is a weakness in the sense that one may be tempted to quickly jump to a complicated simulation model when more formal solutions are available. When that happens, the analyst has wasted time and money in "reinventing the wheel." At the same time he is saddling himself with a solution about which he must make guarded statistical statements when he could easily have generated much more concrete results which could be easily manipulated to demonstrate absolutely what the effects in system design changes would be. The point is that the analyst should simulate only after all reasonable avenues for analytic solutions have been explored.

Simulation of queues is not unlike simulation of any other physical system described by mathematical statements and employing Monte Carlo methods. For that reason we will not further explore such methods in this text but rather refer the reader to one of the many available simulation methods textbooks on the market (see, for example, Schmidt and Taylor, 83).

PROBLEMS

12.1 Cite at least two examples each from your personal experience in which you observed FCFS, LCFS, and RSS queue disciplines. Discuss why these disciplines were being used and suggest what system design modifications would be required to install alternate disciplines in that system.

12.2 Cite at least two examples each from your personal experience in which you observed non-preemptive priority and preemptive priority rules. Discuss why these rules were being used and suggest what impact changing the rules to a simple FCFS policy would have on system operation.

12.3 Prove that the expressions for the second moment of waiting times given by equations (12.6) and (12.7) are valid.

12.4 Consider a single exponential server operating at a rate $\mu = 1$. Plot curves for the variance of waiting time under each of the three disciplines FCFS, LCFS, and RSS as functions of input rate λ in the range $(0 < \lambda < 1)$.

12.5 The stock chaser in a large hardware supply store pulls stock for counter clerks who interface with customers. The clerks write orders which are placed in the order file. The chaser pulls orders from the top of the stack (LCFS). A methods analyst has established that the chaser performs three identifiable tasks. He must 1) check the order against his locator file, which takes an average of thirty seconds; 2) retrieve the item, which takes an average of two minutes; and 3) update the stock records which takes an average of forty-five seconds. The analyst is willing to assume that each of these times is exponentially distributed. Currently customers arrive at the rate of ten per hour. Company policy seeks to assure customers that their orders will not be delayed more than ten minutes.

a) Quantify the probability of meeting that goal under the current order processing system.

b) How would their performance change if they adopted a FCFS system?

c) Suppose that customer arrivals increase to eighteen per hour. How would your answers to (*a*) and (*b*) change?

12.6 Consider a three-class non-preemptive priority system. The arrivals for all classes are Poisson with rates $\lambda_1 = 5$ per hour, $\lambda_2 = 10$ per hour, $\lambda_3 = 15$ per hour. Service time for class 1 consists of a fixed time of three minutes plus a random portion exponentially distributed with mean two minutes. Class-2 service is uniformly distributed between one and two minutes. Class 3 is gamma with mean = 1.1 minutes and variance = 1.21. It costs \$30 per hour, \$24 per hour, and \$36 per hour to queue each unit of class-1, 2, and 3 equipment, respectively.

a) Determine the hourly cost of operating the present priority scheme.

b) Establish an optimum priority system and its associated cost.

c) Estimate the hourly cost for operating this system on a first-come, first-served basis.

12.7 Consider a non-preemptive priority system in which there is a single continuous distribution $b(t)$ for customer service times and Poisson arrivals at rate λ. Priority classes are established by specifying intervals for length of service. A class-j customer is one whose service time t falls in the interval $\tau_{j-1} < t \le \tau_j$. The set of r classes is then defined by the boundary points $(0 = \tau_0 < \tau_1 < \cdots < \tau_r)$. In practice, customers arrive at random but their service time is known before service begins. Those in the queue with the shortest service times are served first.

a) Develop an expression for expected delay of a class-k customer by making the necessary notation changes in equation (12.21).

b) Write an expression for the overall mean delay time under this priority system.

12.8 Consider a non-preemptive priority system with Poisson arrivals and an infinite number of priority classes. There is a common service time distribution $b(t)$. Upon arrival the service time for each customer is known before he enters service. The customer with the minimum service time of all who are waiting is given top priority. Prove that the average delay time in this system can be written as

$$E(W) = \frac{\lambda E(t^2)}{2} \int_0^\infty \left[\frac{b(t)}{\left(1 - \lambda \int_0^t \tau b(\tau)\, d\tau\right)^2} \right] dt$$

12.9 Consider the operation of a computer center in which all batch processing jobs are submitted with an estimate of job processing time attached. Furthermore suppose that the estimates are accurate enough that the analyst is willing to assume that estimated and actual times are equal. Jobs arrive in Poisson fashion at the rate of forty per hour. The processing time is exponentially distributed with a mean of 1.2 minutes. Current practice is to run the system on a FCFS basis.

a) Calculate the average turnaround time for jobs under current policies.

b) Suppose that a two-class priority scheme is introduced such that all jobs with estimated run times of less than two minutes are given top priority. What will be the savings in turnaround time?

c) Suppose that the system always selects the next job on the basis of shortest processing time. What will be the savings in average turnaround time? (Hint: use the results of problem 12.8.)

12.10 Consider a two-class non-preemptive priority system with Poisson arrivals and a single exponential service distribution. Priority classes are determined by the parameter s. All customers whose service time is less than $s\mu$ are class-1 customers; the rest belong in class 2.

a) Write an equation for the average delay time in this system.

b) Find a function for s which will minimize average delay.

c) Plot optimum values of s versus offered load for this system.

12.11 A single telephone booth serves a coeducational dormitory. The demand rate from men is three calls per hour for an average duration of 2.5 minutes. The demand rate from women is five calls per hour with an average duration of ten minutes.

a) In preenlightened days men would always offer to let the ladies go first (on a non-preemptive basis). What was the average delay time before a man could get at the phone?

b) After taking a course in queueing theory one of the residents realizes that the most efficient method would be to let the men go first. What would be the effect on average student delay of switching to this policy?

c) An enlightened woman who studied queues and psychology has suggested that since male conversations were more important perhaps the men should be given preemptive priority rights. What would be the impact of this policy on women's waiting times?

d) The student court is threatening to step in and will demand that everyone be served on a FCFS basis. If that comes to pass, what will be the net gain or loss to each group?

13

MODEL SELECTION AND DATA ANALYSIS

Nearly every model presented in this textbook has been an exact model in the sense that all parameters were assumed to be known. As they stand they are useful as design tools to answer broad questions such as how many service counters make sense or how large a vehicle should be put on the bus route under hypothesized conditions of demand and performance. The questions we now want to consider are ones of implementation. What confidence do you place in your conclusion that the expected line length will seldom exceed ten? How can you justify the use of an (M/M/c) model to describe the tool crib operation in your particular plant? These are typical questions which must be faced by those who apply the models.

MAJOR QUESTIONS

There are two principal statistical problems to be treated. One problem area assumes that we know the general form of the underlying service and demand distributions but must estimate their parameters. From these estimates we then need to make confidence interval statements concerning system performance measures under alternative design configurations. Chapter 14 addresses such problems.

The second major problem area, which really is the more critical of the two, concerns the initial selection of a model to describe an observed system. The problem here is to select probability density functions for service and interarrival times on the basis of observed data. Both problems are extremely important but often neglected in the queueing literature. The reasons may be that 1) a behavioral model is adequate for the analysts' purposes, 2) model building is easier to do than statistical analysis, or 3) the analysts fear that the reality of data may destroy their carefully constructed hypothetical world.

Major problems encountered in the statistical analysis of queues are well outlined by Cox (25). The interested reader should consult that article for further guidance and an excellent bibliography of pre-1965 work in statistical analysis related to queueing theory. The discussion which follows draws heavily on suggestions made by Cox (25) and Epstein (35).

JUSTIFYING THE MODEL

Let us just take the position of an analyst who must construct a model for a proposed modification to a queueing system. He can obtain demand data from the existing stream of customer requests for service. He can obtain service data by experimenting with a single unit of the new service facility to be used. The problem is to identify the probability distribution and stationarity properties of the arrival and service mechanisms taken separately before combining them in a new operating system. Since there are many models available if one can justify the use of Poisson arrivals and an exponential service process, we will concentrate on testing the validity of the exponential distribution as a model for interevent times. An extensive summary of techniques for accomplishing this is given by Epstein (35).

GRAPHICAL METHODS

To gain some intuitive insight into the way a system operates, nothing can replace plotting the data in as many different ways as possible. Taking time to "run your fingers through the data" may save you from performing many more formal but time consuming tests. You may very quickly answer such questions as does the arrival rate change over time and is the distribution obviously something other than exponential?

Suppose, for example, that instruction cards for each arriving job in a machine shop are stamped by a time clock at the time of their arrival. These cards have been collected for one eight-hour day and their log-in times noted in Table 13-1. (All times are in minutes from the origin with zero representing 8:00 AM).

TABLE 13-1
JOB ARRIVAL TIMES

023	167	234	309	450
049	181	242	353	
053	194	245	381	
118	217	283	392	
147	221	308	421	

We might choose to do several things with these data. Let us first make an estimate of arrival rate $\hat{\lambda}$. As an initial approximation we might say that the arrival rate is equal to the number of arrivals divided by the elapsed time period. That is

$$\hat{\lambda} = \frac{N}{T} = \frac{21}{480} = .04375/\text{min.} \tag{13.1}$$

Let us next plot the times of arrivals on a simple time axis. This plot is shown in Figure 13-1. At first glance one might think such a time series is very nonrandom. We seem to have a few clusters of arrival events separated by long periods of inactivity. A moment's reflection, however, should convince the reader that this is not atypical of an exponential distribution for time between events. The exponential has the property of very high probability for short time intervals and smaller but significant probability for long interevent times. This implies that the clustering of several events in a short period is to be expected because of the high frequency of brief interarrival times. However, an occasional long inter-arrival period will leave large gaps with no events in the time plot.

Let us next convert this time series into a table of interarrival times. The length of the first interarrival interval is simply the difference between the clock times at which the first and second arrivals took place. In the general the rth such interval is

$$\tau_r = t_{r+1} - t_r \tag{13.2}$$

The resulting set of interarrival times is then given by Table 13-2

TABLE 13-2
TIME BETWEEN ARRIVAL EVENTS

i	τ_i	i	τ_i	i	τ_i	i	τ_i
1	26	6	14	11	8	16	44
2	4	7	13	12	3	17	28
3	65	8	23	13	38	18	11
4	29	9	4	14	25	19	29
5	20	10	13	15	1	20	29

By calculating the sample mean and standard deviation from these data we find

$$\bar{\tau} = \sum_{i=1}^{20} \tau_i/20 = 21.35 \tag{13.3}$$

and

$$s_\tau = \sqrt{\frac{\sum_{i=1}^{20} (\bar{\tau}_i - \bar{\tau})^2}{19}} = 15.87 \tag{13.4}$$

the median or "middle value" for this sample is 21.5. Since we know an exponential distribution will have the mean and standard deviation equal, these statistics may cast some doubt on the assumption of an exponential interarrival distribution. However, we have not yet quantified that doubt.

Let us next plot the interarrival times in the order of their occurrence with the mean and median values used as reference lines. That plot is shown in Figure 13-2. What we seek here is information concerning the randomness of the process. If the system has strong nonrandom compo-

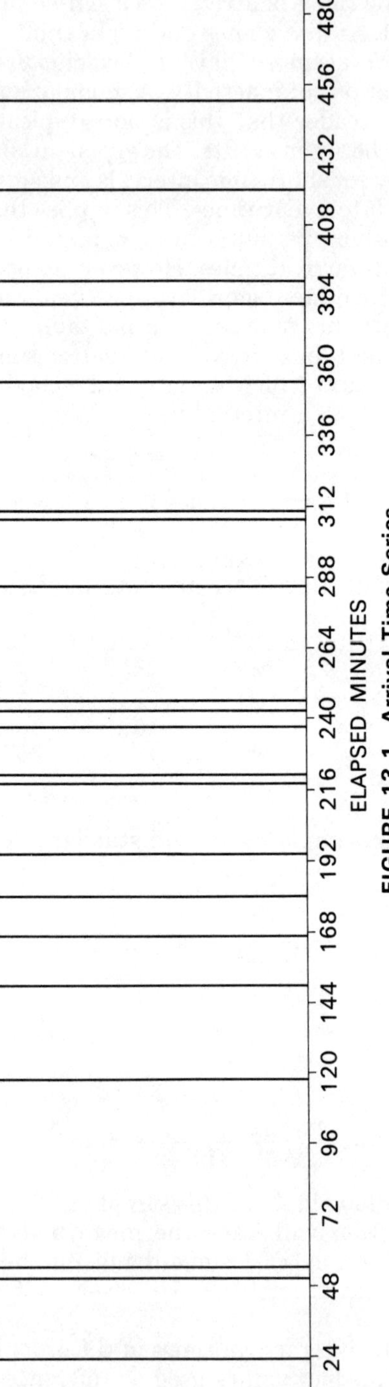

FIGURE 13-1 Arrival Time Series

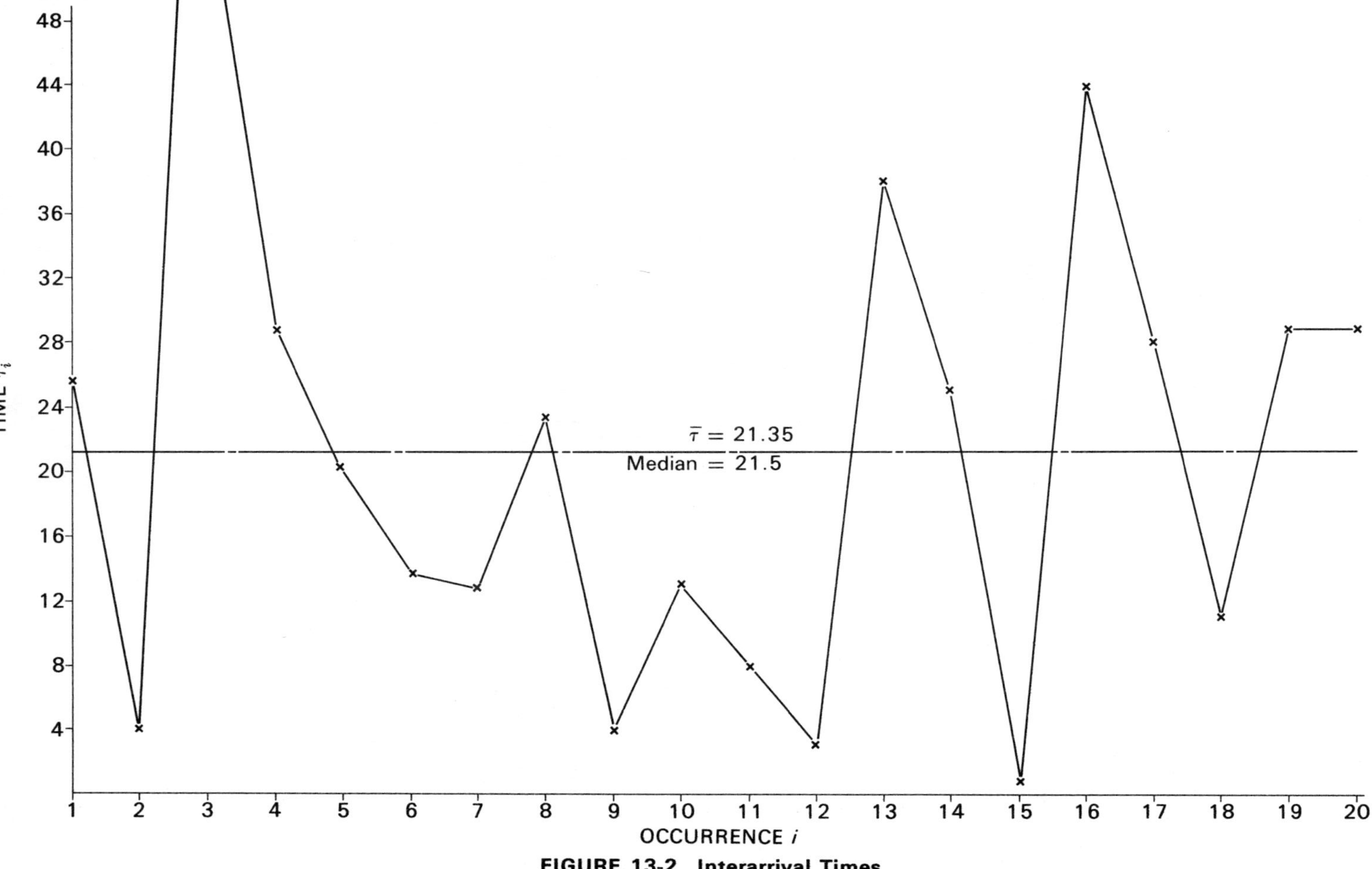

FIGURE 13-2 Interarrival Times

nents, we might for example note a trend up or down in the interarrival times. If the process shifts in step fashion throughout the day, we might note the early interarrival times grouped below the mean and the late ones above the mean, for example. Figure 13-2 shows no obvious indications of such nonrandomness.

There are also more formal methods of determining if there are non-random effects present in this process. One such method involves the sampling distribution of "runs." A discussion of this technique can be found in Duncan, (32). A point falling above the median is considered to belong to one class and a point below the median to another class. A "run above average" then consists of a succession of data points which exceed the median value. Runs below the median value are similarly defined. The test statistic is simply a count of the total number of runs in a given class.

In Figure 13-2 the number of runs above the median can be seen to be

$$
\begin{aligned}
\text{Runs of 1} \ &= 2 \\
\text{Runs of 2} \ &= \underline{4} \\
\text{Total Runs} &= 6
\end{aligned}
$$

Runs below the median are

$$
\begin{aligned}
\text{Runs of 1} \ &= 3 \\
\text{Runs of 2} \ &= 0 \\
\text{Runs of 3} \ &= 1 \\
\text{Runs of 4} \ &= \underline{1} \\
\text{Total Runs} &= 5
\end{aligned}
$$

The total number of runs of both kinds is 11. The number of data points falling above the median is 10 and the number falling below the median is 10. To use Duncan's tables we then define the parameters $r = 10, s = 10$. From those tables (32:928) we find that the probability of finding 6 or less runs is $P = 0.05$. Our data shows a total of 11 runs. Hence from this analysis we cannot conclude that the number of runs is less than would be expected on the assumption of randomness.

Since we have found no reason to suspect that the system is non-stationary, let us now perform a final graphical test by matching observed data against a theoretical curve with the same mean. (Ideally we would like to have more data points, say 50 or more, to perform this test but will use the data of Table 13-2 for purposes of illustration.) If the underlying distribution is exponential, we would expect the cumulative distribution to be given by

$$
F(t) = 1 - e^{-\lambda t}, \ t \geq 0 \tag{13.5}
$$

Although we could plot this function directly, it is more illuminating to use logarithms.

From equation (13.5) it follows that

$$
y = \ln\left(\frac{1}{1 - F(t)}\right) = \lambda t \tag{13.6}
$$

Therefore if we plot y against t, we should get a straight line with slope $\lambda = 1/E(t)$. The sample estimate for $F(t)$ is given by

$$F(t_i) = \frac{i}{n+1} \tag{13.7}$$

where n is the number of orders received and t_i is the length of the ith longest interarrival interval. Converting the data to these terms we have Table 13-3. These data are plotted in Figure 13-3. Note that the observed

TABLE 13-3
CUMULATIVE DISTRIBUTION

i	t_i	$F(t_i)$	$y = \ln\left(\dfrac{21}{21-i}\right)$
1	1	.0476	.0488
2	3	.0952	.1001
3	4	.1429	.1442
4	4	.1905	.2113
5	8	.2381	.2719
6	11	.2857	.3365
7	13	.3333	.4055
8	13	.3810	.4796
9	14	.4286	.5596
10	20	.4762	.6466
11	23	.5238	.7419
12	25	.5714	.8473
13	26	.6190	.9651
14	28	.6667	1.0986
15	29	.7143	1.2523
16	29	.7619	1.4351
17	29	.8095	1.6582
18	38	.8571	1.9459
19	44	.9048	2.3514
20	65	.9524	3.0445

data does fall reasonably close to a straight line passing through the origin and having slope $\hat{\lambda} = 1/E(t) = .0468$. For convenience one might also have plotted $(n+1)/(n+1-i)$ versus t_i on semilog paper to reach the same conclusion. Since we have not found strong evidence that the exponential distribution is an inadequate representation, we may choose at this point to assume that the arrival process is indeed Poisson with a rate of approximately 2.8 customers per hour.

A similar graphical analysis is obviously possible for the service time distribution. In the case of a new or modified facility, however, it may be necessary to experiment with a single server to obtain the necessary data.

METHOD OF MOMENTS

An analyst may sometimes be faced with data from a system for which he has no strong prior indication of the underlying distribution form. In those situations his best strategy may be to

4.0
3.0
2.0
1.0
4 8 12 16 20 24 28 32 36 40 44 48 52 56 60 64 68
$y = \ln\left(\frac{n+1}{n+1-i}\right)$
OBSERVED TIME BETWEEN ARRIVALS

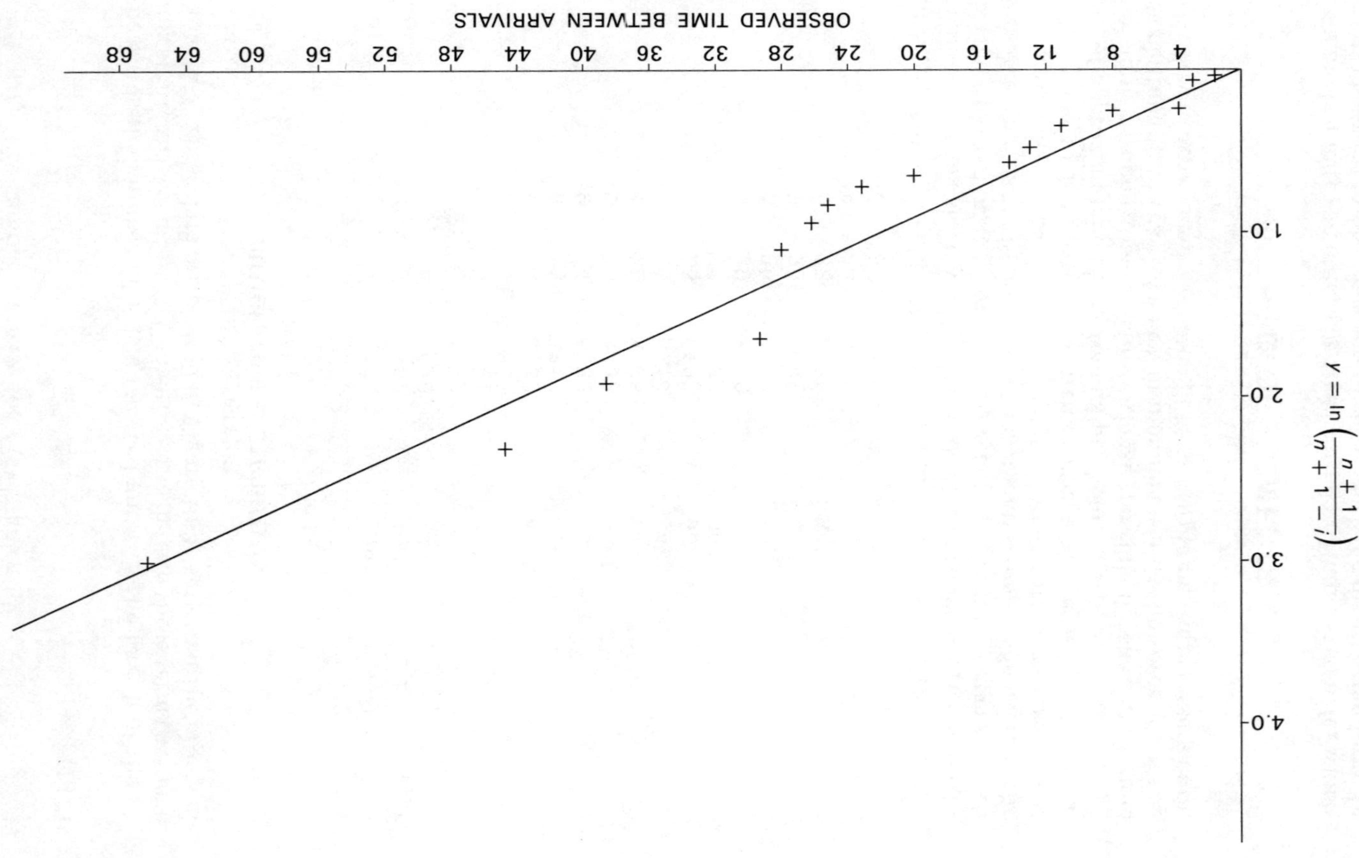

FIGURE 13-3

1. Plot a histogram of the sample data.
2. Select a general class of distributions as possible candidates for describing these data on the basis of the approximate shape of the histogram. (In queueing systems some form of the Gamma distribution is often a good candidate.)
3. Estimate the parameters of the theoretical curve by matching the moments of that curve against corresponding sample moments.
4. Plot the theoretical curve so determined on the same axes as the histogram.
5. Apply the TLAR criterion ("that looks about right") to compare the curve shapes.

Such an approach may lack the rigor of more formal hypothesis tests but it has the clear advantages of ease of application and an intuitive feel for the goodness of fit of candidate distributions.

Suppose that we have a histogram for which we want to find the best fitting member of the gamma family of density functions. The gamma is a two-parameter distribution (a, b) of the form

$$f(t) = \frac{b(bt)^a}{\Gamma(a + 1)}\, e^{-bt}, \qquad t \geq 0 \tag{13.8}$$

whose mean and variance are given by

$$\mu = \frac{a + 1}{b} \tag{13.9}$$

$$\sigma^2 = \frac{a + 1}{b^2} \tag{13.10}$$

The exponential distribution is a member of this family of curves in which $a = 0$.

Solving equations (13.9) and (13.10) for the parameters a and b we find that

$$b = \mu/\sigma^2 \tag{13.11}$$
$$a = (\mu/\sigma)^2 - 1 \tag{13.12}$$

We are now in position to calculate estimates for a and b based on the sample mean and variance obtained from the data.

CURVE FITTING

When one wishes to have a more quantitative justification than the foregoing graphical analysis or method of matching moments for selecting the exponential (or other distribution) as the model for interevent times, he may select from a large body of standard curve fitting techniques available in the literature. Among these techniques are the chi-square goodness of fit test, the F-test, the Kolmogorov-Smirnov test and the Anderson-Darling test. These tests each have their own special advantages and weaknesses. The interested reader should consult a statistics text such as Bowker and Lieberman for the details of these tests

(12). We will limit our discussion to the widely used χ^2 test for general distributions, and a more easily applied test based on the normal distribution for the special case of a Poisson null hypothesis.

The χ^2 Test

The χ^2 goodness of fit test is used to test the hypothesis that the observed relative frequency $f(x)$ equals a particular theoretical relative frequency $f_0(x)$. This technique has some weaknesses, most notable of which are the need for large samples and its dependence upon the choice of intervals into which the time axis is divided. Some analysts may prefer the Kolmogorov-Smirnov test which also tests sample observations against a hypothesized distribution. However, the Kolmogorov-Smirnov test requires that the hypothesized distribution, including all of its parameters, be completely specified. The χ^2 test can be used when the hypothesized distribution form is specified but not its parameters. Since most of our work requires that we estimate parameters as well as distribution form we will use the χ^2 test.

The statistic used for the χ^2 test is

$$\chi^2 = \sum_{i=1}^{k} \frac{(F_i - f_i)^2}{f_i} \tag{13.13}$$

where:

F_i = number of sample observations found in class interval i

f_i = number of observations expected under H_0

k = number of class intervals; k is usually selected so that no interval has $f_i < 5$.

The number of degrees of freedom associated with the test is equal to the number of class intervals, k, minus the number of conditions imposed by the fitting process. If the hypothesized distribution is completely specified, there is a single condition imposed namely

$$\Sigma f_i = \Sigma F_i \tag{13.14}$$

and the degrees of freedom would be $(k - 1)$. For every parameter which must be estimated an additional degree of freedom is lost. For example, if we were testing the hypothesis that interarrival time was exponential, the sample estimate for the single parameter (the mean) of that distribution would be

$$\bar{t} = 1/\lambda = \sum_{i=1}^{n} t_i/n \tag{13.15}$$

where t_i is the length of the ith observed interval and n is the number of observations. In this case our χ^2 test would have $(k - 2)$ degrees of freedom.

To illustrate the use of the χ^2 test let us consider the data of Table 13-4.

TABLE 13-4
OBSERVED INTERARRIVAL TIMES

53	29	13	32	44	33	9	7	2	13
16	34	56	19	12	9	31	4	21	20
19	4	11	34	18	25	24	25	13	2
7	22	22	5	13	1	11	39	42	64
18	13	11	9	2	12	1	5	28	18
23	51	27	32	78	92	138	12	35	6

The mean of those data is $\bar{t} = 24.48$. Let us test the hypothesis that we have an exponential interarrival distribution which implies that our hypothesized distribution is

$$f_0(t) = \frac{1}{24.48} exp(-t/24.48) \tag{13.16}$$

If we observe r interarrival intervals the expected number falling in the interval $(t_i - t_{i-1})$ under the hypothesized distribution will be

$$E(n_i) = f_i = rp_i \tag{13.17}$$

where

$$p_i = \int_{t_{i-1}}^{t_i} \lambda e^{-\lambda t} dt \tag{13.18}$$

Let us divide the time axis into five intervals, count the frequency of observations falling in each interval, and calculate the expected number as follows:

Interval	$t_i - t_{i-1}$	F_i	f_i
1	0–10	15	20.1
2	10+–20	18	13.4
3	20+–30	10	8.9
4	30+–40	8	5.9
5	> 40	9	11.7
	Total	60	60.0

The χ^2 statistic from equation (13.13) is

$$\chi^2 = \frac{(15 - 20.1)^2}{20.1} + \frac{(18 - 13.4)^2}{13.4} + \frac{(10 - 8.9)^2}{8.9} + \frac{(8 - 5.9)^2}{5.9} + \frac{(9 - 11.7)^2}{11.7}$$

$$= 4.38$$

Since we have used sample data to estimate the mean our critical value will be obtained from the χ^2 distribution with $k - 2 = 3$ degrees of freedom. If we use the 5 percent level of significance, the percentage point of the χ^2 distribution with 3 degrees of freedom is 7.815 (see any good book on statistical techniques). Since 4.38 is less than 7.815 we cannot reject the hypothesis that the underlying distribution is exponential.

The reader should note that although we are emphasizing the exponential distribution in this section, the χ^2 goodness-of-fit test is equally well suited for testing hypotheses concerning other distributions.

Normal Test

When we are interested in testing whether an underlying process is Poisson, we can use some special properties of that distribution to devise a quick test which is free of the difficulties attributed to the χ^2 test. Recall from our earlier discussion that if one observes a Poisson process for a fixed length of time T and if r events occur in $[0, T]$ at times $t_1 \leq t_2 \ldots t_r \leq T$ then these times can be considered to be r independent observations on a random variable uniformly distributed over $[0, T]$. In similar fashion it can be shown that if we observe a Poisson process until exactly r events occur (r being a preassigned number) at times $t_1, t_2, \ldots , t_r$, then the $(r - 1)$ random variables $t_1, t_2, \ldots , t_{r-1}$ when unordered will be uniformly distributed on $(0, t_r)$. (See Epstein, 35:96).

With our prior knowledge about the uniform distribution we can now invoke the central limit theorem to argue that the sum of the observed arrival times will be approximately normally distributed. That is, for a predetermined number of events r,

$$S = \sum_{i=1}^{r-1} t_i \tag{13.19}$$

will be approximately normally distributed with

$$\mu = (r - 1) \, t_r/2 \tag{13.20}$$
$$\sigma^2 = (r - 1) \, t_r^2/12 \tag{13.21}$$

Similarly for a predetermined time interval T,

$$S = \sum_{i=1}^{r} t_i \tag{13.22}$$

will be approximately normally distributed with

$$\mu = r \, T/2 \tag{13.23}$$

and

$$\sigma^2 = r \, T^2/12 \tag{13.24}$$

We can then use standard tables for the normal probability distribution to test the hypothesis that the underlying process is Poisson with a constant rate parameter.

Although the number r may be selected for computational convenience, some rules of thumb suggest that it should be greater than ten in order to invoke the central limit theorem.

To illustrate the technique, reconsider the data of Table 13-1. Those 21 observations were made over a 480 minute period. This means that we want to use equations (13.22), (13.23), and (13.24) to calculate the necessary statistics. It then follows that

$$S = 4988$$

$$\mu = 21(480)/2 = 5040$$

$$\sigma = \sqrt{21(480)^2/12} = \sqrt{403200} = 635$$

The 95 percent acceptance interval for this test is then

$$S_A = 5040 \pm 1.96(635) = (3795, 6285)$$

Since the sample statistic $S = 4988$ falls well within this interval we cannot reject the hypothesis that the underlying process is Poisson. We now have a more rigorous test which verifies our earlier graphical analysis of those data.

RATE FLUCTUATIONS

In deciding whether one can use a steady-state model or must go to a time-varying representation of a queueing system it may be desirable to test whether the arrival rate and service rate fluctuate over time. To accomplish this, one could employ a test for homogenity of variance such as Bartlett's test (see Epstein, 35:88).

To apply the test consider kr ordered event times, such as the arrival times of the first kr customers of the day, which have been arranged in k groups each of size r. The assumption is that the mean arrival rate is constant within each group. The object is to test whether it fluctuates from group to group. We must first calculate the total of all interarrival times (t_i) within each group where

$$T_1 = T(t_{1r}) = \sum_{i=1}^{r} t_i \tag{13.25}$$

$$T_j = T(t_{jr} - t_{j(r-1)}) = \sum_{i=r(j-1)+1}^{jr} t_i \tag{13.26}$$

$$T = T(t_{kr}) = \sum_{j=1}^{k} T_j \tag{13.27}$$

The variables T_j are mutually independent and $2\lambda T_j$ is for each group distributed as $\chi^2(2r)$ while $2\lambda T$ is distributed as $\chi^2(2rk)$. [$\chi^2(n)$ refers to a χ^2 distribution with n degrees of freedom.] The test statistic

$$C = 2rk \left\{ \ln\left(\frac{T}{k}\right) - \frac{1}{k} \sum_{i=1}^{k} \ln T_i \right\} \Big/ \left(1 + \frac{k+1}{6rk}\right) \tag{13.28}$$

will be approximately distributed as $\chi^2(k-1)$. The hypothesis that λ is the same over all groups is rejected at the significance level α if $C > \chi^2_{1-\alpha}(k-1)$ where

$$Pr[\chi^2(k-1) > \chi^2_{1-\alpha}(k-1)] = \alpha \qquad (13.29)$$

To illustrate the test consider the arrival times recorded in Table 13-5.

TABLE 13-5
ARRIVAL TIMES

.71	3.77	5.18	6.35	7.98
1.05	3.82	5.24	6.42	8.00
1.22	4.68	5.42	6.48	
1.25	4.70	5.53	7.08	
1.80	4.82	5.84	7.43	
2.95	4.88	5.85	7.71	
3.31	5.02	6.29	7.93	

Suppose we now divide these data into three groups of ten each. Using equations (13.25), (13.26), and (13.27), we can calculate the total interarrival times within each group as

$$T_1 = \sum_{i=1}^{10} t_i = 4.68$$

$$T_2 = \sum_{i=11}^{20} t_i = 1.17$$

$$T_3 = \sum_{i=21}^{30} t_i = 2.15$$

$$T = \sum_{j=1}^{3} T_j = 8.00$$

(The number of groups is somewhat arbitrary, being tempered by the intuitive feel one has for possible underlying reasons for rate changes over time, e.g. if one suspects that the first ten customers may be from a different population than the rest, the groupings of ten provide a reasonable test statistic.)

The test statistic from equation (13.28) is

$$C = \frac{2(30)\left\{\ln\left(\frac{8}{3}\right) - \frac{1}{3}(\ln 4.68 + \ln 1.17 + \ln 2.15)\right\}}{1 + \dfrac{4}{6(30)}}$$

$$= 9.32$$

From the χ^2 tables with 2 degrees of freedom we find the critical point for the 5 percent significance level at

$$\chi^2_{.95}(2) = 5.99$$

Since $C = 9.32$ is greater than 5.99 we must reject the hypothesis that the arrival rate is constant over these three groups. A stationary model would not be appropriate for this system.

CORRELATED ARRIVALS

Cox points out that there are some results reported which suggest that one of the most dangerous departures from the standard renewal assumptions is the occurrence of substantial positive autocorrelations in the series of arrival intervals (25:305). The autocorrelation at lag k refers to the correlation between any two observations in a time series that are k periods apart. That is

$$\rho(k) = \frac{cov(x_t, x_{t+k})}{\sqrt{\sigma_{x_t}^2 \sigma_{x_{t+k}}^2}} \tag{13.30}$$

A plot of $\rho(k)$ versus the lag k is called the autocorrelation function.

If we want to estimate the autocorrelation function from a series of n interarrival times $\{t_j\}$ where j refers to the jth observed arrival, must first estimate the mean

$$\bar{t} = \sum_{j=1}^{n} t_i/n \tag{13.31}$$

This is then used to calculate the autocovariance for as many lags as can be supported by the data. (Box and Jenkins suggests that the maximum lag should not exceed $k_{max} = n/4$ (13:33)).

$$Cov(t_j, t_{j+k}) = \sum_{j=k+1}^{n} y_j y_{j-k}/(n - k - 1) \tag{13.32}$$

where

$$y_i = t_j - \bar{t} \tag{13.33}$$

We can also calculate the sample variance as

$$\hat{\sigma}_{t_j}^2 = \sum_{j=1}^{n} (t_j - \bar{t})^2/(n - 1) \tag{13.34}$$

Since for a stationary process the variance is the same for all j, the estimated values for the autocorrelation function are then obtained from

$$\hat{\rho}(k) = \frac{(n - 1) \sum_{j=k+1}^{n} (t_j - \bar{t})(t_{j-k} - \bar{t})}{(n - k - 1) \sum_{j=1}^{n} (t_j - \bar{t})^2} \tag{13.35}$$

If the $\hat{\rho}(k)$ values approach one for small k, it is not reasonable to use standard Poisson models. However, Cox suggests that it is reasonably safe to use a model which gives nearly the correct variance for the number of arrivals in time periods of lengths up to three or four times the mean queueing time (25:305).

CONSERVATIVE DESIGN

Before leaving the discussion of model selection we should make two observations concerning the widely used Poisson/exponential assumptions. First, the Poisson process is a *mathematical* concept. No real process can be expected to be exactly in agreement with it. What we seek are ways of identifying reasonable agreement with the phenomena under investigation. All of the foregoing tests are attempts to bolster our confidence that an assumed Poisson model can be used as a reasonable surrogate for the real process.

The second observation concerns the choice of the exponential distribution as a model for the service system. This is a conservative choice. If one examines the expected waiting time for alternative choices of service distribution, the exponential will lead to a longer wait than for any other distribution whose coefficient of variation is less than unity. This includes the family of gamma distributions often used to describe service operations. The point is that design decisions made under the assumption that service is exponential, even when it obviously is not, will err on the safe side in terms of predicted line lengths and waiting times.

PROBLEMS

13.1 The following data represent times at which customer arrivals have been observed to occur during an eight-hour shift.

0817	0918	1011	1109	1204	1311	1410	1509
0820	0933	1019	1115	1220	1326	1425	1526
0826	0934	1029	1132	1236	1344	1441	1544
0844	0945	1039	1148	1237	1355	1459	1559
0858	0956	1052		1254			

a) Estimate the arrival rate λ. State all assumptions.

b) Plot the arrival times.

c) Prepare a table of interarrival times.

d) Calculate the mean and variance for observed interarrival times.

e) What is your conclusion about the underlying process distribution?

13.2 Use the sampling distribution of runs to test the data of problem 13.1 for randomness.

13.3 Plot the data of problem 13.1 against a cumulative exponential distribution with the same mean. What conclusions can you draw about the exponential representation?

13.4 Use the method of moments to find an appropriate distribution for the interarrival distribution implied by the data of problem 13.1. State all assumptions. Test the fit of your proposed model by the χ^2 test.

13.5 Use the χ^2 goodness-of-fit test to test the hypothesis that the data in problem 13.1 comes from a Poisson process with a mean of four per hour. State all assumptions.

13.6 Use the normal test to determine if one can assume that the data of problem 13.1 comes from a Poisson process. State all assumptions.

13.7 Consider the following interarrival times recorded for the first fifty customers of the day. Twenty arrived between 7:30 AM and 12:00 noon; the balance arrived after noon.

AM				PM					
7	2	10	20	5	6	7	1	26	1
46	7	1	1	12	2	8	2	13	3
1	7	36	3	2	10	1	1	11	8
26	17	12	18	5	13	15	4	16	7
27	5	4	13	1	10	2	5	10	19

   a) Estimate the arrival rate $\hat{\lambda}$. State all assumptions.

   b) Test whether or not a single Poisson distribution can be used to describe this process.

   c) Use the method of moments to find an appropriate interarrival distribution.

13.8 Reconsider the data of problem 13.7. Suppose that we now have reason to believe that morning customers are different from afternoon customers.

   a) Test this hypothesis.

   b) Fit an appropriate model to both morning and afternoon processes.

13.9 Consider the following service time data taken from three separate tool crib clerks working at the same counter.

Clerk 1		Minutes Clerk 2		Clerk 3	
.3	.9	9.1	3.7	4.8	.5
2.6	14.1	5.1	1.9	4.0	2.7
6.5	2.9	14.4	11.4	.5	3.7
23.0	2.2	2.4	11.0	.1	15.6
24.1	2.3	.9	5.1	2.7	1.0

a) Can you justify modeling this as an (M/M/3) system? Explain.

b) Find appropriate distributions to describe clerk behavior.

13.10 The purchasing department at a large manufacturing plant processes purchase orders for all departments within the plant. The company is willing to assume that processing times are exponentially distributed. A log sheet has been devised so that the number of order requests reaching the department each hour can be recorded. The following data have been collected:

Hour	Requests	Hour	Requests	Hour	Requests	Hour	Requests
1	30	14	24	27	28	39	28
2	29	15	24	28	28	40	26
3	26	16	25	29	26	41	28
4	26	17	27	30	28	42	28
5	26	18	26	31	28	43	27
6	26	19	29	32	29	44	26
7	25	20	28	33	32	45	27
8	20	21	28	34	33	46	27
9	22	22	29	35	31	47	25
10	25	23	31	36	33	48	24
11	24	24	28	37	34	49	20
12	26	25	26	38	30	50	19
13	23	26	25				

a) Is this a Poisson process? Explain.

b) Plot the autocorrelation function for these data.

c) What type of queueing model would you recommend?

13.11 A commercial bakery is about to install a new cake packaging line. At present their decorated cakes are transferred in open pans to their retail shelves. They have observed this operation for thirty minutes and have recorded the following times at which cakes are placed on pans (minutes measured from beginning of observations at $t = 0$).

.54	2.14	9.19	11.31	14.27	23.27	29.74
.76	3.98	9.45	11.94	15.33	25.49	
1.03	5.74	10.54	12.06	15.71	26.97	
1.59	8.42	10.83	12.20	21.76	27.02	
2.00	8.71	10.91	12.25	22.71	27.76	

Their new system will use a single packer to place cakes in boxes. The table upon which the packer works can handle up to thirty cakes. Any arriving when the table is full are set aside to be packed at a later time by the second shift packer. No cakes are decorated on second shift. They have watched a packer simulate placing cakes in boxes with the following observed times:

.82	.46	.18	.42	.26	.62	2.81
.53	.21	1.61	.21	.13	1.77	.46
.22	.52	1.97	.33	.30	.33	2.52
.53	.08	1.51	1.71	.40	.06	1.90
.29	.29	.21	.49	.41	1.14	1.24

Select a model to describe the queue of cakes which will form before the packer. State all assumptions and justify your model recommendation.

14

PARAMETER ESTIMATION AND HYPOTHESIS TESTING

Let us now turn our attention to the parameter estimation problem, the resulting confidence interval statements and tests of hypotheses. In the estimates which follow we will assume that we can collect data for an on-going queueing system. Furthermore we will assume that the general form of the underlying arrival and service processes are known to us. Our tasks are to find efficient ways to estimate parameters of the distributions and test hypotheses concerning proposed model forms. Most of the techniques in this chapter 1) presuppose that the birth and death postulates apply, 2) assume that the system is stationary in the sense that λ_i and μ_i are not functions of time, and 3) are based upon the method of maximum likelihood.

The pioneering work in estimation of this sort was begun by Clarke (19). He was concerned with obtaining the maximum likelihood estimators for the arrival and service parameters in an (M/M/1) queue. Related work of a more general nature concerning statistical methods for Markov processes was performed by Billingsley (10 and 11). Wolff later made extensive use of Billingsley's results to develop a number of very useful estimators and associated hypothesis tests for the general birth and death queueing processes (89). Lilliefors suggested a technique for calculating confidence intervals for performance measures which is closely related to Wolff's work (65). The discussion which follows draws heavily upon the results of those authors as well as summary information from Cox (25).

MAXIMUM LIKELIHOOD ESTIMATES

The method of maximum likelihood is a useful technique for deriving point estimates. Consider first the problem of estimating a single parameter such as the rate parameter in an assumed Poisson arrival process. Suppose that we have a random sample of size k for time between arrivals from a stochastic process which must be used to estimate the parameter λ. Assuming that the observations are independent we can form the joint density function of the sample from the product

$$L(\lambda) = \prod_{i=1}^{k} \lambda e^{-\lambda t_i} \qquad (14.1)$$

where t_i is the ith sample observation. The function $L(\lambda)$ is called the *likelihood* function.

Because of its product form it is often more convenient to manipulate the log of the likelihood function. The maximum likelihood estimator for the single unknown parameter λ is that value of λ which maximizes the joint probability density of occurrence of the observed sample. Equivalently we seek $\widehat{\lambda}$ so that

$$\ln L(\widehat{\lambda}) = \underset{\lambda}{Max}\,[\ln L(\lambda)] \qquad (14.2)$$

From equation (14.1) we have

$$\ln L(\lambda) = k\ln\lambda - \lambda \sum_{i=1}^{k} t_i \qquad (14.3)$$

We can locate the critical value of λ by setting the first derivative equal to zero. That is

$$\frac{d}{d\lambda}\ln L(\lambda) = \frac{k}{\lambda} - \sum_{i=1}^{k} t_i = 0 \qquad (14.4)$$

from which

$$\widehat{\lambda} = \frac{k}{\sum_{i=1}^{k} t_i} \qquad (14.5)$$

The fact that this is the maximum point for $\ln L(\lambda)$ [and equivalently for $L(\lambda)$] is readily apparent since

$$\frac{d^2}{d\lambda^2}\ln L(\lambda) = -k\lambda^{-2}$$

which is always negative as required for strictly convex functions.

The maximum likelihood estimator for λ is intuitively appealing since its reciprocal

$$\frac{1}{\widehat{\lambda}} = \frac{\sum_{i=1}^{k} t_i}{k} \qquad (14.6)$$

which is the estimated mean of the underlying distribution, is simply the mean of the sample observations. This result seems trivial in retrospect, but the technique illustrated can be applied in equally straightforward fashion to much more complicated models. When multiple parameters are involved the optimization step requires solution of a set of simultaneous equations obtained by setting the partial derivatives with respect to each parameter in turn equal to zero. Otherwise nothing different is required.

(M/M/1) Estimates

Now suppose that we can observe a simple single-channel queue over a continuous period of length T. Furthermore we have reason to believe that the underlying interarrival and service distributions are each exponential but with unknown parameters λ and μ. What we seek are the maximum likelihood estimates $\widehat{\lambda}$ and $\widehat{\mu}$ of those parameters.

Let $n(t)$ be the observed state of the system at time t. Furthermore let n_1 be the state of the system at time $t_1 = 0$ and let t_{i+1} be the time at which the ith system state change occurs. State changes occur in step fashion as arrival or service events take place. The number of transitions k which take place during the observation period $[0, T]$ is a random variable.

The likelihood function can be constructed in pieces as follows:

1. When the system is occupied, a transition upward $(n_{i+1} = n_i + 1)$ due to an arrival event at time t_{i+1} requires that an arrival occur in $(\tau_i, \tau_i + d\tau_i)$ and that no service take place in $[0, \tau_i)$, where $\tau_i = t_{i+1} - t_i$ is the time between transitions. The joint density for this event is

$$L_i = \lambda e^{-(\lambda+\mu)\tau_i} \tag{14.7}$$

2. When the system is occupied, a transition downward $(n_{i+1} = n_i - 1)$ due to completion of service at time t_{i+1} requires that a service occur in $(\tau_i, \tau_i + d\tau_i)$ and that no arrivals take place in $[0, \tau_i)$. The joint density for this event is

$$L_i = \mu e^{-(\lambda+\mu)\tau_i} \tag{14.8}$$

3. When the system is empty a transition due to an arrival event at time t_{i+1} contributes

$$L_i = \lambda e^{-\lambda\tau_i} \tag{14.9}$$

4. At the end of the period of observation $(t = T)$ the incomplete interval contributes

$$L_k = e^{-(\lambda+\mu)\tau_k}$$

   or

$$L_k = e^{-\lambda\tau_k} \tag{14.10}$$

depending on whether the system is empty or busy at that time.

The total likelihood function can then be obtained by taking the product of the components L_i described above over all observed transitions $i = 1, 2, \ldots, k$. Let t_e be the time spent in the empty state, and t_b the time spent in any busy state, then $T = t_e + t_b$. Let n_e be the number of arrivals occurring when the system is in the empty state, n_b the number of arrivals occurring when it is in a busy state and n_s the number of services completed. The total number of arrivals is given by $n_a = n_e + n_b$. The likelihood function can then be written as

$$L_c(\lambda, \mu) = \prod_{i=1}^{k} L_i$$

$$= \lambda^{n_a} \mu^{n_s} e^{-(\lambda+\mu)t_b - \lambda t_e} \tag{14.11}$$

Note that this function ignores the probability of beginning the sample function with n_1. For that reason Cox refers to this as the conditional likelihood given the initial state (25:292).

The log of equation (14.11) is

$$\ln L_c(\lambda,\mu) = n_a \ln\lambda + n_s \ln\mu - (\lambda + \mu)t_b - \lambda t_e \tag{14.12}$$

from which it is easy to establish the maximum likelihood estimators as

$$\widehat{\lambda}_c = \frac{n_a}{t_e + t_b} \tag{14.13}$$

$$\widehat{\mu}_c = \frac{n_s}{t_b} \tag{14.14}$$

Again these estimates have intuitive appeal. Equation (14.13) suggests that we count the total number of arrival events in $[0, T)$ and divide by the length of the observation period to estimate the average arrival rate. Equation (14.14) in similar fashion suggests that we count the number of services completed in $[0, T)$ and divide by the total time during that period in which the server was busy to obtain an estimate for the average service rate.

If one wishes to account for the initial state n_1, he must first make some assumption about its distribution. Clarke did this by assuming that $(\rho = \lambda/\mu < 1)$ and the queue was in an equilibrium condition (19). It therefore follows that n_1 is distributed by the geometric distribution for an (M/M/1) queue as

$$p_{n_1}(0) = \rho^{n_1}(1 - \rho) \tag{14.15}$$

By including this term in the likelihood function we are led to the unconditional log likelihood

$$\ln L(\lambda, \mu) = n_1 \ln(\lambda/\mu) + \ln(1 - \lambda/\mu) + \ln L_c(\lambda, \mu) \tag{14.16}$$

Maximizing this function leads to a quadratic equation in $\widehat{\lambda}$.

As an alternative Cox suggests that we treat the additional terms as a small correction to the conditional likelihood. If we subtract an estimate of the mean system size from the initial condition, then divide by the time of observation, we can remove the effect of differences of n_1 from the mean. It therefore follows that it is approximately true that

$$\widehat{\lambda} \approx \frac{n_a + [n_1 - \widehat{\rho}/(1 - \widehat{\rho})]}{T} \tag{14.17}$$

and

$$\widehat{\mu} \approx \frac{n_s - [n_1 - \widehat{\rho}/(1 - \widehat{\rho})]}{t_b} \tag{14.18}$$

where

$$\widehat{\rho} = \frac{\widehat{\lambda}_c}{\widehat{\mu}_c} = \frac{n_a t_b}{n_s T} \tag{14.19}$$

CONFIDENCE INTERVALS FOR RATES

In addition to the point estimates for arrival rate and service rate, it is to the designer's advantage if he can make confidence interval statements about the true value of those parameters. For example, rather than state that our estimate for the average arrival rate is $\widehat{\lambda} = 15$ per hour we would rather make statements such as "we are 95 percent confident that the true mean rate falls in the interval $12 < \lambda < 18$."

The meaning of confidence interval statements must be carefully interpreted. For an excellent discussion of confidence intervals, see Mood (68:Ch.11). Suppose for example we make our confidence interval estimate on the basis of a sample of size k. The statement that we are 95 percent confident that the true mean falls between a and b implies that if we were to repeat our sample of k a number of times and calculate a new interval for each sample, 95 percent of those intervals would be expected to contain the true mean. The interval $[a, b]$ is a random interval.

The technique for constructing confidence intervals requires that we find a function, say y, of the sample observations and the parameter to be estimated which has a distribution independent of the parameter and any other parameters. Once that distribution is identified then probability statements of the form

$$P(a < y < b) = \alpha \tag{14.20}$$

can be converted to confidence interval statements about the parameter. Any points a and b which define α portion of the area under the density $f(y)$ can be used to designate a confidence interval. However, the points defining the interval of shortest length are to be preferred as they give more precise information about the location of the true parameter.

Consider again a sample of k interarrival times from a Poisson process which will be used to estimate the mean interarrival time and consequently the arrival rate. The maximum likelihood estimate for the mean interarrival time is

$$1/\widehat{\lambda} = \frac{1}{k}[t_1 + t_2 + \cdots + t_k] \tag{14.21}$$

Each observation t_i is a random sample drawn from an exponential distribution with the unknown parameter λ. Suppose that we define a new random variable y such that

$$y = \frac{2\lambda k}{\widehat{\lambda}} = 2\lambda[t_1 + t_2 + \cdots + t_k] \tag{14.22}$$

The new variable y consists of a sum of independent exponentially distributed random variables t_i multiplied by a constant 2λ.

We can easily find the distribution for the sum of random variables by using the convolution properties of Laplace transforms as discussed in Appendix A. That is if the transform of the distribution for t_i is

$$\mathcal{L}[f(t_i)] = \int_0^\infty e^{-st_i}\lambda e^{-\lambda t_i}\,dt_i$$

$$= \frac{\lambda}{s + \lambda} \tag{14.23}$$

then the transform for the distribution of the sum of k such variables is

$$\mathcal{L}\left[g\left(\sum_{i=1}^{k} t_i\right)\right] = \left(\frac{\lambda}{s + \lambda}\right)^k \tag{14.24}$$

Furthermore the variable y for which we seek a distribution is simply the random variable $T = \sum_{i=1}^{k} t_i$ multiplied by a constant (2λ). We can therefore use another property of Laplace transforms outlined in Appendix A which says that the transform of the distribution of a random variable multiplied by a constant can be obtained by replacing the transform variable s by the product of s times that constant. It therefore follows that the transform for the distribution sought is

$$\mathcal{L}[h(y)] = \mathcal{L}[g(T)]\Big|_{s=2\lambda s}$$

$$= \left(\frac{\lambda}{2\lambda s + \lambda}\right)^k = \left(\frac{\frac{1}{2}}{s + \frac{1}{2}}\right)^k \tag{14.25}$$

From the table of transform pairs it follows that the random variable y is chi-square distributed with $2k$ degrees of freedom. That is

$$h(y) = \frac{1}{2^k \Gamma(k)} y^{k-1} e^{-y/2}, \, y > 0 \tag{14.26}$$

Note that this distribution is independent of the parameter λ as required for our confidence interval estimates.

Now suppose that we wish to define a 95 percent confidence interval for the true arrival rate from a sample of $k = 10$ arrivals observed during a period of $T = 2$ hours. We first need to determine values for a and b such that

$$P(a < y < b) = .95 \tag{14.27}$$

From any χ^2 table we can find that $a = 9.591$ and $b = 34.170$ are values which satisfy this statement when y is distributed as $\chi^2(20)$. By using equation (14.22) we can rewrite this expression as

$$P(a < y < b) = P\left(\frac{a\widehat{\lambda}}{2k} < \lambda < \frac{b\widehat{\lambda}}{2k}\right)$$

$$= P\left(\frac{a}{2T} < \lambda < \frac{b}{2T}\right) = \alpha \qquad (14.28)$$

Therefore from the data reported it follows that a 95 percent confidence interval for λ is

$$P(2.40 < \lambda < 8.54) = .95 \qquad (14.29)$$

Note that from the limited data available it is possible that the true value of λ may differ markedly from our point estimate of $\widehat{\lambda} = 5$ per hour.

It is obvious that we could develop confidence interval statements for the service rate μ in like fashion. In this case the observation period T would simply be a sum of busy periods and the number of events k would be the number served during those busy periods.

In summary, to find the $100\,\alpha$ percent confidence interval for the rate parameter λ for an exponentially distributed interevent time process:

1. Sum all observed interevent times

$$T = \sum_{i=1}^{k} t_i$$

2. Use the tables for the χ^2 distribution with $2k$ degrees of freedom to locate values for a and b such that

$$P(a < \chi^2(2k) < b) = \alpha$$

3. Convert to a confidence interval statement by

$$P\left(\frac{a}{2T} < \lambda < \frac{b}{2T}\right) = \alpha$$

LOAD FACTOR CONFIDENCE INTERVALS

A technique for calculating confidence intervals for various performance measures of simple queues has been developed by Lilliefors (65). His method depends upon finding the proper distribution for the sample estimate of the offered load

$$\widehat{a} = \widehat{\lambda}/\,\widehat{\mu} \qquad (14.30)$$

(In Lilliefors notation traffic intensity $\widehat{\rho}$ is used in place of $\widehat{a}$. The two are equivalent in the notation of this textbook only for single channel systems.) Once the distribution for $\widehat{a}$ is known one can then determine upper and lower confidence limits for any monotonic function of a (or $\rho = a/c$) by simply replacing the a variable by its appropriate limiting values.

It may be enlightening, for example, to determine an appropriate interval estimate for the mean number of customers present in an (M/M/1) queue. We know from previous modeling efforts that the expected number in the system can be expressed as

$$L = (\lambda/\mu)/(1 - \lambda/\mu) = \rho/(1 - \rho) \tag{14.31}$$

But this expression is exact only if we know λ and μ exactly. The problem arises because we most often have only statistical estimates, $\widehat{\lambda}$ and $\widehat{\mu}$, of these parameters. The question, then, is what sort of confidence interval statement can we make concerning our calculated value of L when we are uncertain about the exact location of the true values of λ and μ.

Consider first only those systems with Poisson arrivals and an exponential service process. Following the techniques of the last section we can determine a distribution for the variable $y_1 = 1/\widehat{\theta}$, where $\widehat{\theta}$ is an estimate of the rate parameter θ for an exponential distribution. Further suppose that we have separate estimates for both $\widehat{\lambda}$ and $\widehat{\mu}$ where $\widehat{\lambda}$ is estimated from k observations and $\widehat{\mu}$ is estimated from r observations. Here we are assuming the simplest form of the estimates, e.g.,

$$1/\widehat{\mu} = \sum_{i=1}^{r} \tau_i/r \tag{14.32}$$

The maximum likelihood estimator for the ratio $a = \lambda/\mu$ can then be expressed as

$$\widehat{a} = \frac{1/\widehat{\mu}}{1/\widehat{\lambda}} = \widehat{\lambda}/\widehat{\mu} \tag{14.33}$$

Again using the convolution properties of the Laplace transform we find that the transform for the distribution for the mean interevent time, as determined from a sample of n observations from a common exponential distribution is

$$\mathcal{L}[g(1/\widehat{\theta})] = \{\mathcal{L}[f(t)]\}^n \Big|_{s=s/n}$$
$$= \left(\frac{n\theta}{s + n\theta}\right)^n \tag{14.34}$$

The inverse of this transform is a special form of the gamma distribution known as the nth Erlang distribution. That is

$$g(1/\widehat{\theta}) = \frac{(n\theta)^n}{\Gamma(n)} (1/\widehat{\theta})^{n-1} e^{-n\theta/\widehat{\theta}} \tag{14.35}$$

From this analysis it follows that the distribution of our sample mean interarrival time must be k^{th} Erlang with parameter λ. In like fashion the distribution for the sample mean service time must be r^{th} Erlang with parameter μ. The ratio

$$\widehat{a} = \frac{\overline{t}_s}{\overline{t}_a} = \frac{1/\widehat{\mu}}{1/\widehat{\lambda}}$$

is then the ratio of two gamma (or Erlang) distributed random variables.

By suitable scaling, namely

$$X = \frac{2\lambda k}{\widehat{\lambda}}$$

and

$$Y = \frac{2\mu r}{\widehat{\mu}}$$

We can develop two new random variables X and Y which are χ^2 distributed with $2k$ and $2r$ degrees of freedom, respectively. Let us now form a ratio

$$U = \frac{X/2k}{Y/2r} = \frac{r \sum_{i=1}^{k} t_i}{k \sum_{i=1}^{r} \tau_i} \left(\frac{\lambda}{\mu}\right) = \frac{\overline{t}_a}{\overline{t}_s} a \tag{14.36}$$

Since this ratio is formed from χ^2 distributed variates, it is well known that U will be F-distributed with $2k$ and $2r$ degrees of freedom. See, for example, Bowker and Lieberman (12:120). That is

$$P(U \leq c) = F(2k, 2r) \tag{14.37}$$

We are now in position to use standard F-tables to determine an appropriate confidence interval for the offered load $a = \lambda/\mu$. We must first find c and d from $F(2k, 2r)$ such that

$$P\left(c < \frac{\overline{t}_a}{\overline{t}_s} a < d\right) = \alpha \tag{14.38}$$

It then follows that the 100α percent confidence interval for the offered load can be written as

$$P\left(c \frac{\overline{t}_s}{\overline{t}_a} < a < d \frac{\overline{t}_s}{\overline{t}_a}\right) = \alpha \tag{14.39}$$

or

$$P(c\widehat{a} < a < d\widehat{a}) = \alpha \tag{14.40}$$

where $\widehat{a}$ is our sample estimate of the offered load a.

Example. Suppose that data from an (M/M/1) queueing system has been collected to estimate the arrival and service rates. The average of twenty observed interarrival times has been calculated as $\overline{t}_a = 5$ minutes. The average of fifteen observed service times has been calculated as $\overline{t}_s = 4$ minutes. The point estimate for the offered load is therefore $\widehat{a} = 4/5 = 0.8$. To determine a 95 percent confidence interval for a we can find from the tables for $F(40, 30)$ that

$$P(0.52 < F < 2.01) = .95$$

It therefore follows from equation (14.40) that the corresponding confidence interval statement is

$$P(0.42 < a < 1.61) = .95$$

Given the length of the confidence interval and the fact that a large portion of that interval exceeds unity one might be well advised to collect more data before attempting to apply a steady state model which assumes that $a < 1$.

PERFORMANCE MEASURE CONFIDENCE INTERVALS

Consider now typical measures of performance one calculates from the usual steady-state queueing models. For example from Chapter 6 we find for the (M/M/1) system that

$$L = a/(1 - a) \qquad\qquad (14.41)$$
$$L_q = a^2/(1 - a) \qquad\qquad (14.42)$$

and

$$p_0 = 1 - a \qquad\qquad (14.43)$$

Each of these measures is a monotonic function of the offered load a. From Lilliefors' (65) argument it follows that one can obtain upper and lower confidence limits for each. This is accomplished by first determining the upper and lower limits for a, say a_U and a_L, then using them each in turn in the appropriate statement. The ratio of offered load a to its estimate $\widehat{a}$ is of course F-distributed as demonstrated in the previous section.

Suppose for example that we have ten observations on interarrival times and fifteen on service times from which we estimate $\widehat{a} = 0.2$. Applying equation (14.40) for 95 percent confidence we can establish the upper and lower limits for a by using the F-tables for $F(20, 30)$ as

$$a_U = d\widehat{a} = 2.20(0.2) = 0.44$$
$$a_L = c\widehat{a} = 0.4255(0.2) = 0.085$$

Our 95 percent confidence limits for the expected number of customers in the system is then obtained from

$$P\left(\frac{.085}{1 - .085} < L < \frac{.44}{1 - .44}\right) = .95$$

or

$$P(0.093 < L < 0.786) = .95$$

These limits can then be compared with the point estimate

$$\widehat{L} = \frac{\widehat{a}}{1 - \widehat{a}} = 0.25$$

to determine whether or not more data is needed. Limits for L_q and p_0 can be calculated in similar fashion.

The standard multi-channel configuration also offers opportunities for easy determination of confidence intervals for some typical measures. Again referring to Chapter 6 we see that for a c-channel exponential service, Poisson arrival system one can calculate

$$p_0 = \left[\sum_{n=0}^{c-1} a^n/n! + \frac{a^c}{c!(1 - a/c)} \right]^{-1} \tag{14.44}$$

$$L_q = \frac{a^{c+1} p_0}{(c + 1)!(1 - a/c)^2} \tag{14.45}$$

and

$$L = L_q + a \tag{14.46}$$

The upper and lower confidence limits for each of these quantities can be determined in straightforward fashion from the upper and lower limits on the offered load $a = \lambda/\mu$ as previously demonstrated. Extension to other forms of the birth and death equations is also possible.

One can use the same approach to handle limited cases for more general service times. Suppose that the service time distribution is n^{th} Erlang with parameter μ. This implies that the service facility can be mathematically simulated by the n-fold convolution of identical exponential distribution with parameter $n\mu$. Such a distribution is often used to approximate more general service distributions to facilitate model construction.

With this assumed model for service times our sample estimate of the mean service time

$$\frac{1}{\widehat{\mu}} = \sum_{i=1}^{r} \tau_i/r \tag{14.47}$$

will be nr^{th} Erlang distributed with parameter μ. The truth of this statement can be easily established by transform arguments similar to those used in equation (14.34). The Laplace transform for the nth Erlang distribution of service times is

$$\mathcal{L}[f(t_i)] = \left(\frac{n\mu}{s + n\mu} \right)^n \tag{14.48}$$

It follows that the sample mean, which is a sum of r variables t_i each multiplied by $(1/r)$, will have the transformed distribution

$$\mathcal{L}[g(1/\widehat{\mu})] = \left(\frac{n\mu}{s + n\mu} \right)^{nr} \Big|_{s=s/r}$$

$$= \left(\frac{(nr)\mu}{s + nr\mu} \right)^{nr} \tag{14.49}$$

The inverse of this transform is the nr^{th} Erlang

$$g(1/\widehat{\mu}) = \frac{(nr\mu)^{nr}}{\Gamma(nr)} \left(\frac{1}{\widehat{\mu}} \right)^{nr-1} e^{-nr\mu/\widehat{\mu}} \tag{14.50}$$

If we assume that the arrival process is Poisson with the estimated mean interarrival time obtained from k observations we can perform the scaling

$$X = \frac{2\lambda k}{\widehat{\lambda}}$$

$$Y = \frac{2n\mu r}{\widehat{\mu}}$$

which yields two χ^2 distributed random variables with $2k$ and $2nr$ degrees of freedom. It follows that the ratio

$$\frac{X/2k}{Y/2nr} = a/\widehat{a} \tag{14.51}$$

will be F-distributed with $2k$ and $2nr$ degrees of freedom. That distribution can be used to establish confidence interval estimates for a which are used in turn to establish confidence intervals for L and L_q.

HYPOTHESIS TESTS

It is possible for us to use the information developed in the calculation of confidence intervals to construct some simple hypothesis tests concerning values of the parameters. For instance the variable

$$y = 2\lambda \sum_{i=1}^{k} t_i$$

where the t_i are identically exponentially distributed random variables with parameter λ, was shown in equation (14.26) to be χ^2 distributed with $2k$ degrees of freedom. Suppose that we propose the null hypothesis that

$$H_0 \colon \lambda = \lambda_0$$

against the alternative

$$H_1 \colon \lambda = \lambda_1 > \lambda_0$$

Further suppose that we want the risk of a Type I error to be α and the risk of a Type II error to be β. What we now need to determine is a proper sample size and a critical region.

Under the hypothesis that the true rate is $\lambda = \lambda_0$, the test statistic

$$y = 2\lambda_0 T \tag{14.52}$$

where $T = \sum_{i=1}^{k} t_i$ can be compared to the upper α point on the χ^2 $(2k)$ curve. Under the hypothesis that $\lambda = \lambda_1$ the test statistic

$$y = 2\lambda_1 T \tag{14.53}$$

can be compared to the lower β point on the χ^2 $(2k)$ curve. These two values can then be used to specify the critical point T_c for total time observed and the required sample size k. The solution to the implied pair of simultaneous equations is most easily accomplished by "hunt and try" techniques in the χ^2 tables.

For example suppose that

$$H_0: \lambda = 5$$
$$H_1: \lambda = 10$$
$$\alpha = .05, \beta = 0.2$$

Trying a test value $k = 20$ we find from the χ^2 tables that

$$\chi^2_{.05}(40) = 26.5$$

From equation (14.52) it then follows that

$$26.5 = 2(5)T_c$$

from which $T_c = 2.65$. This implies that the null hypothesis should be rejected if $\sum_{i=1}^{20} t_i < 2.65$. Applying this critical value against the risk of a Type II error we have

$$y_c = 2\lambda_1 T_c$$
$$= 2(10)(2.65) = 53$$

From the χ^2 (40) table the probability of finding a value greater than 53 is .075. It follows that the risk of a β error is approximately .075. If we try a sample size of $k = 15$, we find that for the null hypothesis $\lambda = 5$

$$\chi^2_{.05}(30) = 18.49$$

from which $T_c = 1.849$ from the Type I error requirements. If the alternative hypothesis that $\lambda = 10$ is true it follows that

$$y_c = 2\lambda_1 T_c = 36.98$$

From the $\chi^2(30)$ tables the probability of exceeding 36.98 is approximately 0.19. Therefore this test closely matches our design goals. To summarize:

1. Observe the arrival process until fifteen arrivals have occurred.
2. If the elapsed time is less than 1.849 hours, reject $H_0: \lambda_0 = 5$ for a Type I error risk of .05.
3. Considering the alternative $\lambda_1 = 10$, the risk of a type II error is 0.19.

It should be obvious to the reader that similar hypothesis tests could be designed for the mean service rate and the offered load since the distribution for each of those parameters is well known. It is also worth noting that as a practical matter one is often faced with a predetermined number of observations. In that case the critical value T_c can be easily calculated for any desired significance level α. The test is completely specified in this case by T_c and α. The β value is whatever fraction turns up when the critical time value T_c is used under conditions of the alternative hypothesis. Alternatively one may have fixed the observation interval T and simply counted the number of events in that interval. In this case T_c is known and the critical number is how many arrival events can reasonably be expected to occur in that interval. Here one might create a crude test by searching for a value of k, say k_c, such that

$$\chi_\alpha^2(2k_c) = 2\lambda_0 T \qquad (14.54)$$

The null hypothesis is rejected if $k > k_c$ events occurred during T. This test is crude in that the observation interval does not necessarily end with an arrival event as would be required for equation (14.54) to be exactly true.

MULTIPLE PARAMETER SYSTEMS

Let us now examine more general birth and death processes in which the rates λ_n and μ_n are functions of the state of the system but do not vary with time. Here we seek to make inferences about a finite number of such parameters based upon continuous observation of the system over some time period of length T. The techniques are due to Wolff (89) who made specific use of more general theorems by Billingsley (11).

The class of models to be treated can be described by a finite number of parameters (i.e., arrival and service rates) satisfying the following restrictions:

1. $\mu_0 = 0$
2. $\mu_j > 0, j = 1, 2, \ldots$
3. $\lambda_j > 0$, for all j when system capacity is unlimited
4. For systems with finite capacity k,
5. $a)$ $\lambda_j > 0$ for all $j < k$
   $b)$ $\lambda_j = 0$ for $j = k, k + 1, \ldots$
6. The mean recurrence time for each state is finite.

The sample function is similar to the one we previously analyzed for the (M/M/1) system. The state of the system at time t is $n(t)$, $t_1 = 0$ and t_{i+1} is the time at which the ith state change occurs. The number of transitions k which take place during the observation period $[0, T)$ is a random variable.

The likelihood function for a realization of that sample function is constructed in a manner parallel to that demonstrated in equations (14.7) through (14.12). The only difference is that each piece of the likelihood function must now be written as a function of the state of the system at the time a transition occurs.

1. Suppose the system is in state $j > 0$ and a transition upward occurs on the ith change of state. The probability density for this event is

$$L(i \mid j) = \lambda_j e^{-(\lambda_j + \mu_j)\tau_i} \qquad (14.55)$$

where $\tau_i = t_{i+1} - t_i$ is the time between transitions.

2. If the system is in state $j > 0$ and the ith transition is downward, the appropriate density is

$$L(i\,|\,j) = \mu_j e^{-(\lambda_j + \mu_j)\tau_i} \tag{14.56}$$

3. If the system is empty, a transition due to an arrival event at time t_{i+1} contributes

$$L(i\,|\,0) = \lambda_0 e^{-\lambda_0 \tau_i} \tag{14.57}$$

4. At the end of the period of observation $(t = T)$ the incomplete interval contributes

$$L(k\,|\,j) = e^{-(\lambda_j + \mu_j)\tau_k} \text{ if } j > 0 \tag{14.58}$$

or

$$L(k\,|\,0) = e^{-\lambda_0 \tau_k} \text{ if } j = 0 \tag{14.59}$$

Ignoring the initial state n_1, the total likelihood function can now be expressed as

$$L_c(\boldsymbol{\theta}) = \prod_{i=1}^{k} L(i\,|\,j)$$

$$= \prod_{j=1}^{\infty} \lambda_j{}^{a_j} \mu_j{}^{s_j} e^{-(\lambda_j + \mu_j)\overline{t}_j} \lambda_0{}^{a_0} e^{-\lambda_0 \overline{t}_0} \tag{14.60}$$

where

a_j = number of arrivals observed when the system was in state j
s_j = number of services observed when the system was in state j
$\overline{t}_j$ = total time spent in state j
$\boldsymbol{\theta} = (\lambda_0, \lambda_1, \ldots, \lambda_m, \mu_1, \mu_2, \ldots, \mu_m)$ = vector of parameters

The natural log of the likelihood function yields

$$\ln L_c(\boldsymbol{\theta}) = \sum_{j=0}^{\infty} a_j \ln\lambda_j + \sum_{j=1}^{\infty} s_j \ln\mu_j - \sum_{j=1}^{\infty} \overline{t}_j(\lambda_j + \mu_j) - \overline{t}_0 \lambda_0 \tag{14.61}$$

To use the log likelihood function it is necessary that we specify a finite number of parameters for the model to be estimated, take derivatives with respect to each of those parameters in turn, set the derivatives equal to zero, and solve the resulting set of equations. For example, suppose that we have a truncated single-channel system with capacity k in which we feel that service rate may vary with the number of customers present. The parameter vector for this model has $(k + 1)$ elements, $\boldsymbol{\theta} = (\lambda, \mu_1, \mu_2, \ldots, \mu_k)$. The corresponding log likelihood function is

$$\ln L_c(\boldsymbol{\theta}) = \ln \lambda \sum_{j=0}^{k-1} a_j + \sum_{j=1}^{k} s_j \ln \mu_j - \sum_{j=1}^{k-1} \overline{t}_j(\lambda + \mu_j) - \overline{t}_k \mu_k - \overline{t}_0 \lambda \tag{14.62}$$

Taking derivatives with respect to λ leads to

$$\frac{d}{d\lambda} \ln L_c(\boldsymbol{\theta}) = \frac{1}{\lambda} \sum_{j=0}^{k-1} a_j - \sum_{j=0}^{k-1} \overline{t}_j = \frac{1}{\lambda} \sum_{j=0}^{k-1} a_j - (T - \overline{t}_k) = 0$$

from which

$$\widehat{\lambda} = \sum_{j=0}^{k-1} a_j / (T - \overline{t}_k) \tag{14.63}$$

Taking derivatives with respect to μ_j leads to

$$\frac{d}{d\mu_j} \ln L_c(\boldsymbol{\theta}) = \frac{s_j}{\mu_j} - \overline{t}_j = 0$$

from which

$$\widehat{\mu}_j = s_j / \overline{t}_j \tag{14.64}$$

The maximum likelihood estimators in this case have great intuitive appeal. To find the arrival rate simply record the number of arrival events and divide by the time during which the system was unsaturated. To find the state dependent service rate μ_j simply record the number of services observed while the system occupied state j and divide by the total time it was in that state.

The most important part of the analysis goes one step further where we can use these results to design hypothesis tests concerning our vector of parameters. Suppose that we hypothesize that some vector $\boldsymbol{\theta}^0 = (\lambda^0, \mu_1^0, \mu_2^0, \ldots, \mu_k^0)$ is the true representation of the parameters. Wolff has shown that if H_0: $\boldsymbol{\theta} = \boldsymbol{\theta}^0$ is true then

$$2[\max_{\theta} L_c(\boldsymbol{\theta}) - \ln L_c(\boldsymbol{\theta}^0)] \xrightarrow{L} \chi^2(M) \tag{14.65}$$

where $\chi^2(M)$ is Chi-Square distributed with M degrees of freedom and M is the number of parameters being estimated. The symbol $\xrightarrow{L}$ indicates convergence of the function. The hypothesis test is then designed by determining a critical region such that

$$P(\chi^2(M) \geq c) = \alpha \tag{14.66}$$

If the number calculated on the left side of equation (14.65) as a result of sample observations exceeds the critical number c, the null hypothesis is rejected.

Wolff also shows that one can determine the power of the test as a function of the null hypothesis $\boldsymbol{\theta}^0$ and an alternative $\boldsymbol{\theta}^1$. This requires calculation of a vector

$$\mathbf{h} = k^{1/2}(\boldsymbol{\theta}^0 - \boldsymbol{\theta}^1) \tag{14.67}$$

the variance-covariance matrix $\boldsymbol{\sigma}(\boldsymbol{\theta})$ and a parameter

$$\Delta^2 = \mathbf{h}\boldsymbol{\sigma}(\boldsymbol{\theta}^1)\mathbf{h}'$$

where Δ is the noncentrality parameter of the noncentral χ^2 distribution. The statistic is

$$2[\max_{\boldsymbol{\theta}} \ln L_c(\boldsymbol{\theta}) - \ln L_c(\boldsymbol{\theta}^0)] \xrightarrow{\;L\;} \chi^2(M, \Delta) \qquad (14.68)$$

The power of the test is then determined from the noncentral χ^2 as

$$P(\chi^2(M, \Delta) > c) = 1 - \beta \qquad (14.69)$$

The variance-covariance matrix for some general classes of models are calculated in detail by Wolff (89).

Example. To return to our state dependent service rate, truncated model suppose that the maximum capacity is $k = 4$ and that the following data have been collected:

$$\sum_{j=0}^{3} a_j = 96, \; T = 480 \text{ minutes}$$

$$s_1 = 45, \; \overline{t}_1 = 270$$
$$s_2 = 18, \; \overline{t}_2 = 100$$
$$s_3 = 12, \; \overline{t}_3 = 40$$
$$s_4 = 16, \; \overline{t}_4 = 45$$

The implied idle time is

$$\overline{t}_0 = T - \sum_{i=1}^{4} t_i = 25 \text{ minutes}$$

Our null hypothesis is that

$$\boldsymbol{\theta}^0 = (\lambda^0, \mu_1^0, \mu_2^0, \mu_3^0, \mu_4^0) = (0.2, 0.2, 0.2, 0.25, 0.25)$$

The maximum likelihood estimates for each of these five parameters are

$$\widehat{\lambda} = 96/435 = 0.22$$
$$\widehat{\mu}_1 = 45/270 = 0.17$$
$$\widehat{\mu}_2 = 18/100 = 0.18$$
$$\widehat{\mu}_3 = 12/40 = 0.30$$
$$\widehat{\mu}_4 = 16/45 = 0.36$$

Our test statistic is

$$2[\max_{\boldsymbol{\theta}} \ln L_c(\boldsymbol{\theta}) - \ln L_c(\boldsymbol{\theta}^0)]$$

$$= 2[\ln \widehat{\lambda} \sum_{j=0}^{k-1} a_j + \sum_{j=1}^{k} s_j \ln \widehat{\mu}_j - \sum_{j=1}^{k-1} \overline{t}_j(\widehat{\lambda} + \widehat{\mu}_j) - \overline{t}_k \widehat{\mu}_k - \overline{t}_0 \widehat{\lambda}$$

$$- (\ln \lambda^0 \sum_{i=0}^{k-1} a_j + \sum_{j=1}^{k} s_j \ln \mu_j^0 - \sum_{j=1}^{k-1} \overline{t}_j(\lambda^0 + \mu_j^0) - \overline{t}_k \mu_k^0 - \overline{t}_0 \lambda^0)]$$

$$= 2[\ln (\widehat{\lambda}/\lambda^0) \sum_{j=0}^{k-1} a_j + \sum_{j=1}^{k} s_j \ln(\widehat{\mu}/\mu^0) + \sum_{j=1}^{k-1} \overline{t}_j(\lambda^0 + \mu_j^0) + \overline{t}_k \mu_k^0 + \overline{t}_0 \lambda^0$$

$$- \sum_{j=0}^{k} (a_j + s_j)]$$

Substituting the observed sample values we have

$$S = 2[\ln\left(\frac{.22}{.2}\right)(96) + 45\ln\left(\frac{.17}{.2}\right) + 18\ln\left(\frac{.18}{.2}\right) + 12\ln\left(\frac{.30}{.25}\right)$$

$$+ 16\ln\left(\frac{.36}{.25}\right) + 270(.4) + 100(.4) + 40(.45) + 45(.25) + 25(.2) - 187]$$

$$= 16.4$$

From the $\chi^2(5)$ tables we find that

$$P(\chi^2(5) > 11.07) = 0.05$$

Therefore since the sample value $S = 16.4$ is considerably larger than the critical value $c = 11.07$, we must reject the null hypothesis at the 5 percent significance level.

NESTED MODELS

Wolff has posed another test which is useful for testing restricted models under the assumption that the original model of a more general class is true (89). Suppose that ϕ is a vector of k' parameters, where $k' < k$, obtained from the components of the original model vector θ by restricting some of them. Assume that the original model is true, then if the restricted model is true the statistic S' will be Chi-Square distributed by

$$S' = 2[\max_{\theta} \ln L_c(\theta) - \max_{\phi} \ln L_c(\phi)] \xrightarrow{L} \chi^2(k - k') \quad (14.70)$$

This technique requires that we find maximum likelihood estimators for both the general and the restricted models before the test statistic can be formed.

As an example of when such a test might be useful, consider that a system in which $\mu_1 = \mu_2 = \mu_3 = \mu_4 = \mu$ is nested within our last example where the μ_i's were assumed to be unique. The vector of parameters for the restricted model is $\phi = (\lambda, \mu)$. The log likelihood for the restricted model is

$$\ln L_c(\phi) = \ln\lambda \sum_{j=0}^{3} a_j + \ln\mu \sum_{j=1}^{4} s_j - \sum_{j=1}^{3} \overline{t_j}\,(\lambda + \mu) - \overline{t_4}\mu - \overline{t_0}\lambda \quad (14.71)$$

The maximum likelihood estimator for λ is unchanged. The maximum likelihood estimator for μ becomes

$$\widehat{\mu} = \sum_{j=1}^{4} s_j/(T - t_0) \quad (14.72)$$

The test statistic (equation 14.70) can now be written from equations (14.62), (14.63), (14.64), (14.71), and (14.72) as

$$S' = 2\left[\sum_{j=1}^{4} s_j \ln\left(\widehat{\mu_j}\right) - \sum_{j=1}^{4} s_j \ln\left(\widehat{\mu}\right)\right] \quad (14.73)$$

Using the data from the last example we have

$$\widehat{\mu} = 91/455 = 0.2$$
$$\widehat{\mu}_1 = 0.17, \widehat{\mu}_2 = 0.18, \widehat{\mu}_3 = 0.3, \widehat{\mu}_4 = 0.36$$
$$s_1 = 45, s_2 = 18, s_3 = 12, s_4 = 16$$

from which

$$S' = 2\left[45 \ln\left(\frac{.17}{.2}\right) + 18 \ln\left(\frac{.18}{.2}\right) + 12 \ln\left(\frac{.3}{.2}\right) + 16 \ln\left(\frac{.36}{.2}\right)\right]$$

$$= 10.12$$

The critical value is obtained from

$$P(\chi^2(5 - 2) > 7.81) = 0.05$$

Therefore since $S' = 10.12$ exceeds 7.81 we must reject the hypothesis that $\mu_1 = \mu_2 = \mu_3 = \mu_4$ at the 5 percent level of significance. Note that this suggests that we do in fact have unique values for the μ_j parameters, although they are not necessarily the ones we posed in our previous null hypothesis. Tests of other hypotheses such as $\mu_1 = \mu_2 = 0.2$ and $\mu_3 = \mu_4 = 0.3$ might now be in order. Execution of such tests is left as an exercise for the reader.

DATA COLLECTION

Before leaving the subject of statistical analyses of queues let us pause for a moment's reflection on the data collection problem. The numbers necessary to perform all of the analyses suggested in these last two chapters must be collected by someone. Often data collection must be performed by an analyst with a stop-watch. In those cases, especially when human operators are involved, it pays for the analyst to carefully observe good time-study techniques. In other cases one may be able to profitably employ work sampling or fixed interval observation techniques. We will not take room to explore such topics here but do raise the caveat that a serious practitioner should not neglect more mundane tools such as time-study. His final models depend heavily upon accurate data which is seldom handed to him on a sheet of paper.

PROBLEMS

14.1 Consider a queueing system in which the service time distribution is Erlang of the form

$$b(t) = \frac{\mu(\mu t)^{k-1}e^{-\mu t}}{(k - 1)!}$$

Develop maximum likelihood estimates for the parameters μ and k.

14.2 Consider the log likelihood function of equation (14.16).

    a) Find an expression for $\widehat{\lambda}$ and $\widehat{\mu}$ from this equation.

b) Compare numerical values for the estimates made in this manner with estimates made using the approximations of equations (14.17) and (14.18) for $n_1 = 0, n_1 = E(n_1)$ and $n_1 = 2E(n_1)$ when $n_a = 50$, $n_s = 45, t_b = 40$, and $T = 60$.

14.3 Consider a single-channel queueing system in which there are state dependent exponential service rates, say

$$\mu_n = n^\gamma \mu, \qquad n = 1, 2, \ldots$$

and Poisson arrivals. Develop maximum likelihood estimators for the parameters, μ, μ_n, γ, and λ.

14.4 An arrival process in an (M/M/1) queueing system has been observed for three hours. During that time twenty arrivals occurred.

a) Compute a 90 percent confidence interval for the arrival rate λ.

b) Test the null hypothesis that $\lambda = 6$ per hour against the alternative that $\lambda > 6$.

14.5 Consider an (M/M/1) system in which we have observed fifty arrivals and forty-five services in a one-hour interval. During that time the server was idle for twenty minutes.

a) Estimate a 95 percent confidence interval for λ.

b) Estimate a 95 percent confidence interval for μ.

c) Estimate a 95 percent confidence interval for the traffic intensity.

d) Would you reject the hypothesis that traffic intensity is less than one? Explain.

14.6 Design an hypothesis test for a Poisson arrival process to test the hypothesis that $\lambda_0 = 9$ against the alternative $\lambda_1 = 12$ where $\alpha = .05, \beta = .10$.

14.7 Consider an (M/M/1) system in which twelve arrivals were observed over a period of one hour, and twenty services have been recorded with a mean time of three minutes.

a) Calculate a 95 percent confidence interval for the offered load.

b) Calculate a 90 percent confidence interval for expected queue length.

c) Quantify your confidence that the expected number in the system will be less than five.

14.8 Consider an (M/M/c) system in which it took two hours to accumulate thirty arrivals. There were three servers on duty and sixty customers were discharged in a separate four-hour period. During the four-hour discharge period there were no servers busy for one hour, one for twenty minutes, two for forty minutes, and three for two hours.

a) Find a 95 percent confidence for the offered load.

b) Find a 95 percent confidence interval for the utilization factor.

c) Find a 95 percent confidence interval for the expected number of customers in the queue.

14.9 Consider a single channel system $(M/E_3/1)$ where service is assumed to follow a three-Erlang distribution. The average service time has been estimated to be three minutes on the basis of twenty observations. The average interarrival time is estimated to be four minutes from a sample of thirty observations.

a) Find a 95 percent confidence interval for the offered load.

b) Find a 95 percent confidence interval for the expected number in the queue.

14.10 Consider an $(M/M/1)$ system in which the maximum likelihood estimate for the number of customers in the system remains constant at $\hat{L} = 2$. Suppose that the same number of observations are taken for both service times and interarrival times.

a) Prepare a plot of the 95 percent confidence band for L as a function of sample size.

b) What conclusions can you draw?

14.11 Consider a bank teller who is observed during an eight-hour banking day. During that time 120 customers arrive and are duly served. The teller was idle for a total of one hour and 30 minutes during this period. An $(M/M/1)$ model is proposed.

a) Estimate a 95 percent confidence interval for teller idle time in an eight-hour day.

b) Find a 90 percent confidence interval for mean queue length.

14.12 Consider a state-dependent system in which it is felt that service rate and arrival rate vary with number of customers present in the following manner:

n	μ_n	λ_n
0	0	λ_0
1	μ_1	λ_1
2	μ_1	λ_1
3	μ_2	λ_1
4	μ_2	λ_2
5	μ_3	λ_2
6	μ_3	0

a) Find the maximum likelihood estimators for $(\lambda_0, \lambda_1, \lambda_2, \mu_1, \mu_2, \mu_3)$.

b) Suppose that one hypothesizes that $\mu_1 = 1, \mu_2 = 2, \mu_3 = 3, \lambda_0 = \lambda_1 = 1, \lambda_2 = 0.5$. Further suppose that the following data are available:

n	a_n	s_n	$\bar{t}_n$ (min)
0	5	0	30
1	10	10	40
2	12	12	30
3	6	8	20
4	4	3	10
5	2	3	5
6	0	3	5

Test the hypothesis that H_0 is true.

  c) If you reject H_0 in part (*b*), propose another null hypothesis which appears plausible on the basis of observed data and test that one.

14.13 Reconsider the data of problem 14.12(*b*). Suppose that we now hypothesize the $\lambda_0 = \lambda_1 = \lambda_2 = \lambda$ and $\mu_1 = \mu_2 = \mu_3 = \mu$. Test the hypothesis that this is a proper model.

14.14 Reconsider the data from the nested model example in the text. Test the hypothesis that $\mu_1 = \mu_2 = 0.2$ and $\mu_3 = \mu_4 = 0.3$.

14.15 Prepare a paper which will offer guidelines to a fledgling analyst on when one might employ each of the following data collection techniques in preparing to analyze a general birth-death queueing system in a manufacturing plant:

  a) Make histograms from production control records.

  b) Use stop-watch time-study techniques.

  c) Use work sampling procedures.

  d) Collect data from a simulation study.

APPENDIX A

TRANSFORM
TECHNIQUES

The purpose of Appendix A is to provide the reader with a few elementary tools from operational mathematics which may expedite the solution of queueing models. The presentation which follows tends to be "cook-book" in nature. No attempt is made to provide exhaustive proofs or theoretical justification for the techniques presented. For the theory of operational mathematics one should study a text such as Churchill (18). For a more complete presentation of the probability implications of transforms, see Giffin (40).

WHY TRANSFORMS?

Transforms are used in the context of queueing models for the same reasons that logarithms are used in elementary algebra. They make complicated formulations easier to manipulate. In systems of algebraic equations one can often replace multiplication by addition of logs, division by subtraction, and exponentiation by multiplication. In systems of differential or integral equations one can often accomplish similar reduction in complexity by replacing derivatives and integrals with simple algebraic functions of transforms. The resulting algebraie expressions in transforms can then be more easily solved than the original differential or integral equations. Similar reductions in complexity can also be made for difference equations and summations which arise in systems of equations with integer arguments. We will use Laplace transforms for functions of continuous variables, geometric transforms for functions of discrete variables, and both for mixed functions such as the differential-difference equations which describe most queueing systems.

The transforms have an added bonus in the context of probability modeling. In many cases we may perform a series of calculations which lead us to an expression for the transform of a probability distribution. If our main interest is in the moments of that distribution, we can obtain the moments directly from the transform without going through the sometimes arduous task of inverting the transform.

DEFINITIONS

The "operational method" is really nothing new to anyone who has studied elementary calculus. One could think of the act of differentiation as a transformation. That is

$$D(f(t)) = f'(t) \tag{A.1}$$

implies that the operator D has "transformed" the function $f(t)$ into the function $f'(t)$. In this context the function $3t^2$ is the transform of t^3. In similar fashion the operation

$$I(f(t)) = \int_0^t f(\tau)\, d\tau \tag{A.2}$$

transforms the function $f(t)$ into a functional $F(t)$. In each case the inverse transformation exists. That is, if we know $F(t)$, we can recover $f(t)$. Furthermore, the transforms and their inverses are unique. There is only one function $f'(t)$ which can be obtained by applying the operator D to $f(t)$ and only one function $f(t)$ which can be obtained by applying the inverse transformation (antiderivative) D^{-1} to the function $f'(t)$.

The modern concept of operational mathematics is based on functional transformations. Here we will be concerned with operators such as $\mathcal{L}$ which transforms the function $f(t)$ into a function in the complex variable s by

$$\mathcal{L}(f(t)) = \int_0^\infty e^{-st} f(t)\, dt = F(s) \tag{A.3}$$

This is the Laplace transform of the function $f(t)$. Like the "D" and "I" operators from basic calculus this transformation is unique and does possess an inverse.

Unfortunately there is a certain amount of confusion generated by giving new names to old tools. This is especially true in elementary presentations of probability and statistics. Terms such as "characteristic function" and "moment generating" function are used without any hint that these are simply transforms which have long been widely understood in classical mathematics. A brief guide to terminology is given in Table A-1.

All of the transforms in Table A-1 can be used in solving queueing models. However, we will limit our attention to Laplace transforms for functions of continuous variables and geometric transforms for functions of discrete variables. These transforms are widely tabled, easy to manipulate, and illustrate techniques common to the other transforms listed.

LAPLACE TRANSFORM

Since most of the probability distributions of interest to us in the queueing theory context are zero to the left of the origin, we will limit our

TABLE A-1
TRANSFORM TERMINOLOGY FOR DENSITIES $f(t)$

Probability Theory	Classical Term	Definition
Characteristic Function	Fourier Transform	$C(s) = \displaystyle\int_{-\infty}^{\infty} e^{ist} f(t)\,dt$
Moment Generating Function	Exponential Transform	$M(s) = \displaystyle\int_{-\infty}^{\infty} e^{st} f(t)\,dt$
	Laplace Transform	$F(s) = \displaystyle\int_{0}^{\infty} e^{-st} f(t)\,dt$
Generating Function	Geometric Transform	$G(z) = \displaystyle\sum_{i=0}^{\infty} z^i f(i)$
	z-Transform	$Z(z) = \displaystyle\sum_{i=0}^{\infty} z^{-i} f(i)$

attention to the unilateral Laplace transform. The unilateral Laplace transform of a function $f(t)$ is defined as

$$\mathcal{L}(f(t)) = F(s) = \int_{0}^{\infty} f(t) e^{-st}\,dt \tag{A.4}$$

The inverse transformation, which requires complex variable theory, is defined as

$$F^{-1}(s) = f(t) = \frac{1}{2\pi i} \lim_{b \to \infty} \int_{a-ib}^{a+ib} F(s) e^{st}\,ds \tag{A.5}$$

The variable s is a complex variable of the form

$$s = a + i\omega \tag{A.6}$$

Fortunately we are seldom required to perform the contour integration implied by equation (A.5). Instead we can employ a small set of theorems and basic tables of transform pairs to perform the inversion. This table look-up philosophy is the same one used in solving basic calculus problems. Once the terms are defined, one rarely goes back to the formal limit arguments to integrate a function, for example. Rather one attempts to break the function into its simplest components which are then integrated by finding appropriate pairs of functions in a table of integrals. In like manner we will break functions and their transforms into simple components which can be transformed or inverted by finding appropriate pairs in a table of transform pairs.

In most of our calculations the complex variable s is treated as a dummy variable. When we are required to evaluate particular transforms we will normally restrict our attention to values of s in the interval $[0, 1]$. Not every function has a Laplace transform. Those which do are said to be of exponential order; that is, they do not grow more rapidly than ce^{at}

where c is a constant and the real portion of s exceeds a. The probability distributions of interest to us are of exponential order.

PROPERTIES OF LAPLACE TRANSFORMS

Linearity

A fundamental property of Laplace transforms, and indeed of all transforms used in this text is that they are linear operators. This means that if $f_1(t)$ and $f_2(t)$ are functions to be transformed by the operator T, then for any arbitrary constants c_1 and c_2 it is true that

$$T(c_1 f_1(t) + c_2 f_2(t)) = c_1 T(f_1(t)) + c_2 T(f_2(t)) \qquad \text{(A.7)}$$

This is an extremely important property which often enables us to find transforms and inverse transforms by decomposing a complicated function into a linear combination of easily recognized components.

For example, suppose that one wants to find the transform of

$$f(t) = 5t + 3\sin(6t) \qquad \text{(A.8)}$$

Because of the linear property of Laplace transforms we can write this as

$$\mathcal{L}(f(t)) = 5\mathcal{L}(t) + 3\mathcal{L}(\sin(6t)) \qquad \text{(A.9)}$$

As this point we can either compute the component transforms from the definition of equation (A.4) or refer to a table of transform pairs. In either event we find

$$\mathcal{L}(t) = \frac{1}{s^2} \text{ and } \mathcal{L}(\sin(6t)) = \frac{6}{s^2 + 36}$$

It therefore follows that

$$\mathcal{L}(f(t)) = \frac{5}{s^2} + \frac{18}{s^2 + 36} \qquad \text{(A.10)}$$

Conversely, suppose that we have derived a transform which we want to invert, say

$$F(s) = \frac{s + 1}{s^2 + 4s} \qquad \text{(A.11)}$$

This can be rewritten as

$$F(s) = \frac{1}{4}\left(\frac{1}{s}\right) + \frac{3}{4}\left(\frac{1}{s + 4}\right) \qquad \text{(A.12)}$$

The inverse $f(t) = F^{-1}(s)$ can then be determined either by contour integration for each of the components via equation (A.5) or by table look-up. By either method

$$F^{-1}(s) = \frac{1}{4}t + \frac{3}{4}e^{-4t} \qquad \text{(A.13)}$$

Tables of transform pairs for simple functions can be easily generated by carrying out the integration implied by the definition. For example in queueing theory one frequently encounters the exponential distribution

$$f(t) = \mu e^{-\mu t}, \qquad t > 0 \tag{A.14}$$

The Laplace transform of this distribution is

$$\mathcal{L}(f(t)) = \int_0^\infty \mu e^{-\mu t} e^{-st}\, dt$$

$$= \frac{\mu e^{-(\mu+s)t}}{-(\mu+s)}\bigg|_{t=0}^{\infty} = \frac{\mu}{s+\mu} \tag{A.15}$$

Having once performed the integration we can place this pair of functions in our table and at all times thereafter solve for the transform and inverse transform of any exponential distribution by simple table look-up.

Partial Fractions

It is important to have techniques for decomposing complicated functions into linear combinations of simpler components which we can locate in the tables. One way to do this is by the method of *partial fraction expansion.*

We often encounter transforms which take the form of a quotient of polynomials, say

$$F(s) = \frac{A(s)}{B(s)} \tag{A.16}$$

where $A(s)$ and $B(s)$ are polynomials in s with the degree of the numerator less than the degree of the denominator. If the denominator can be factored into a product of distinct linear terms, we know from algebra that the fraction can be expressed as a sum of separate terms each having as its denominator a factor of the original denominator. That is

$$F(s) = \frac{A(s)}{(s-a_1)(s-a_2)\cdots(s-a_k)} = \frac{c_1}{s-a_1} + \frac{c_2}{s-a_2} + \cdots \frac{c_k}{s-a_k} \tag{A.17}$$

The roots of the denominator are called the *poles* of the transform. If there are k distinct roots, the transform is said to have k *simple* or *first order poles.* When a factor is repeated n times the transform is said to have an nth order pole. The partial fraction expansion for such a transform then contains n terms for each nth order pole. For example,

$$F(s) = \frac{A(s)}{(s-a_1)^n(s-a_2)\cdots(s-a_k)}$$

$$= \frac{c_{11}}{(s-a_1)^n} + \frac{c_{12}}{(s-a_1)^{n-1}} + \cdots + \frac{c_{1n}}{(s-a_1)} + \frac{c_2}{s-a_2}$$

$$+ \cdots + \frac{c_k}{s-a_k} \tag{A.18}$$

where a_1 is an nth order pole and all other a_i's are first order poles.

The coefficients in the partial fraction expansion are most easily calculated from the following limit arguments. For a first order pole at $s = a_i$

$$c_i = \lim_{s \to a_i} (s - a_i) F(s) \tag{A.19}$$

For an nth order pole at $s = a_i$, there are n coefficients

$$c_{ij} = \frac{1}{(j-1)!} \lim_{s \to a_i} \frac{d^{j-1}}{ds^{j-1}} (s - a_i)^n F(s) \tag{A.20}$$

where

$$0 < j \leq n$$

For example suppose that the transform to be inverted is

$$F(s) = \frac{s+1}{s(s+2)^2} \tag{A.21}$$

This transform has a first order pole at $s = 0$ and a second order pole at $s = -2$. From the partial fraction expansion we know

$$F(s) = \frac{c_1}{s} + \frac{c_{21}}{(s+2)^2} + \frac{c_{22}}{s+2} \tag{A.22}$$

where

$$c_1 = \lim_{s \to 0} s \left[\frac{s+1}{s(s+2)^2} \right] = \frac{1}{4}$$

$$c_{21} = \lim_{s \to -2} (s+2)^2 \left[\frac{s+1}{s(s+2)^2} \right] = \frac{1}{2}$$

$$c_{22} = \lim_{s \to -2} \frac{d}{ds} \left[(s+2)^2 \left(\frac{s+1}{s(s+2)^2} \right) \right] = -\frac{1}{4}$$

It therefore follows that

$$F(s) = \frac{1}{4} \left(\frac{1}{s} \right) + \frac{1}{2} \left(\frac{1}{(s+2)^2} \right) - \frac{1}{4} \left(\frac{1}{s+2} \right) \tag{A.23}$$

The inverse is readily determined by table look-up as

$$f(t) = \frac{1}{4} + \frac{1}{2} t e^{-2t} - \frac{1}{4} e^{-2t} \tag{A.24}$$

Differentiation And Integration

Laplace transforms are especially valuable in simplifying certain operations involving derivatives and integrals. The transform of the first derivative of a function $f(t)$ can be written as

$$\mathcal{L}(f'(t)) = s\mathcal{L}(f(t)) - f(0) \tag{A.25}$$

where $f(0)$ represents the boundary condition at $t = 0$. The transform for higher order derivatives generalizes to

$$\mathcal{L}(f^{(n)}(t)) = s^n \mathcal{L}(f(t)) - s^{n-1}f(0) \cdots - sf^{(n-2)}(0) - f^{(n-1)}(0) \qquad \text{(A.26)}$$

where

$$f^{(n)}(t) = \frac{d^n f(t)}{dt^n}$$

and there are n boundary conditions; i.e., values of derivatives of the function at $t = 0$.

Certain types of integration can also be performed quickly by transforms. For example, the transform of the cumulative probability distribution is related to the transform of the density by

$$\mathcal{L}\left(\int_0^t f(\tau)\, d\tau\right) = \frac{1}{s}\,\mathcal{L}(f(t)) \qquad \text{(A.27)}$$

A second type of integral frequently encountered in queueing problems is the convolution integral. The convolution operation can be replaced by a simple product of transforms as

$$\mathcal{L}\left(\int_0^t f(\tau)\, g(t - \tau)\, d\tau\right) = \mathcal{L}(f(t))\mathcal{L}(g(t)) \qquad \text{(A.28)}$$

From the viewpoint of probability modeling, this is perhaps the most important property of transforms. We will have much more to say about such integrals in future sections.

To demonstrate the power of these techniques consider the following integro-differential equation which arises in the analysis of certain classes of circuits.

$$L\frac{dI(t)}{dt} + RI(t) + \frac{1}{C}\int_0^t I(\tau)\, d\tau = k \qquad \text{(A.29)}$$

The problem is to find the function $I(t)$ which satisfies the equation. We must first form the subsidiary equation by transforming equation (A.29). To do this we will use the properties of linearity (A.7), transform of derivatives (A.26), and transform of integrals (A.27). The subsidiary equation then becomes

$$L[s\mathcal{L}(I(t)) - I(0)] + R\mathcal{L}(I(t)) + \frac{1}{Cs}\mathcal{L}(I(t)) = \frac{k}{s} \qquad \text{(A.30)}$$

Assuming that the system is initially at rest, $I(0) = 0$, we now have a simple algebraic equation in $\mathcal{L}(I(t))$. Solving the subsidiary equation yields

$$\mathcal{L}(I(t)) = \frac{kC}{LCs^2 + RCs + 1} \qquad \text{(A.31)}$$

The desired function $I(t)$ can now be recovered by the usual inversion techniques.

Other operations including shifting, multiplication, division, and limits are also possible to perform with Laplace transforms. These operations are summarized in Table A-4.

GEOMETRIC TRANSFORMS

Geometric transforms provide the same capabilities for functions of discrete variables that Laplace transforms provide for functions of continuous variables. We will again limit our attention to those functions which are zero to the left of the origin. The geometric transform of the function $f(n)$ is then defined by the sum

$$G(f(n)) = G_n(z) = \sum_{n=0}^{\infty} f(n)z^n \tag{A.32}$$

The transform variable z, like the complex variable s in the Laplace transform, will be treated as a dummy. When we are required to evaluate particular functions of the transform we normally limit our attention to real values of z which fall in the interval $[0, 1]$.

Whenever the sum in equation (A.32) has a finite upper limit, the series does converge and consequently the transform exists. When the number of terms is infinite, not all functions $f(n)$ can be transformed. One can test for convergence by using the ratio test. That is, given the series $\sum_{i=0}^{\infty} u$, find the limit

$$L = \lim_{n \to \infty} \frac{u_{n+1}}{u_n} \tag{A.33}$$

If $L < 1$, the series is absolutely convergent. If $L > 1$, the series is divergent. If $L = 1$, no conclusion can be drawn. For most probability functions encountered in queueing models, convergence can be assumed without test. Only when difficulties in finding identities are encountered will we resort to explicit tests for convergence.

Inverting a geometric transform, at least for low-order terms, may be an easier task than inverting a Laplace transform. One can often recover the first few terms of the function $f(n)$ by long division or by Maclaurin series expansion of the transform. If a general formula for $f(n)$ is desired, we can use the same techniques of partial fraction expansion and table look-up suggested in the previous section for Laplace transforms.

For example, suppose that we have developed the transform

$$G_n(z) = \frac{0.2}{1 - 0.8z} \tag{A.34}$$

Since this is a ratio of polynomials in z we can divide the numerator by the denominator to obtain

$$
\begin{array}{r}
0.2 + 0.16z + 0.128z^2 + \cdots \\[4pt]
\hline
1 - 0.8z \overline{)\,0.2} \\
0.2 - 0.16z \\
\hline
0.16z \\
0.16z - 0.128z^2 \\
\hline
0.128z^2 \\
0.128z^2 - .1024z^3 \\
\hline
\cdots
\end{array}
$$

From the definition of $G_n(f(n))$ we know that $f(n)$ must be the coefficient of z^n in the summation developed by long division above. It must therefore follow that $f(0) = 0.2$, $f(1) = 0.16$, $f(2) = 0.128\ldots$. In this manner we can recover as much of the function $f(n)$ as we have patience for doing long division. One can verify these results by consulting the tables to find that the general term for this particular function is

$$G_n^{-1}(z) = f(n) = 0.2(0.8)^n \tag{A.35}$$

As a second example for which long division is *not* helpful consider the transform

$$G_n(z) = e^{-5(1-z)} \tag{A.36}$$

We can evaluate low-order terms of $f(n)$ for this function by the Maclaurin series.

$$G_n(z) = \sum_{n=0}^{\infty} \frac{1}{n!} \frac{d^n G(z)}{dz^n} \bigg|_{z=0} z^n \tag{A.37}$$

Again appealing to the definition of $G_n(z)$ we know that $f(n)$ should appear as the coefficient of z^n in its power series expansion. That is

$$f(n) = \frac{1}{n!} \frac{d^n G_n(z)}{dz^n} \bigg|_{z=0} \tag{A.38}$$

For the problem at hand we have

$$f(0) = e^{-5(1-z)} \bigg|_{z=0} = e^{-5}$$

$$f(1) = \frac{5e^{-5(1-z)}}{1} \bigg|_{z=0} = 5e^{-5}$$

$$f(2) = \frac{25e^{-5(1-z)}}{2} \bigg|_{z=0} = \frac{25}{2} e^{-5}$$

In this manner we can recover as much of the function $f(n)$ as we have patience for taking derivatives. These results can be verified by checking the tables where we find that the general term is

$$G_n^{-1}(z) = f(n) = \frac{(5)^n}{n!} e^{-5} \tag{A.39}$$

PROPERTIES OF GEOMETRIC TRANSFORMS

Geometric transforms exhibit properties which closely parallel those of Laplace transforms. To begin, the geometric transform is a linear operator. This means that

$$G(c_1 f_1(n) + c_2 f_2(n)) = c_1 G(f_1(n)) + c_2 G(f_2(n)) \qquad \text{(A.40)}$$

It follows that one can often find transforms and inverse transforms by decomposing complicated functions into simpler components which can be located in the tables.

The entries in the tables are developed by applying the definition to simple functions. For example suppose that we want the geometric transform of the distribution

$$f(n) = 0.3(0.7)^n, \; n = 0, 1, \ldots \qquad \text{(A.41)}$$

By definition

$$G_n(z) = \sum_{n=0}^{\infty} 0.3(0.7z)^n \qquad \text{(A.42)}$$

But from the calculus of finite differences we know that

$$\sum_{n=a}^{b} r^n = \left. \frac{r^n}{r-1} \right|_{n=a}^{b+1} \qquad \text{(A.43)}$$

Therefore for a proper range of z we have

$$G_n(z) = 0.3 \lim_{b \to \infty} \left[\frac{(0.7z)^{b+1}}{0.7z - 1} - \frac{(0.7z)^0}{0.7z - 1} \right]$$

$$= \frac{0.3}{1 - 0.7z} \qquad \text{(A.44)}$$

This transform pair can now be entered in the tables for future reference so that we never need to recalculate this particular sum or its inverse.

Differences And Sums

The discrete analogs of derivatives and integrals are differences and sums. The first forward difference of the function $f(n)$ is defined as

$$\Delta f(n) = f(n + 1) - f(n) \qquad \text{(A.45)}$$

The geometric transform of this function can be shown to be

$$G(\Delta f(n)) = (z^{-1} - 1)G(f(n)) - z^{-1}f(0) \qquad \text{(A.46)}$$

In similar fashion the transform for differences involving higher order shifts can be written as a function of $G(f(n))$ as

$$G(f(n + r)) = z^{-r} \left[G(f(n)) - \sum_{i=0}^{r-1} g(i)z^i \right] \qquad \text{(A.47)}$$

Summations, which are the discrete analogs of integrals can also be expressed as functions of transforms. For example, the transform of the cumulative probability distribution for a discrete random variable is related to the transform of its probability mass function by

$$G\left(\sum_{i=0}^{n} f(i)\right) = \frac{1}{1-z} G(f(n)) \qquad (A.48)$$

However, a more important summation from the point of view of probability modeling is the convolution sum. This summation is easily manipulated with geometric transforms since

$$G\left[\sum_{i=0}^{n} f(i)g(n-i)\right] = G(f(n))G(g(n)) \qquad (A.49)$$

That is we can write the transform of the convolution sum as a product of transforms of the functions being convolved.

Differential-Difference Equations

Many of the models of queueing systems are written in the form of an infinite set differential-difference equation. For example,

$$\frac{dp_0(t)}{dt} = -\lambda p_0(t)$$

$$\frac{dp_n(t)}{dt} = \lambda p_{n-1}(t) - \lambda p_n(t), \qquad n \geq 1 \qquad (A.50)$$

is a set of equations containing both derivatives and differences which is continuous in time, t, and discrete in space, n. Such equations are called differential-difference equations. A common solution strategy is to reduce the infinite set of equations to a single partial differential equation in transforms, solve that subsidiary equation, then invert twice to find the unknown function $p_n(t)$.

Applying this strategy to the set (A.50), let us first multiply each equation containing a derivative whose subscript is n by z^n and add over all n. That yields

$$\sum_{n=0}^{\infty} \frac{dp_n(t)}{dt} z^n = \lambda \sum_{n=0}^{\infty} p_n(t)z^{n+1} - \lambda \sum_{n=0}^{\infty} p_n(t) \qquad (A.51)$$

Defining the geometric transform of the function $p_n(t)$ as

$$G(z, t) = \sum_{n=0}^{\infty} p_n(t)z^n \qquad (A.52)$$

we can recognize that

$$\frac{\partial G(z, t)}{\partial t} = \sum_{n=0}^{\infty} \frac{dp_n(t)}{dt} z^n \qquad (A.53)$$

Therefore equation (A.51) can be rewritten as

$$\frac{\partial G(z, t)}{\partial t} = \lambda(z - 1)G(z, t) \tag{A.54}$$

Taking Laplace transforms of both sides yields

$$s\mathcal{L}(G(z, t)) - G(z, 0) = \lambda(z - 1)\mathcal{L}(G(z, t)) \tag{A.55}$$

from which

$$\mathcal{L}(G(z, t)) = \frac{G(z, 0)}{s - \lambda(z - 1)} \tag{A.56}$$

Assuming that the system is initially empty we know that

$$G(z, 0) = 1 \tag{A.57}$$

Therefore equation (A.56) becomes

$$\mathcal{L}(G(z, t)) = \frac{1}{s - \lambda(z - 1)} \tag{A.58}$$

Inverting the Laplace transform gives

$$G(z, t) = e^{\lambda(z-1)t} \tag{A.59}$$

Inverting the resulting geometric transform gives

$$p_n(t) = \frac{(\lambda t)^n}{n!} e^{-\lambda t} \tag{A.60}$$

This is a reasonable solution procedure for many time-varying stochastic processes which can be described by infinite sets of differential-difference equations.

MOMENTS OF DISTRIBUTIONS

An important property of transforms of probability distributions is that the moments of the distribution can be recovered directly from the transform by taking derivatives. It is not necessary to first invert the transform and then operate on the resulting distribution.

Consider first a continuous random variable, t, distributed by $f(t)$. The Laplace transform of the density function is defined as

$$\mathcal{L}(f(t)) = F_t(s) = \int_0^\infty e^{-st} f(t) \, dt \tag{A.61}$$

If this transform has been obtained by extensive algebraic manipulation, it is sometimes convenient to check whether the resulting function in s can possibly be the transform of a legitimate probability function. If we evaluate $\mathcal{L}(f(t)) = F_t(s)$ at $s = 0$, we should have

$$F_t(0) = \int_0^\infty e^{0t} \cdot f(t) \, dt = 1 \tag{A.62}$$

If the function being considered does not go to unity as s goes to zero, it cannot possibly be the transform of a density function. This property is also useful when one has a transform completely specified except for one unknown. It may be possible to set $s = 0$ and solve the resulting expression $F_t(0) = 1$ for the unknown parameter.

If we take the first derivative of $F_t(s)$ then evaluate at $s = 0$, we have

$$\frac{dF_t(s)}{ds} = \int_0^\infty - te^{-st} f(t)\, dt \Big|_{s=0}$$

$$= - \int_0^\infty tf(t)\, dt$$

$$= -E(t) \tag{A.63}$$

We have thereby "generated" the first moment of the distribution. Continuing to higher order moments, it is easy to show that the rth moment can be obtained from

$$\frac{d^r F_t(s)}{ds^r} \Big|_{s=0} = (-1)^r E(t^r) \tag{A.64}$$

For example suppose that we are given the transform

$$F_t(s) = \left(\frac{3}{s+3}\right)^2 \tag{A.65}$$

First check whether it satisfies the necessary condition that $F_t(0) = 1$.

$$F_t(0) = \left(\frac{3}{0+3}\right)^2 = 1$$

Next find the mean of the distribution.

$$E(t) = (-1)\frac{dF_t(s)}{ds}\Big|_{s=0}$$

$$= (-1)\left[2\left(\frac{3}{s+3}\right)\left(\frac{-3}{(s+3)^2}\right)\right]\Big|_{s=0}$$

$$= 2/3$$

Similar calculations are also possible for geometric transforms of discrete distributions. Consider the integer valued random variable n distributed by $p(n)$. The geometric transform of $p(n)$ is described by

$$G(p(n)) = G_n(z) = \sum_{n=0}^\infty z^n p(n) \tag{A.66}$$

If it is a proper transform $G_n(z)$ evaluated at $z = 1$ should be unity, that is

$$G_n(1) = \sum_{n=0}^\infty 1^n p(n) = 1 \tag{A.67}$$

If we take the first derivative of $G_n(z)$ evaluated at $z = 1$, we have

$$\left.\frac{dG_n(z)}{dz}\right|_{z=1} = \sum_{n=0}^{\infty} nz^n p(n)\,\bigg|_{z=1}$$
$$= E(n) \tag{A.68}$$

Thus we can "generate" the mean of the distribution by taking the derivative with respect to z.

Higher order moments require slightly more computation with geometric transforms than with Laplace. For example, taking the second derivative of $G_n(z)$ we have

$$\left.\frac{d^2G_n(z)}{dz^2}\right|_{z=1} = \sum_{n=0}^{\infty} n(n-1)z^{n-2}p(n)\,\bigg|_{z=1}$$
$$= E(n(n-1)) \tag{A.69}$$

This is the second factorial moment of the distribution. However, it is apparent that the usual power moment can be readily recovered by

$$E(n^2) = E(n(n-1)) + E(n)$$
$$= \left.\frac{d^2G_n(z)}{dz^2}\right|_{z=1} + E(n) \tag{A.70}$$

If we define x, r-factorial to be

$$x^{(r)} = \frac{x!}{(x-r)!} \tag{A.71}$$

we can find the rth factorial moment by

$$\left.\frac{d^rG_n(z)}{dz^r}\right|_{z=1} = E(n^{(r)}) \tag{A.72}$$

If power moments are needed they can be recovered by algebraically expanding the resulting factorial expression.

ADDING RANDOM VARIABLES

Consider the problem of determining the distribution for the sum of two independent random variables, say

$$y = x_1 + x_2 \tag{A.73}$$

If x_1 and x_2 are continuously distributed by $f_1(x_1)$ and $f_2(x_2)$, it follows that the distribution of y is given by the convolution integral

$$h(y) = \int_0^y f_1(x)f_2(y-x)\,dx \tag{A.74}$$

If x_1 and x_2 are discrete, the distribution for y is given by the convolution sum

$$h(y) = \sum_{x=0}^{y} f_1(x)f_2(y-x) \tag{A.75}$$

We noted earlier that these convolution operations could be replaced by simple products in transform domain. It therefore follows that

$$T(h(y)) = T(f_1(x_1))T(f_2(x_2)) \tag{A.76}$$

where the "T" operator is either $\mathcal{L}$ or G depending upon whether continuous or discrete variables are involved. If there are n terms in the summation (A.73) and all random variables are identically distributed by $f(x)$, we can write

$$T(h(y)) = [T(f(x))]^n \tag{A.77}$$

There is an interesting extension to this problem which occurs when the number of terms in the sum is itself a random variable. Let

$$y = x_1 + x_2 + \cdots + x_n \tag{A.78}$$

be such a random sum of random variables where x_i is distributed by $f(x)$ for all i and n is distributed by $p(n)$. The variable x can be either continuous or discrete but n must be discrete. We could then treat equation (A.77) as a conditional transform, i.e., $T(h(y)|n)$ is the transform of $h(y)$ for a fixed n. The unconditional transform can then be obtained by weighting each term by $p(n)$ and summing over all n. That is

$$T(h(y)) = \sum_{n=0}^{\infty} T(h(y)|n)p(n)$$

$$= \sum_{n=0}^{\infty} [T(f(x))]^n p(n) \tag{A.79}$$

But the resulting function looks similar to the geometric transform for $p(n)$ in which the z argument has been replaced by the transform $T(f(x))$. We can therefore represent the transform for a random sum of random variables as

$$T(h(y)) = G_n(z) \Big|_{z = T(f(x))}$$

$$= G_n[T(f(x))] \tag{A.80}$$

For example, suppose that x is exponentially distributed such that

$$T(f(x)) = \frac{5}{s + 5}$$

and n is Poisson distributed such that

$$G_n(z) = exp(-15(1 - z))$$

It then follows that $y = x_1 + x_2 + \cdots + x_n$ is distributed such that

$$T(h(y)) = \mathcal{L}(h(y)) = exp\left(-15\left(1 - \frac{5}{s + 5}\right)\right)$$

Once the transforms of complicated functions of random variables are known one can then compute the moments of the distributions by taking the appropriate derivatives. Fortunately one can solve for the mean and variance in general terms for random sums from equation (A.80). The result is that

$$\mu_y = \mu_n \mu_x \qquad\qquad (A.81)$$

$$\sigma_y^2 = \mu_n \sigma_x^2 + \mu_x^2 \sigma_n^2 \qquad\qquad (A.82)$$

This means that one can compute the mean and variance for a random sum directly from the moments of the component variables in that sum. One does not need to know the transform or even the probability distribution to obtain these important statistics.

SUMMARY

We have attempted to highlight some of the basic properties of operational methods as they apply to models of queueing systems. The subject is much too broad to think that this short appendix makes one an expert. However, the author does feel that this brief introduction, careful examination of the identities and theorems in Tables A-2, A-3, A-4, and A-5, plus practice, should give the reader an adequate grasp to follow most of the discussion in this textbook. The theme throughout is to use the tables. Transforms are useful only as long as they reduce computational burden.

TABLE A-2
TRANSFORMS OF COMMON PROBABILITY DISTRIBUTIONS

Distribution	Equation	Transform	Type
Exponential	$\mu e^{-\mu t},\ t > 0$	$\dfrac{\mu}{s + \mu}$	Laplace
Rectangular	$(b - a)^{-1},\ a \le t \le b$	$\dfrac{e^{-as} - e^{-bs}}{s(b - a)}$	Laplace
Chi Square	$\dfrac{1}{2^k \Gamma(k)}\, t^{k-1} e^{-t/2},\ t > 0$	$\left(\dfrac{1/2}{s + 1/2}\right)^k$	Laplace
k-Erlang	$\dfrac{(k\mu)^k (t)^{k-1}}{\Gamma(k)}\, e^{-k\mu t},\ t > 0$	$\left(\dfrac{k\mu}{s + k\mu}\right)^k$	Laplace
Gamma	$\dfrac{t^{a-1} e^{-t/b}}{\Gamma(a) b^a},\ t > 0$	$\left(\dfrac{b^{-1}}{s + b^{-1}}\right)^a$	Laplace
Constant	$\delta(t - a),\ t = a$	e^{-sa}	Laplace
Geometric	$p(1 - p)^n,\ n = 0, 1, 2 \ldots$	$\dfrac{p}{1 - (1 - p)z}$	Geometric
Poisson	$\dfrac{\mu^n e^{-\mu}}{n!},\ n = 0, 1, 2 \ldots$	$e^{-\mu(1-z)}$	Geometric
Binomial	$\binom{k}{n} p^n (1 - p)^{k-n},$ $n = 0, 1, \ldots, k$	$[pz + (1 - p)]^k$	Geometric
Negative Binomial	$\binom{n + r - 1}{r - 1} p^r (1 - p)^n,$ $n = 0, 1, 2 \ldots$	$\left[\dfrac{p}{1 - (1 - p)z}\right]^r$	Geometric

TABLE A-3
PROBABILITY IMPLICATIONS
T = GENERAL TRANSFORM OPERATOR; $G_n(z)$ = GEOMETRIC TRANSFORM; $F_t(s)$ = LAPLACE TRANSFORM

Operation	Transform Computation
1. $y = x_1 + x_2$ x_1 and x_2 independent	$T(g(y)) = T(f_1(x_1))T(f_2(x_2))$
2. $y = \sum_{i=1}^{n} x_i$ x_i's independent	$T(g(y)) = \prod_{i=1}^{n} T(f_i(x_i))$
3. $y = x_1 + x_2 + \cdots + x_n$ x_i's independent and identically distributed; n discrete	$T(g(y)) = G_n(T(f(x)))$
4. $y = ax$ a constant; x continuous	$F_y(s) = F_x(s)\Big|_{s=as} = F_x(as)$
5. $y = ax + b$ a and b constant; x continuous	$F_y(s) = e^{-sb}F_x(as)$
6. $y = a_1x_1 + a_2x_2 + \cdots + a_nx_n + b$ a_i's and b constant; x_i's continuous and independent	$F_y(s) = e^{-sb}[F_{x_1}(a_1s)F_{x_2}(a_2s) \ldots F_{x_n}(a_ns)]$
7. $pr(x \leq k) = \int_0^k f(x)\,dx$ x continuous	$\mathcal{L}\left(\int_0^x f(r)\,dr\right) = \frac{1}{s}F_x(s)$
8. $pr(n \leq k) = \sum_{n=0}^{k} p(n)$ n discrete	$G\left(\sum_{i=0}^{n} p(i)\right) = \frac{1}{1-z}G_n(z)$
9. $h(n) = \int_0^{\infty} p(n\,|\,t)f(t)\,dt$ $n\,|\,t$ Poisson distributed with mean λt; t continuous	$G_n(z) = F_t(\lambda(1-z))$
10. $E(x^r)$ x continuous	$E(x^r) = (-1)^r \frac{d^r F_x(s)}{ds^r}\Big|_{s=0}$
11. $E(n^{(r)})$ n discrete; factorial moments	$E(n^{(r)}) = \frac{d^r G_n(z)}{dz^r}\Big|_{z=1}$
12. σ_n^2 n discrete	$\sigma_n^2 = G_n''(1) + G_n'(1) - (G_n'(1))^2$
13. $E(y = x_1 + x_2 + \cdots + x_n)$ x_i's independent and identically distributed; n discrete	$E(y) = \mu_n\mu_x$
14. $\mathrm{Var}\,(y = x_1 + x_2 + \cdots + x_n)$ x_i's independent and identically distributed; n discrete	$\sigma_y^2 = \mu_n\sigma_x^2 + \mu_x^2\sigma_n^2$

TABLE A-4
LAPLACE TRANSFORM THEOREMS AND IDENTITIES

$f(t)$	$\mathcal{L}(f(t)) = F(s) = \int_0^\infty f(t)e^{-st}\,dt$
1. $ag(t) + bh(t)$	$aG(s) + bH(s)$
2. $\dfrac{d^n g(t)}{dt^n}$	$s^n G(s) - s^{n-1}g(0) - s^{n-2}g^{(1)}(0) \ldots$ $\quad - g^{(n-1)}(0)$ where $g^{(i)}(t) = \dfrac{d^i g(t)}{dt^i}$
3. $\displaystyle\int_0^t g(r)\,dr$	$\dfrac{1}{s}G(s)$
4. $\displaystyle\int_0^t g(r)h(t-r)\,dr$	$G(s)H(s)$
5. $e^{at}g(t)$	$G(s-a)$
6. $g(t-a)U(t-a)$ where $U(t)$ is the unit step function	$e^{-as}G(s)$
7. $t^n g(t)$	$(-1)^n \dfrac{d^n}{ds^n}G(s)$
8. $\dfrac{1}{t}g(t)$	$\displaystyle\int_s^\infty G(s)\,ds$
9. $g(ct),\ c > 0$	$\dfrac{1}{c}G\left(\dfrac{s}{c}\right)$
10. $\lim\limits_{t\to 0} f(t)$	$\lim\limits_{s\to\infty} sF(s)$
11. $\lim\limits_{t\to\infty} f(t)$	$\lim\limits_{s\to 0} sF(s)$ if all poles for $F(s)$ lie in left half of the s-plane
12. $\delta(t)$ where $\displaystyle\int_{-\infty}^\infty \delta(t)\,dt = 1$ and $\delta(t) = 0,\ t \neq 0$	1
13. c	$\dfrac{c}{s}$
14. $t^n,\ n = 1, 2, 3, \ldots$	$\dfrac{n!}{s^{n+1}}$
15. $t^a,\ a > 0$	$\dfrac{\Gamma(a+1)}{s^{a+1}}$
16. e^{at}	$\dfrac{1}{s-a}$
17. te^{at}	$\dfrac{1}{(s-a)^2}$
18. $\dfrac{t^{n-1}e^{at}}{\Gamma(n)}$	$\dfrac{1}{(s-a)^n}$
19. $U(t-a) = \begin{cases} 0 & t < a \\ 1 & t \geq a \end{cases}$	$\dfrac{e^{-as}}{s}$
20. $[g_{ij}(t)]$, a matrix	$[G_{ij}(s)]$
21. $\sin(at)$	$\dfrac{a}{s^2 + a^2}$
22. $\cos(at)$	$\dfrac{s}{s^2 + a^2}$

TABLE A-5
GEOMETRIC TRANSFORM THEOREMS AND IDENTITIES

$p(n)$	$G(p(n)) = G_n(z) = \sum\limits_{n=0}^{\infty} p(n)z^n$
1. $ag(n_1) + bh(n_2)$	$aG_{n_1}(z) + bG_{n_2}(z)$
2. $\Delta g(n)$	$(z^{-1} - 1)G_n(z) - z^{-1}g(0)$
3. $\sum\limits_{i=0}^{n} g(i)$	$\dfrac{1}{1-z}G_n(z)$
4. $\sum\limits_{i=0}^{n} g(i)h(n-i)$	$G(g(n))G(h(n))$
5. $g(n+r)$	$z^{-r}\left(G_n(z) - \sum\limits_{i=0}^{r-1} g(i)z^i\right)$
6. $g(n-r)U(n-r)$ where $U(n)$ is the unit step function	$z^r G_n(z)$
7. $k^{rn}g(n)$	$G_n(k^r z)$
8. $\lim\limits_{n\to 0} g(n)$	$G_n(z)\Big|_{z=0}$
9. $\lim\limits_{n\to\infty} g(n)$	$\lim\limits_{z\to 1}(1-z)G_n(z)$ if all poles for $G_n(z)$ lie outside the unit circle
10. 1	$\dfrac{1}{1-z}$
11. c	$\dfrac{c}{1-z}$
12. c^n	$\dfrac{1}{1-cz}$
13. n	$\dfrac{z}{(1-z)^2}$
14. n^r	$\left(z\dfrac{d}{dz}\right)^r \dfrac{1}{1-z}$
15. $n^{(r)}$	$\dfrac{r!z^r}{(1-z)^{r+1}}$
16. $\dfrac{c^n}{n!}$	e^{cz}
17. nc^n	$\dfrac{cz}{(1-cz)^2}$
18. $\sin(cn)$	$\dfrac{z\sin c}{z^2 + 1 - 2z\cos c}$
19. $\cos(cn)$	$\dfrac{1 - z\cos c}{z^2 + 1 - 2z\cos c}$
20. $[g_{ij}(n)]$, a matrix	$[G(g_{ij}(n))]$

APPENDIX B

STEADY-STATE BIRTH-DEATH PROGRAM

The computer program presented in this section was originally developed by James J. Mager as a part of his graduate program conducted under the direction of the author.[1]

The program will compute the steady state probability distribution for any queueing system which can be described by the general model for birth-death processes, namely

$$p_n = \frac{\displaystyle\prod_{i=0}^{n-1} \lambda_i}{\displaystyle\prod_{i=1}^{n} \mu_i} p_0$$

$$p_0 = \left[1 + \sum_{n=1}^{\infty} \frac{\displaystyle\prod_{i=0}^{n-1} \lambda_i}{\displaystyle\prod_{i=1}^{n} \mu_i} \right]^{-1}$$

Specific applications for this model are presented in Chapter 6.

The program print capability is limited to a maximum of $NN = 501$ states. If there is a significant probability that the number of customers in the system exceeds 500 the distribution is truncated. If the user wishes to expand the set of states to be printed, the dimension statements at the beginning of the program must be changed.

The user can define the problem to be solved in one of three ways; standard format, parameter tables, or parameter functions. The inputs required for each method are listed as follows:

Standard Format. Standard format refers to those systems which can be defined in terms of the ORSA notation (M/M/c/n/m). The model

---

[1] Adapted with permission of the author from James J. Mager, *A General Purpose Solution Program for (M/M) Queueing Systems,* Unpublished project report for ISE 693, The Ohio State University, 1976.

requires specification of a constant demand rate λ, the service rate per channel μ, the number of channels c, the truncation point for number of customers in the system n, and the size of the calling population m. In the program these parameters are identified as LAMBDA, MU, C, N and M respectively. If any of the parameters C, N or M are to be infinite, the user must enter 9999 in that data block.

Two data cards appended to the program deck will set-up the model for this class of problems. The input is contained in the following columns of those cards:

Card	Column(s)	Contents
1	1	1
2	1–10	LAMBDA in real format (decimal point must be included)
2	11–20	MU in real format (decimal point must be included)
2	21–25	C in integer format (right justified)
2	26–30	N in integer format (right justified)
2	31–35	M in integer format (right justified)

A typical application for this method of problem description is the limited population, multi-channel model given by equation (6.83).

Parameter Tables. Parameter tables are used when the analyst wishes to explicitly account for state dependent arrival rates, service rates, and operating channels. Here the user must specify a maximum number of states, NCAP, and for every state, I, which is less than capacity, an arrival rate LAMBDA (I), service rate MU(I), and operating channel count C(I). The states of the system are $(0, 1, ---, NCAP - 1)$. Note that NCAP is the number of states not the physical capacity of the system.

The number of data cards required depends upon the value of NCAP selected, the total being NCAP plus two. The input is contained in the following columns of those cards:

Card	Column(s)	Contents
1	1	2
2	1–5	NCAP in integer format (right justified)
3	1–10	LAMBDA(0) in real format (decimal point must be included)
3	11–20	MU(0) in real format (decimal point must be included)
3	21–25	C(0) in real format (decimal point must be included)
4	1–10	LAMBDA(1)
4	11–20	MU(1)

4	21–25	C(1)
⋮		
NCAP + 2	1–10	LAMBDA(NCAP − 1)
NCAP + 2	11–20	MU(NCAP − 1)
NCAP + 2	21–25	C(NCAP − 1)

A typical application for this method of problem description is the channel control problem represented by equation (6.107).

Parameter Functions. Parameter functions are used when there are well-defined functional relationships between the parameters and the state of the system. The objective is the same as that expressed for parameter tables except that now the user must specify LAMBDA(I), MU(I) and C(I) in functional form for the first NCAP states of the system.

Input is accomplished by a combination of two data cards and FORTRAN statements inserted in the main program deck. The data cards contain the following:

Card	Column(s)	Contents
1	1	3
2	1–5	NCAP in integer format (right justified)

The FORTRAN statements are inserted behind the comment card reading

****PLACE FUNCTION CARDS HERE FOR MODEL=3****

A minimum of three statements, one for each function, is required. Because there is no zero index in a FORTRAN array and to avoid mixing real and integer variables in arithmetic statements it is necessary to define a new variable $S = I - 1$. The variable S then appears on the righthand side of all function definitions as noted in the examples:

Fortran Statement	Function implied
LAMBDA(I) = 5 − S	$\lambda_i = 5 - i,\ i = S = 0, 1, \text{---}$ $I = 1, 2, \text{---}$
MU(I) = S**2	$\mu_i = i^2,\ i = S = 0, 1, \text{---}$ $I = 1, 2, \text{---}$
IF(S.LE.4.) C(I) = S IF(S.GT.4.) C(I) = 4	$c_i = \begin{cases} i, i = S \le 4 \\ 4, i > 4 \end{cases}$

A typical application for this method of problem description is the pressure coefficient problem represented by equation (6.100).

Regardless of the method of input used, the program computes and prints the following information:

$$L = \text{expected number in the system, including the units being served}$$

$$LQ = \text{expected number in the queue, excluding the units being served}$$

$$W = \text{expected time spent in the system by a customer, including service time}$$

$$WQ = \text{expected time spent in the system by a customer, excluding service time}$$

$$RHO = \text{traffic intensity}$$

$$= \frac{\text{(arrival rate)}}{\text{(Number of servers)(service rate per server)}}$$

For all I in $[0, J]$ where J is the minimum of

1. the smallest K such that $Pr(I > K) < 10^{-5}$
2. NN set by the user

the program prints:

$$P(I) = \text{probability of I units in the system}$$
$$PCUM(I) = \text{probability of I or less units in the system}$$
$$LAMBDA(I) = \text{arrival rate as a function of number in the system}$$
$$MU(I) = \text{service rate as a function of number in the system}$$
$$RHO(I) = \text{traffic intensity as a function of number in the system}$$
$$C(I) = \text{operating channels as a function of number in the system}$$

When MU(I) is very large, underflow conditions may result for RHO(I). However, in test runs made to date this has never affected the accuracy of results when execution continues ignoring the RHO(I) calculation. (Note that MU(I) represents the *total* service rate when there are I customers in the systems.)

If the ratio LAMBDA(I)/MU(I) is very large overflow may occur which will prevent the program from completing execution. The program, which is generously laden with comment cards to aid the user, is written in Fortran IV.

PROGRAM LISTING

```
C     **** ADJUST NEXT TWO CARDS FOR NN, IF NECESSARY ****
      NN=501
      REALLAMBDA(501),MU(501),P(501),PCUM(501),RHO(501),C(501),PROD(501)
      REAL L,LQ,W,WQ,IFACT,LAMEFF
      READ (5,10) MODEL
   10 FORMAT (I1)
      KEY=0
      MES=0
      IF (MODEL.EQ.1) GO TO 100
      IF (MODEL.EQ.2) GO TO 240
      IF (MODEL.NE.3) GO TO 400
      READ (5,245) NCAP
      DO 90 I=1,NCAP
      S=I-1
C     **** PLACE FUNCTION CARDS HERE FOR MODEL=3 ****
   90 CONTINUE
      DO 95 I=2,NCAP
      RHO(I)=LAMBDA(I)/MU(I)
   95 CONTINUE
      GO TO 247
  100 CONTINUE
      READ (5,105) XLAMBD,XMU,NC,N,M
  105 FORMAT (F10.0,F10.0,I5,I5,I5)
      C(1)=NC
      LAMBDA(1)=XLAMBD
      MU(1)=XMU
      RHO(1)=LAMBDA(1)/MU(1)
      XRHO=RHO(1)/NC
      PCUM(1)=0.
      IF (N.LT.9999) GO TO 180
      IF (M.LT.9999) GO TO 180
      KEY=1
      IF (C(1).GT.1) GO TO 130
C
C     STANDARD RESULTS FOR THE (M/M/1) QUEUE
C
      IF (RHO(1).GE.1.0) MES=1
      IF (RHO(1).GE.1.0) GO TO 300
      DO 110 I=1,NN
      K=I
      KK=I-1
      P(I)=(1.-RHO(1))*(RHO(1)**(I-1))
      PCUM(1)=P(1)
      IF (I.EQ.1) GO TO 110
      PCUM(I)=PCUM(I-1)+P(I)
      IF (PCUM(I).GT.0.99999) GO TO 115
  110 CONTINUE
  115 L =RHO(1)/(1.-RHO(1))
      LQ=(RHO(1)**2.)/(1.-RHO(1))
      W =1./(MU(1)*(1.-RHO(1)))
      WQ=RHO(1)/(MU(1)*(1.-RHO(1)))
      DO 120 I=2,NN
      LAMBDA(I)=LAMBDA(I-1)
      MU(I)=MU(I-1)
      RHO(I)=RHO(I-1)
      C(I)=C(I-1)
  120 CONTINUE
      MU(1)=0.
      C(1)=0.
```

```
      GO TO 300
  130 CONTINUE
C
C     STANDARD RESULTS FOR THE (M/M/C) QUEUE
C
      IF (C(1).EQ.9999.) GO TO 160
      IF (RHO(1).GE.C(1)) MES=1
      IF (RHO(1).GE.C(1)) GO TO 300
      SUM=0.
      CFACT=1.0
      DO 135 I=1,NC
      IFACT=1.0
      DO 133 J=1,I
      IF (J.EQ.1) GO TO 133
      IFACT=IFACT*(J-1)
  133 CONTINUE
      SUM=SUM+(RHO(1)**(I-1)/IFACT)
      CFACT=CFACT*I
  135 CONTINUE
      P(1)=1.0/(SUM+(RHO(1)**C(1)/(CFACT*(1.0-(RHO(1)/C(1))))))
      DO 145 I=2,NN
      K=I
      KK=I-1
      CC=C(1)+1.0
      IFACT=1.0
      IF (I.GT.CC) GO TO 143
      DO 142 J=1,I
      IF (J.EQ.1) GO TO 142
      IFACT=IFACT*(J-1)
  142 CONTINUE
      P(I)=P(1)*(RHO(1)**(I-1))/IFACT
      GO TO 144
  143 P(I)=P(1)*(RHO(1)**(I-1))/((C(1)**(I-1-NC))*CFACT)
  144 PCUM(1)=P(1)
      IF (I.EQ.1) GO TO 145
      PCUM(I)=PCUM(I-1)+P(I)
      IF (PCUM(I).GT.0.99999) GO TO 150
  145 CONTINUE
  150 ND=NC+1
      LQ=P(ND)*(C(1)*RHO(1))/((C(1)-RHO(1))**2)
      L =LQ+RHO(1)
      WQ=LQ/LAMBDA(1)
      W =WQ+1.0/MU(1)
      DO 155 I=2,NN
      LAMBDA(I)=LAMBDA(I-1)
      MU(I)=MU(1)*C(1)
      C(I)=C(I-1)
      RHO(I)=LAMBDA(I)/MU(I)
  155 CONTINUE
      DO 158 I=2,NC
      MU(I)=MU(1)*(I-1)
      C(I)=I-1
      RHO(I)=LAMBDA(I)/MU(I)
  158 CONTINUE
      MU(1)=0.0
      C(1)=0.0
      GO TO 300
C
C     STANDARD RESULTS FOR THE (M/M/INFINITY) QUEUE
C
  160 CONTINUE
      NC=9999
```

```
      DO 170 I=1,NN
      K=I
      KK=I-1
      IFACT=1.0
      DO 165 J=1,I
      IF (J.EQ.1) GO TO 165
      IFACT=IFACT*(J-1)
  165 CONTINUE
      P(I)=(EXP(-RHO(1))*(RHO(1)**(I-1))/IFACT)
      PCUM(1)=P(1)
      IF (I.EQ.1) GO TO 170
      PCUM(I)=PCUM(I-1)+P(I)
      IF (PCUM(I).GT.0.99999) GO TO 175
  170 CONTINUE
  175 L =RHO(1)
      LQ=0.0
      W =1.0/MU(1)
      WQ=0.0
      XRHO=0.0
      DO 179 I=2,NN
      LAMBDA(I)=LAMBDA(1)
      MU(I)=MU(1)*(I-1)
      RHO(I)=LAMBDA(I)/MU(I)
      C(I)=I-1
  179 CONTINUE
      MU(1)=0.0
      MU(1)=0.0
      C(1)=0.0
      GO TO 300
C
C     MU(I) AND C(I) FOR THE (M/M/C/N) AND (M/M/C/N/M) QUEUES
C
  180 CONTINUE
      KEY=2
      NCAP=N+1
      IF (N.GE.M) NCAP=M+1
      C(1)=0.0
      MU(1)=0.0
      NCC=NC+1
      DO 185 I=2,NCC
      C(I)=I-1
      MU(I)=C(I)*XMU
  185 CONTINUE
      IF (NC.GT.N) GO TO 191
      IF (NC.GT.M) GO TO 191
      DO 190 I=NCC,NCAP
      C(I)=NC
      MU(I)=NC*XMU
  190 CONTINUE
  191 IF (M.LT.9999) GO TO 210
C
C     LAMBDA(I) AND RHO(I) FOR THE (M/M/C/N) QUEUE
C
      DO 200 I=2,NCAP
      LAMBDA(I)=XLAMBD
      RHO(I)=LAMBDA(I)/MU(I)
  200 CONTINUE
      LAMBDA(NCAP)=0.0
      GO TO 260
C
C     LAMBDA(I) AND RHO(I) FOR THE (M/M/C/N/M) QUEUE
C
```

```
  210 CONTINUE
      KEY=3
      DO 220 I=1,NCAP
      LAMBDA(I)=XLAMBD*(M-I+1)
  220 CONTINUE
      DO 230 I=2,NCAP
      RHO(I)=LAMBDA(I)/MU(I)
  230 CONTINUE
      GO TO 260
C
C     READ NCAP, LAMBDA(I), MU(I), AND C(I) FOR MODEL=2
C
  240 CONTINUE
      READ (5,245) NCAP
  245 FORMAT (I5)
      READ (5,246) (LAMBDA(I),MU(I),C(I),I=1,NCAP)
  246 FORMAT (F10.0,F10.0,F5.0)
      RHO(1)=0.10000E+10
C
C     CHECK FOR VALID MU(I) AND C(I)
C
  247 CONTINUE
      DO 250 I=2,NCAP
      K=I
      KK=I-1
      IF (MU(I).LE.0.0) MES=2
      IF (MU(I).LE.0.0) GO TO 300
      IF (C(I).LE.0.0) MES=3
      IF (C(I).LE.0.0) GO TO 300
      RHO(I)=LAMBDA(I)/MU(I)
  250 CONTINUE
C
C     STANDARD RESULTS USING LAMBDA(I), MU(I), AND C(I) AS INPUT
C
  260 CONTINUE
      SUM=0.0
      ND=NCAP-1
      DO 270 I=1,ND
      PROD(I)=1.0
      DO 265 J=1,I
      PROD(I)=PROD(I)*(LAMBDA(J)/MU(J+1))
  265 CONTINUE
      SUM=SUM+PROD(I)
  270 CONTINUE
      P(1)=1.0/(1.0+SUM)
      PCUM(1)=P(1)
      IF (PCUM(1).GT.0.99999) GO TO 285
      DO 280 I=2,NCAP
      K=I
      KK=I-1
      P(I)=P(1)*PROD(I-1)
      PCUM(I)=PCUM(I-1)+P(I)
      IF (PCUM(I).GT.0.99999) GO TO 285
  280 CONTINUE
  285 CONTINUE
      L =0.0
      LQ=0.0
      LAMEFF=0.0
      DO 290 I=1,K
      L =L+(I-1)*P(I)
      J=I-1
      LAMEFF=LAMEFF+P(I)*LAMBDA(I)
```

```
      IF (J.LE.C(I)) GO TO 290
      LQ=LQ+P(I)*(J-C(I))
  290 CONTINUE
      W =L/LAMEFF
      WQ=LQ/LAMEFF
C
C     WRITE THE HEADINGS AND RESULTS
C
  300 CONTINUE
      IF (MODEL.EQ.2) WRITE (6,304)
  304 FORMAT (1H1,14X,'STATE DEPENDENT DATA READ FROM CARDS',/)
      IF (MODEL.EQ.3) WRITE (6,305)
  305 FORMAT (1H1,15X,'INPUT DATA READ IN FUNCTIONAL FORM',/)
      IF (MODEL.EQ.2) GO TO 315
      IF (MODEL.EQ.3) GO TO 315
      IF (KEY.EQ.1) WRITE (6,310) NC
  310 FORMAT (1H1,30X, '(M/M/',I4,')',/)
      IF (KEY.EQ.2) WRITE (6,311) NC,N
  311 FORMAT (1H1,30X, '(M/M/',I4,'/',I4,')',/)
      IF (KEY.EQ.3) WRITE (6,312) NC,N,M
  312 FORMAT (1H1,30X, '(M/M/',I4,'/',I4,'/',I4,')',/)
      WRITE (6,314) XLAMBD,XMU,XRHO
  314 FORMAT(1H ,10X,'LAMBDA=',E12.5,5X,'MU=',E12.5,5X,'RHO=',E12.5,/)
  315 CONTINUE
      IF (MES.EQ.1) GO TO 410
      IF (MES.EQ.2) GO TO 420
      IF (MES.EQ.3) GO TO 430
  316 WRITE (6,317) L,LQ
  317 FORMAT (1H ,14X,' L=',E12.5,5X,'LQ=',E12.5,/)
      WRITE (6,320) W,WQ
  320 FORMAT (1H ,14X, ' W=',E12.5,5X, 'WQ=',E12.5,/)
      WRITE (6,330)
  330 FORMAT (1H0,5X, 'I',5X, 'P(I)',5X,'PCUM(I)',5X, 'LAMBDA(I)',5X, 'M
     1U(I)',9X,'RHO(I)',6X,'C(I)',/)
      WRITE (6,335) P(1),PCUM(1),LAMBDA(1),MU(1),C(1)
  335 FORMAT (1H ,5X,'0',3X,F7.5,3X,F7.5,3X,E12.5,2X,E12.5,3X,'*********
     1**',2X,F4.0)
      DO 350 I=2,K
      J=I-1
      WRITE (6,340) J,P(I),PCUM(I),LAMBDA(I),MU(I),RHO(I),C(I)
  340 FORMAT (1H ,2X,I4,3X,F7.5,3X,F7.5,3X,E12.5,2X,E12.5,2X,E12.5,2X,F4
     1.0)
  350 CONTINUE
      IF (K.EQ.NN) GO TO 499
      IF (K.EQ.NCAP) GO TO 499
      WRITE (6,360) KK
  360 FORMAT (1H0,2X, '*** PROB(I>',I3,') IS LESS THAN 0.00001 ***')
      GO TO 499
C
C     WRITE THE ERROR MESSAGES
C
  400 WRITE (6,405)
  405 FORMAT (1H1,15X,'INVALID DATA--MODEL IS NOT EQUAL TO 1,2, OR 3')
      GO TO 499
  410 WRITE (6,415)
  415 FORMAT (1H ,15X,'INVALID DATA--RHO IS GREATER THAN OR EQUAL TO 1')
      GO TO 499
  420 WRITE (6,425) KK
  425 FORMAT (1H ,15X,'INVALID DATA--MU(',I4,') IS NOT POSITIVE')
      GO TO 499
  430 WRITE (6,435) KK
  435 FORMAT (1H ,15X,'INVALID DATA--C(',I4,' IS NOT POSITIVE')
```

```
  499 CONTINUE
      STOP
      END
/*
```

```
1
7.          12.              2 9999 9999
/*
//
```

SAMPLE OUTPUT

(M/M/   1)

LAMBDA= 0.30000E+01   MU= 0.80000E+01   RHO= 0.37500E+00

L= 0.60000E+00   LQ= 0.22500E+00

W= 0.20000E+00   WQ= 0.75000E-01

I	P(I)	PCUM(I)	LAMBDA(I)	MU(I)	RHO(I)	C(I)
0	0.62500	0.62500	0.30000E+01	0.0	************	0.
1	0.23438	0.85938	0.30000E+01	0.80000E+01	0.37500E+00	1.
2	0.08789	0.94727	0.30000E+01	0.80000E+01	0.37500E+00	1.
3	0.03296	0.98022	0.30000E+01	0.80000E+01	0.37500E+00	1.
4	0.01236	0.99258	0.30000E+01	0.80000E+01	0.37500E+00	1.
5	0.00463	0.99722	0.30000E+01	0.80000E+01	0.37500E+00	1.
6	0.00174	0.99896	0.30000E+01	0.80000E+01	0.37500E+00	1.
7	0.00065	0.99961	0.30000E+01	0.80000E+01	0.37500E+00	1.
8	0.00024	0.99985	0.30000E+01	0.80000E+01	0.37500E+00	1.
9	0.00009	0.99994	0.30000E+01	0.80000E+01	0.37500E+00	1.
10	0.00003	0.99998	0.30000E+01	0.80000E+01	0.37500E+00	1.
11	0.00001	0.99999	0.30000E+01	0.80000E+01	0.37500E+00	1.

*** PROB(I> 11) IS LESS THAN 0.00001 ***

(M/M/   2)

LAMBDA= 0.10500E+02   MU= 0.14000E+02   RHO= 0.37500E+00

L= 0.87273E+00   LQ= 0.12273E+00

W= 0.83117E-01   WQ= 0.11688E-01

I	P(I)	PCUM(I)	LAMBDA(I)	MU(I)	RHO(I)	C(I)
0	0.45455	0.45455	0.10500E+02	0.0	************	0.
1	0.34091	0.79545	0.10500E+02	0.14000E+02	0.75000E+00	1.
2	0.12784	0.92330	0.10500E+02	0.28000E+02	0.37500E+00	2.
3	0.04794	0.97124	0.10500E+02	0.28000E+02	0.37500E+00	2.
4	0.01798	0.98921	0.10500E+02	0.28000E+02	0.37500E+00	2.
5	0.00674	0.99595	0.10500E+02	0.28000E+02	0.37500E+00	2.
6	0.00253	0.99848	0.10500E+02	0.28000E+02	0.37500E+00	2.
7	0.00095	0.99943	0.10500E+02	0.28000E+02	0.37500E+00	2.
8	0.00036	0.99979	0.10500E+02	0.28000E+02	0.37500E+00	2.
9	0.00013	0.99992	0.10500E+02	0.28000E+02	0.37500E+00	2.
10	0.00005	0.99997	0.10500E+02	0.28000E+02	0.37500E+00	2.
11	0.00002	0.99999	0.10500E+02	0.28000E+02	0.37500E+00	2.
12	0.00001	1.00000	0.10500E+02	0.28000E+02	0.37500E+00	2.

*** PROB(I> 12) IS LESS THAN 0.00001 ***

(M/M/   1/   20)

LAMBDA= 0.30000E+01   MU= 0.40000E+01   RHO= 0.75000E+00

L= 0.29499E+01   LQ= 0.22005E+01

W= 0.98409E+00   WQ= 0.73409E+00

I	P(I)	PCUM(I)	LAMBDA(I)	MU(I)	RHO(I)	C(I)
0	0.25060	0.25060	0.30000E+01	0.0	************	0.
1	0.18795	0.43854	0.30000E+01	0.40000E+01	0.75000E+00	1.
2	0.14096	0.57950	0.30000E+01	0.40000E+01	0.75000E+00	1.
3	0.10572	0.68522	0.30000E+01	0.40000E+01	0.75000E+00	1.
4	0.07929	0.76451	0.30000E+01	0.40000E+01	0.75000E+00	1.
5	0.05947	0.82398	0.30000E+01	0.40000E+01	0.75000E+00	1.
6	0.04460	0.86858	0.30000E+01	0.40000E+01	0.75000E+00	1.
7	0.03345	0.90203	0.30000E+01	0.40000E+01	0.75000E+00	1.
8	0.02509	0.92712	0.30000E+01	0.40000E+01	0.75000E+00	1.
9	0.01882	0.94594	0.30000E+01	0.40000E+01	0.75000E+00	1.
10	0.01411	0.96005	0.30000E+01	0.40000E+01	0.75000E+00	1.
11	0.01058	0.97063	0.30000E+01	0.40000E+01	0.75000E+00	1.
12	0.00794	0.97857	0.30000E+01	0.40000E+01	0.75000E+00	1.
13	0.00595	0.98452	0.30000E+01	0.40000E+01	0.75000E+00	1.
14	0.00447	0.98899	0.30000E+01	0.40000E+01	0.75000E+00	1.
15	0.00335	0.99234	0.30000E+01	0.40000E+01	0.75000E+00	1.
16	0.00251	0.99485	0.30000E+01	0.40000E+01	0.75000E+00	1.
17	0.00188	0.99673	0.30000E+01	0.40000E+01	0.75000E+00	1.
18	0.00141	0.99815	0.30000E+01	0.40000E+01	0.75000E+00	1.
19	0.00106	0.99921	0.30000E+01	0.40000E+01	0.75000E+00	1.
20	0.00079	1.00000	0.0	0.40000E+01	0.75000E+00	1.

```
              INPUT DATA READ IN FUNCTIONAL FORM
          L= 0.80172E+00      LQ= 0.21521E-02

          W= 0.10022E+00      WQ= 0.26901E-03


  I      P(I)       PCUM(I)      LAMBDA(I)       MU(I)            RHO(I)        C(I)

  0    0.44916     0.44916     0.80000E+01     0.0            ***********      0.
  1    0.35933     0.80848     0.80000E+01     0.10000E+02     0.80000E+00     1.
  2    0.14373     0.95221     0.80000E+01     0.20000E+02     0.40000E+00     2.
  3    0.03833     0.99054     0.80000E+01     0.30000E+02     0.26667E+00     3.
  4    0.00767     0.99821     0.80000E+01     0.40000E+02     0.20000E+00     4.
  5    0.00147     0.99967     0.80000E+01     0.41826E+02     0.19127E+00     4.
  6    0.00027     0.99994     0.80000E+01     0.43379E+02     0.18442E+00     4.
  7    0.00005     0.99999     0.80000E+01     0.44737E+02     0.17882E+00     4.

*** PROB(I> 7) IS LESS THAN 0.00001 ***


              STATE DEPENDENT DATA READ FROM CARDS
          L= 0.20000E+01      LQ= 0.35294E+00

          W= 0.10000E+01      WQ= 0.17647E+00


  I      P(I)       PCUM(I)      LAMBDA(I)       MU(I)            RHO(I)        C(I)

  0    0.05882     0.05882     0.40000E+01     0.0            ***********      0.
  1    0.23529     0.29412     0.30000E+01     0.10000E+01     0.30000E+01     1.
  2    0.35294     0.64706     0.20000E+01     0.20000E+01     0.10000E+01     2.
  3    0.35294     1.00000     0.10000E+01     0.20000E+01     0.50000E+00     2.


                    (M/M/   5/   2/   5)

     LAMBDA= 0.50000E+01     MU= 0.10000E+02     RHO= 0.10000E+00

             L= 0.12500E+01      LQ= 0.0

          W= 0.66667E-01      WQ= 0.0


  I      P(I)       PCUM(I)      LAMBDA(I)       MU(I)            RHO(I)        C(I)

  0    0.16667     0.16667     0.25000E+02     0.0            ***********      0.
  1    0.41667     0.58333     0.20000E+02     0.10000E+02     0.20000E+01     1.
  2    0.41667     1.00000     0.15000E+02     0.20000E+02     0.75000E+00     2.
```

APPENDIX C

TIME VARYING BIRTH-DEATH PROGRAM

The computer programs presented in this section are extensions of programs originally developed by Michael G. Hartman as a part of his M.S. Thesis written under the direction of the author.[1] The original programs have since been modified and extended by a succession of graduate students working with the author. The most notable added contributions were made by Chi-Houng Eddie Lu and Anderson H. Walters.

These programs are designed to handle (M/M/c/k) systems with nonstationary input. They calculate time-dependent probability distributions for multichannel, truncated systems having time-varying arrival and service rates. The programs require that the user provide 96 values each for the arrival rate and service rate per channel on data cards appended to the program decks. These values represent the step function approximations to the continuously varying rate functions. The number 96 was selected to represent the 96 fifteen-minute time periods in a 24-hour day. If a different number of periods is desirable the user must change the corresponding dimension statements. The data card format is F3.0 with 24 values placed on each card for a total of eight data cards. A ninth data card carries the specification for KI (the number of parallel channels in the system), MM, (the maximum number of customers permitted in the system) and TIMPER (the total length of the time period being divided into 96 subperiods). The format is 2I2, F4.0. In the current program the truncation point MM must be less than 60.

The program solves the finite set of linear differential equations for each subperiod implied by the data cards using Runge-Kutta numerical methods. It assumes that the system is initially empty, i.e., $p_0(0) = 1$. The time-varying solution over the first subperiod, usually taken to be fifteen minutes, is obtained from the constant coefficient differential equations where the arrival rate and service rate are the step values read in for that subperiod. At the end of the first period various operating statistics are

---

[1] Adapted with permission of the author from Michael G. Hartman, "The Truncated, Time Varying Queue as a Model of an Air Traffic Control System," Unpublished M.S. Thesis, The Ohio State University, 1971, pp. 70–73.

calculated and printed along with the entire state probability distribution at that point in time. The distribution at the end of the first subperiod becomes the initial value vector for the next subperiod. These values are then used together with new constant arrival and service rates to find the time-varying solution for the next subperiod. The solution proceeds in this manner until the entire 96 subperiods in TIMPER have been evaluated.

The statistics printed at the end of each subperiod include:

1. SUM	The sum of all values contained in the numerical approximation for the probability distribution at the end of the interval. This should be equal to one but may vary slightly at times due to the approximations inherent in the numerical solution procedure.
2. EQ	The expected number of customers in the queue, exclusive of those in service, at the end of the interval.
3. SDQ	The standard deviation for customers in the queue.
4. ES	The expected number of customers in the system, including those in service, at the end of the interval.
5. SDS	The standard deviation for customers in the systems.
6. ARR	The arrival rate in effect during the interval.
7. SER	The service rate per channel in effect during the interval.
8. MX	The state with the largest probability of occurrence. The probability of this state may have been altered from that obtained from the numerical solution to guarantee a probability vector which will sum to unity. The magnitude of that alteration can be determined by subtracting one from the amount printed for SUM.
9. PW	Probability that all service channels are full at the end of the interval. Equivalently this is the probability that a customer who arrives then will be forced to wait for service.
10. EHOLD	The expected number of customer-minutes holding time, excluding service time, accumulated during that interval.
11. THOLD	The summation of all expected holding times from the origin through the end of the current interval.
12. CUM RJECTS	An approximation for the number of customers rejected due to system saturation from the origin to the end of the current interval.
13. P(N)	The probability that there are N customers

present at the end of the current interval. (Printed for all possible state values.)

The program, which is written in FORTRAN IV, is generously laden with comment cards to facilitate understanding on the part of the user.

PROGRAM LISTING

```
C THIS PROGRAM CALCULATES TRANSIENT STATE PROBABILITIES AND THE RELATED
C     BEHAVIORAL CHARACTERISTICS FOR A KI SERVER TRUNCATED QUEUEING
C     SYSTEM WITH VARIABLE POISSON ARRIVAL AND EXPONENTIAL SERVICE
C     RATES.  A TABULAR STEP-FUNCTION IS USED TO APPROXIMATE THE MEAN
C     ARRIVAL AND SERVICE RATES.
C
C
C SUBROUTINE RKGS, FROM THE IBM SCIENTIFIC SUBROUTINE PACKAGE, IS USED
C     TO SOLVE THE SIMULTANEOUS DIFFERENTIAL EQUATIONS WHICH DESCRIBE
C     THE SYSTEM.  THIS SUBROUTINE USES THE RUNGE-KUTTA METHOD FOR THE
C     SOLUTION OF INITIAL VALUE PROBLEMS.  FOR A COMPLETE DESCRIPTION
C     OF THIS SUBROUTINE SEE:  SYSTEM/360 SCIENTIFIC SUBROUTINE PACKAGE,
C     VERSION 3
C
C
C THE QUEUE STATISTICS CALCULATED ARE:
C     1. EXPECTED NUMBER IN THE QUEUE = EQ
C     2. STD. DEV. OF NUMBER IN THE QUEUE = SDQ
C     3. EXPECTED TOTAL HOLDING TIME FOR A SUB-DIVISION = EHOLD
C     4. SUMMATION OF HOLDING TIMES = THOLD
C     5. APPROXIMATE # OF CUMULATIVE REJECTS = RJECTS
C     6. EXPECTED NUMBER IN SYSTEM = ES
C     7. STD. DEV. OF NUMBER IN SYSTEM = SDS
C     8. PROBABILITY OF DELAY = PW
C
C
C THE VARIABLES USED ARE:
C     TIMPER= THE TOTAL LENGTH OF TIME FOR WHICH THE EQUATIONS ARE TO BE
C               SOLVED
C     KI=NUMBER OF SERVERS IN THE SYSTEM
C     MM=MAXIMUM NUMBER ALLOWED IN THE SYSTEM
C     M= MM+1= # OF POSSIBLE STATES OF THE SYSTEM
C     N= THE # OF SUB-DIVISIONS OF TIMPER FOR BOTH DATA INPUT AND OUTPUT
C        OF RESULTS
C     DT= THE LENGTH OF THE TIME-PERIOD SUB-DIVISION
C     Z= AN ERROR WEIGHT USED BY RKGS
C     R(I)= INPUT VECTOR OF MEAN ARRIVAL RATES FOR THE TIME PERIOD UNDER
C             CONSIDERATION.   I= 1,2,...N
C     L(I)=INPUT VECTOR OF MEAN SERVICE RATE PER CHANNEL FOR PERIOD
C           UNDER CONSIDERATION. I=1,2,.....N
C     P(J)= PROBABILITY OF J-1 IN THE SYSTEM AT A GIVEN TIME; J=1,2,...M
C     DERY(J)= INPUT VECTOR OF ERROR WTS., DESTROYED BY RKGS; LATER ON
C               DERY IS THE VECTOR OF DERIVATIVES OF P AT A POINT T.
C     PRMT(1)= LOWER BOUND OF THE SOLUTION INTERVAL FOR ONE ITERATION
C     PRMT(2)= UPPER BOUND OF THE SOLUTION INTERVAL FOR ONE ITERATION
C     PRMT(3)= INITIAL STEP-SIZE OF INTEGRATION
C     PRMT(4)= UPPER ERROR BOUND ACCEPTABLE ON VALUES OF P
C     AUX= A STORAGE MATRIX USED BY RKGS ; AUX MUST CONTAIN 8 ROWS
C     DQ = EXPECTED NUMBER IN THE QUEUE AT THE END OF CURRENT
C           SUB-DIVISION
C     DQL = EXPECTED NUMBER IN THE QUEUE AT THE END OF THE LAST
C            SUB-DIVISION
C     MX = STATE WITH LARGEST PROBABILITY . MAY BE ADJUSTED TO
C          ACCOMODATE NUMERICAL ERRORS
C
C
C NOTE:  THE PROGRAM IS COMPLETELY GENERAL WITH THE FOLLOWING EXCEPTIONS
```

```
C         1. MM MUST BE LESS THAN 60
C         2. IF N IS CHANGED, THE DIMENSIONS ON R & L MUST BE CHANGED
C         3. THE CALCULATION OF EXPECTED HOLDING TIME FINDS THE AVERAGE
C            EXPECTED NUMBER IN THE QUEUE FOR A SUB-DIVISION AND MULTIPLIES
C            THIS BY THE NUMBER OF MINUTES IN A SUB-DIVISION TO OBTAIN THE
C            TOTAL EXPECTED HOLDING TIME FOR THAT SUB-DIVISION.
C
      EXTERNAL FCT,OUTP
      DIMENSION P(60),DERY(60),PRMT(5),AUX(8,60)
      REAL L(96)
      COMMON I,R(96),L,M,MM,TIMLIM,DT,RJECTS,TSAVE,KI
      COMMON DQ,DQL,THOLD
      N=96
      DQ=0.
       DQL=0.
      THOLD=0.
      READ(5,100)(R(I),I=1,N)
      READ(5,100)(L(I),I=1,N)
      READ(5,50) KI,MM,TIMPER
   50 FORMAT(2I2,F4.0)
      TSAVE=0.
      M=MM+1
      RJECTS=0.
      DT=TIMPER/FLOAT(N)
      Z=1./FLOAT(M)
      PRINT 101,MM,KI
C
C INITIALIZE P(I) AT TIME= 0
C
      DO 2 J=1,60
    2 P(J)=0.
      P(1)=1.
C
C ESTABLISH STEP-SIZE FOR INTEGRATION AND ERROR BOUND
C
      PRMT(3)=.03
      PRMT(4)=.001
C
C SOLVE THE SYSTEM OF EQUATIONS FOR THE N TIME PERIODS IN TIMPER
C
      DO 1 I=1,N
      TIMLIM=DT*FLOAT(I)
      PRMT(1)=TIMLIM-DT
      PRMT(2)=TIMLIM
      DO 3 J=1,M
    3 DERY(J)=Z
    1 CALL RKGS(PRMT,P,DERY,M,IHLF,FCT,OUTP,AUX)
  100 FORMAT(24F3.0)
  101 FORMAT('-',5X,'RESULTS FOR THE TRUNCATED KI SERVER QUEUEING SYSTEM
     1'    ,/,10X,'TIME VARYING POISSON ARRIVALS AND EXPONENTIAL SERVICE'
     2,/,15X,'MAXIMUM QUEUE SIZE=',I2,5X,'NUMBER OF SERVERS=',I2)
      STOP
      END
      SUBROUTINE FCT(T,P,DERY)
C
C THIS SUBROUTINE EVALUATES THE RIGHT HAND SIDE OF THE DIFFERENTIAL
C     EQUATIONS WHICH DESCRIBE THE SYSTEM
C
C T= TIME, THE INDEPENDENT VARIABLE
C
C NEITHER T NOR P SHOULD BE CHANGED BY THIS SUBROUTINE
C
```

```
      DIMENSION P(60),DERY(60)
      REAL L(96)
      COMMON I,R(96),L,M,MM,TIMLIM,DT,RJECTS,TSAVE,KI
      COMMON DQ,DQL,THOLD
      KK=KI+1
      DERY(1)=L(I)*P(2)-R(I)*P(1)
      DERY(M)=R(I)*P(M-1)-FLOAT(KI)*L(I)*P(M)
      DO 1 J=2,KI
      Q=FLOAT(J)
    1 DERY(J)=Q*L(I)*P(J+1)+R(I)*P(J-1)-((Q-1.0)*L(I)+R(I))*P(J)
      DO 2 J=KK,MM
      QQ=FLOAT(KI)
    2 DERY(J)=QQ*L(I)*P(J+1)+R(I)*P(J-1)-(QQ*L(I)+R(I))*P(J)
      RETURN
      END
      SUBROUTINE OUTP(T,P,DERY,K,NDIM,PRMT)
C
C
C THIS SUBROUTINE CALCULATES QUEUE STATISTICS AND OUTPUTS THE RESULTS
C     AT THE END OF EACH OF THE N TIME PERIODS.
C
C THE VARIABLES USED INCLUDE:
C     1. SUM= THE SUM FROM 1 TO M OF THE STATE PROBABILITIES,P,
C              (SHOULD EQUAL 1.0)
C     2. EQ = E(# IN THE QUEUE) AT TIME = T
C     3. SDQ= STD . DEV. OF # IN THE QUEUE
C     4. EHOLD= E(WAITING TIME), NOT INCLUDING SERVICE TIME
C     5. THOLD = CUMULATIVE TOTAL EXPECTED HOLDING TIME
C     6. RJECTS= APPROXIMATE # OF CUMULATIVE REJECTS
C     7. ES=E(# IN THE SYSTEM) AT T
C     8. SDS=STD DEV OF # IN THE SYSTEM
C     9. PW= PROBABILITY THAT A CUSTOMER WILL BE DELAYED
C
      DIMENSION PRMT(5),P(60),DERY(60)
      REAL L(96)
      COMMON I,R(96),L,M,MM,TIMLIM,DT,RJECTS,TSAVE,KI
      COMMON DQ,DQL,THOLD
C
C OUTP IS CALLED BY RKGS AT EACH INCREMENT OF INTEGRATION. TO AVOID
C     UNNECESSARY OUTPUT, CONTROL IS RETURNED TO RKGS, UNLESS THE END
C     OF ONE OF THE N SUB-DIVISIONS OF TIMPER IS REACHED.  IF ADDITIONAL
C     OUTPUT IS DESIRED, THIS 'IF' STATEMENT MAY BE CHANGED ACCORDINGLY.
C
      PM=P(M)
      IF(PM.LT.0.)PM=0.
      RJECTS=RJECTS+R(I)*PM*(T-TSAVE)
      TSAVE=T
      IF(T+.001.LT.TIMLIM)RETURN
      SUM=0.
      PMAX=0.
      MX=0
      DO 4 J=1,M
      IF(P(J).LT.0.)P(J)=0.
      IF(P(J).LT.PMAX)GO TO 4
      PMAX=P(J)
      MX=J
    4 SUM=SUM+P(J)
      IF(SUM.EQ.1.)GO TO 8
      DELT=SUM-1.
      P(MX)=P(MX)-DELT
    8 CONTINUE
      PPW=0.
```

```
      DO 10 J = 1,KI
   10 PPW = PPW + P(J)
      PW =1.-PPW
      KI2=KI+2
      EQ=0.
      VQ=0.
      SDQ=0.
      DO 12 J=KI2,M
      EQ=EQ+FLOAT(J-1-KI)*P(J)
   12 VQ=VQ+((FLOAT(J-1-KI))**2)*P(J)
      IF(VQ.LT.EQ**2)GO TO 13
      SDQ=SQRT(VQ-EQ**2)
   13 DQ=EQ
      EHOLD=30*DT*(DQ+DQL)
      DQL=DQ
      THOLD=THOLD+EHOLD
      ES=0.
      VS=0.
      SDS=0.
      DO 14 J=2,M
      ES=ES+FLOAT(J-1)*P(J)
   14 VS=VS+((FLOAT(J-1))**2)*P(J)
      IF(VS.LT.ES**2)GO TO 15
      SDS=SQRT(VS-ES**2)
   15 PRINT 101,T,SUM,EQ,SDQ,ES,SDS,R(I),L(I),MX
      PRINT 111,PW,EHOLD,THOLD,RJECTS
      PRINT102,(J,J=1,14),(P(J),J=1,15)
      PRINT103,(J,J=15,29),(P(J),J=16,30)
      PRINT103,(J,J=30,44),(P(J),J=31,45)
      PRINT103,(J,J=45,59),(P(J),J=46,60)
  101 FORMAT('-',1X,'TIME=',F6.3,3X,'SUM=',F6.3,3X,'EQ=',F7.4,3X,
     1'SDQ=',F7.4,3X,'ES=',F7.4,3X,'SDS=',F7.4,
     23X,'ARR=',F3.0,3X,'SER=',F3.0,3X,'MX=',I2/)
  111 FORMAT('-',15X,'PW=',F8.6,3X,'EHOLD=',F6.2,'MIN',3X,'THOLD=',F8.2,
     13X,'CUM RJECTS=',F6.2,///)
  102 FORMAT(' ',1X,'STATE # :',9X,'0',14I7,/,1X,'P(N) : ',6X,15F7.2)
  103 FORMAT(' ',1X,'STATE # :',3X,15I7,/,1X,'P(N) :',7X,15F7.2)
      RETURN
      END
/*



  5. 5. 5. 5. 5. 5. 5. 3. 2. 1. 1. 1. 2. 2. 2. 2. 2. 2. 2. 3. 3. 3. 3. 4.
  4. 4. 9.13.19.25.31.37.43.50.50.50.50.50.50.50.51.53.51.49.47.45.45.46.
 47.47.48.48.49.50.51.52.53.54.55.56.58.60.62.61.61.60.61.62.64.66.65.63.
 61.59.57.55.54.52.50.48.46.45.43.41.39.37.35.34.32.30.26.22.17.13. 9. 5.
 45.45.45.45.45.45.45.45.45.45.45.45.45.45.45.45.45.45.45.45.45.45.45.45.
 45.45.45.45.45.45.45.45.45.45.45.45.45.45.45.45.45.45.45.45.45.45.45.45.
 45.45.45.45.45.45.45.45.45.45.45.45.45.45.45.45.45.45.45.45.45.45.45.45.
 45.45.45.45.45.45.45.45.45.45.45.45.45.45.45.45.45.45.45.45.45.45.45.45.
 12524.
/*
//
```

SAMPLE OUTPUT

```
RESULTS FOR THE TRUNCATED KI SERVER QUEUEING SYSTEM
    TIME VARYING POISSON ARRIVALS AND EXPONENTIAL SERVICE
        MAXIMUM QUEUE SIZE=25      NUMBER OF SERVERS= 1


TIME= 0.250    SUM= 1.000    EQ= 0.0138    SDQ= 0.1307    ES= 0.1249    SDS= 0.3747    ARR= 5.    SER=45.    MX= 1


              PW=0.111058    EHOLD=  0.10MIN    THOLD=     0.10    CUM RJECTS=   0.00


   STATE # :       0        1        2        3        4        5        6        7        8        9       10       11       12       13       14
   P(N) :       0.89     0.10     0.01     0.00     0.00     0.00     0.00     0.00     0.00     0.00     0.00     0.00     0.00     0.00     0.00
   STATE # :      15       16       17       18       19       20       21       22       23       24       25       26       27       28       29
   P(N) :       0.00     0.00     0.00     0.00     0.00     0.00     0.00     0.00     0.00     0.00     0.00     0.0      0.0      0.0      0.0
   STATE # :      30       31       32       33       34       35       36       37       38       39       40       41       42       43       44
   P(N) :       0.0      0.0      0.0      0.0      0.0      0.0      0.0      0.0      0.0      0.0      0.0      0.0      0.0      0.0      0.0
   STATE # :      45       46       47       48       49       50       51       52       53       54       55       56       57       58       59
   P(N) :       0.0      0.0      0.0      0.0      0.0      0.0      0.0      0.0      0.0      0.0      0.0      0.0      0.0      0.0      0.0


TIME= 0.500    SUM= 1.000    EQ= 0.0139    SDQ= 0.1310    ES= 0.1250    SDS= 0.3750    ARR= 5.    SER=45.    MX= 1


              PW=0.111110    EHOLD=  0.21MIN    THOLD=     0.31    CUM RJECTS=   0.00


   STATE # :       0        1        2        3        4        5        6        7        8        9       10       11       12       13       14
   P(N) :       0.89     0.10     0.01     0.00     0.00     0.00     0.00     0.00     0.00     0.00     0.00     0.00     0.00     0.00     0.00
   STATE # :      15       16       17       18       19       20       21       22       23       24       25       26       27       28       29
   P(N) :       0.00     0.00     0.00     0.00     0.00     0.00     0.00     0.00     0.00     0.00     0.00     0.0      0.0      0.0      0.0
   STATE # :      30       31       32       33       34       35       36       37       38       39       40       41       42       43       44
   P(N) :       0.0      0.0      0.0      0.0      0.0      0.0      0.0      0.0      0.0      0.0      0.0      0.0      0.0      0.0      0.0
   STATE # :      45       46       47       48       49       50       51       52       53       54       55       56       57       58       59
   P(N) :       0.0      0.0      0.0      0.0      0.0      0.0      0.0      0.0      0.0      0.0      0.0      0.0      0.0      0.0      0.0


TIME= 0.750    SUM= 1.000    EQ= 0.0139    SDQ= 0.1310    ES= 0.1250    SDS= 0.3750    ARR= 5.    SER=45.    MX= 1


              PW=0.111111    EHOLD=  0.21MIN    THOLD=     0.52    CUM RJECTS=   0.00


   STATE # :       0        1        2        3        4        5        6        7        8        9       10       11       12       13       14
   P(N) :       0.89     0.10     0.01     0.00     0.00     0.00     0.00     0.00     0.00     0.00     0.00     0.00     0.00     0.00     0.00
   STATE # :      15       16       17       18       19       20       21       22       23       24       25       26       27       28       29
   P(N) :       0.00     0.00     0.00     0.00     0.00     0.00     0.00     0.00     0.00     0.00     0.00     0.0      0.0      0.0      0.0
   STATE # :      30       31       32       33       34       35       36       37       38       39       40       41       42       43       44
   P(N) :       0.0      0.0      0.0      0.0      0.0      0.0      0.0      0.0      0.0      0.0      0.0      0.0      0.0      0.0      0.0
   STATE # :      45       46       47       48       49       50       51       52       53       54       55       56       57       58       59
   P(N) :       0.0      0.0      0.0      0.0      0.0      0.0      0.0      0.0      0.0      0.0      0.0      0.0      0.0      0.0      0.0


TIME= 1.000    SUM= 1.000    EQ= 0.0139    SDQ= 0.1310    ES= 0.1250    SDS= 0.3750    ARR= 5.    SER=45.    MX= 1
```

APPENDIX D

TIME-VARYING BULK-SERVICE PROGRAM WITH RANDOM ARRIVALS AND DETERMINISTIC SERVICE

The computer programs presented in this section, like those in Appendix C, are extensions of programs originally developed by Michael G. Hartman as a part of his M.S. Thesis written under the direction of the author.[1] The programs have been extensively modified to enhance their general classroom usage.

These programs are designed to calculate time-dependent probability distributions for bulk service, truncated queueing systems having time-varying arrival and service rates. The assumptions are:

1. Arrivals within a period of length DTl, commonly taken to be fifteen minutes, occur in Poisson fashion at a constant rate.
2. Arrival rates may vary from period to period.
3. Service occurs at the start of a service epoch, whose length is defined as $1 \div$ service rate, and will accomodate up to a prespecified batch size of customers. Service rate within each period of length DTl is constant. Service time is deterministic.
4. Service rates may vary from period to period.
5. Service opportunities occur in conveyor fashion. If customers arrive to an empty system they must still wait until the beginning of the next service epoch before they can begin service.

The model operates as a simple finite Markov chain in which state probabilities at time $(t + \bar{t})$ are calculated as a function of the state

---

[1] Adapted with permission of the author from Michael G. Hartman, "The Truncated, Time Varying Queue as a Model of an Air Traffic Control System," Unpublished M.S. Thesis, The Ohio State University, 1971, pp. 80–83.

probabilities at time t and a single-stage, time-dependent transition matrix. That is

$$\mathbf{P}(t + \bar{t}) = \mathbf{P}(t)\mathbf{T}(t)$$

where $\bar{t}$ is the current length of a service epoch and $\mathbf{T}(t)$ is a matrix of the form described in equation (10.95) of Chapter 10. The notation used is the same as that employed for the exponential service time model of Appendix C.

The user must provide 96 values each for the arrival rate and service rate per batch on data cards appended to the program decks. These values represent step function approximations to the continuously varying time functions. The number 96 was selected to represent the 96 fifteen-minute time periods in a 24-hour day. If a different number of periods is desired, the user must change the corresponding dimension statements. The data card format is F3.0 with twenty-four values placed on each card for a total of eight cards. A ninth data card carries the specification for KI, (the maximum batch size), MM, (the maximum number of customers permitted in the system) and TIMPER, (the total length of the time period being divided into 96 subperiods). The format is 2I2, F4.0. In the current program capacity MM must be less than 57 since 56! is the largest factorial which can be calculated and stored without overflow.

For programming convenience, statistics and probabilities are printed only at the end of each subperiod, nominally taken to be fifteen minutes in a standard 24-hour experiment. The system is assumed to be empty at time zero. The distribution at the end of the first subperiod becomes the initial value vector for the next subperiod. These values together with a new transition matrix, whose elements are calculated from current arrival and service rates, are then used to find the time-varying solution for the next subperiod. The solution proceeds in this manner until the entire 96 subperiods in TIMPER have been evaluated.

The statistics printed at the end of each subperiod include:

1.	EQ	The expected number of customers in the queue at the instant just prior to service taking place.
2.	SDQ	The standard deviation for customers in the queue.
3.	EHOLD	The expected number of customer-minutes holding time accumulated during the current subperiod.
4.	THOLD	The summation of all expected holding times from time zero to the end of the current subperiod.
5.	CUM RJECTS	An approximation for the number of customers rejected due to system saturation from time zero to the end of the current subperiod.
6.	ARR	The arrival rate in effect during the current subperiod. (Parameter of a Poisson distribution.)
7.	SER	The service rate per batch in effect during the

<table>
<tr><td></td><td>current subperiod. (Deterministic service times).</td></tr>
<tr><td>8. SUM</td><td>The sum of all values contained in the numerical approximation for the probability distribution at the end of the interval. With no rounding the sum should be equal to one.</td></tr>
<tr><td>9. MX</td><td>The state with the largest probability of occurrence. The probability of this state may have been altered to guarantee a probability vector which will sum to unity.</td></tr>
<tr><td>10. P(N)</td><td>The probability that there were N customers present at the instant just prior to the last service epoch. (Printed for all possible state values.)</td></tr>
</table>

The program, which is written in FORTRAN IV, is generously laden with comment cards to facilitate understanding on the part of the reader.

PROGRAM LISTING

```
C THIS PROGRAM CALCULATES TRANSIENT STATE PROBABILITIES AND THE RELATED
C     BEHAVIORAL CHARACTERISTICS FOR A BULK SERVICE TRUNCATED QUEUEING
C     SYSTEM WITH VARIABLE POISSON ARRIVALS AND DETERMINISTIC SERVICE
C     TIME.
C     A TABULAR STEP-FUNCTION IS USED TO APPROXIMATE THE MEAN ARRIVAL
C     AND SERVICE RATES OVER TIME.
C
C SOLUTION IS CARRIED OUT BY MEANS OF IMBEDDED MARKOV CHAIN ARGUMENTS,
C     WITH THE FOLLOWING MAJOR ASSUMPTION:  SERVICE CAN BE INITIATED
C     ONLY AT THE BEGINNING OF AN EPOCH; THE LENGTH OF AN EPOCH IS EQUAL
C     TO THE LENGTH OF TIME THAT A SERVICE OPERATION TAKES.
C
C
C THE QUEUE STATISTICS CALCULATED ARE:
C     1. EXPECTED # IN THE SYSTEM = EQ
C     2. STD. DEV. OF # IN THE SYSTEM = SDQ
C     3. EXPECTED TOTAL HOLDING TIME FOR A SUB-DIVISION = EHOLD
C     4. SUMMATION OF HOLDING TIME = THOLD
C     5. APPROXIMATE # OF CUMULATIVE REJECTS = RJECTS
C
C
C THE VARIABLES USED ARE:
C     TIMPER = TOTAL LENGTH OF TIME FOR EXPERIMENT
C     KI = BATCH SIZE
C     MM = MAXIMUM NUMBER ALLOWED IN SYSTEM = M-1
C     N= THE # OF SUB-DIVISIONS OF THE TIME PERIOD TO BE SOLVED FOR
C     DT1= THE LENGTH OF A SUB-DIVISION
C     M= THE # OF POSSIBLE STATES OF THE SYSTEM
C     TIME = THE CURRENT TIME OF DAY IN THE PROGRAM
C     TIMLIM= THE TIME AT THE END OF THE CURRENT SUB-DIVISION
C     R(I)= THE ARRIVAL RATE FOR THE I-TH SUB-DIVISION; I=1,2,...N
C     L(I)= THE SERVICE RATE FOR THE I-TH SUB-DIVISION; I=1,2,...N
C     P(J)= THE DEPENDENT VARIABLE DENOTING PROB(J-1 IN THE SYSTEM);
C           J=1,2,...M
C     FACT(I)= (I-1)!; IN DOUBLE PRECISION TO HANDLE LARGE VALUES OF I-1
C     DT= THE LENGTH OF A TRANSITION EPOCH
C     TR(I,J)= THE TRANSITION MATRIX DENOTING THE PROB(J-1 IN SYSTEM AT
C              T+DT|I-1 IN SYSTEM AT T)
C     X(I)= A VECTOR USED TO SAVE NEW VALUES OF P(I) UNTIL ALL M VALUES
C           OF P(I) HAVE BEEN CALCULATED
C     SUM= THE SUM OF P(J), J=1,M; SHOULD EQUAL 1.0
C     DQ = EXPECTED NUMBER IN QUEUE AT CURRENT SERVICE EPOCH
C     DQL = EXPECTED NUMBER IN QUEUE AT LAST SERVICE EPOCH
C
C
C NOTE:  THE PROGRAM IS COMPLETELY GENERAL WITH THE FOLLOWING EXCEPTIONS
C     1. M MUST BE LESS THAN 58
C     2. IF N IS CHANGED, THE DIMENSIONS ON R & L MUST BE CHANGED
C     3. THE CALCULATION OF EXPECTED HOLDING TIME FINDS THE AVERAGE
C        EXPECTED NUMBER IN THE QUEUE FOR A SERVICE INTERVAL AND
C        MULTIPLIES THIS BY THE NUMBER OF MINUTES IN THAT INTERVAL TO
C        OBTAIN THE TOTAL EXPECTED CUSTOMER-MINUTES DELAY FOR THAT
C        INTERVAL.
C
C
      REAL L(96)
      DIMENSION R(96),P(60),X(60)
      COMMON TR(60,60),FACT(60),M,KI
```

```
       DOUBLE PRECISION FACT
       N=96
       READ(5,100)(R(I),I=1,N)
       READ(5,100)(L(I),I=1,N)
       READ(5,50)KI,MM,TIMPER
    50 FORMAT(2I2,F4.0)
       DT1 = TIMPER/FLOAT(N)
       DQ = 0.
       DQL = 0.
       THOLD = 0.
       RJECTS=0.
       TIME=0.
       M = MM+1
       PRINT101,MM,KI
C
C INITIALIZE P(J) AT TIME= 0
C
       DO 10 I=1,60
    10 P(I)=0.
       P(1)=1.
C
C CALCULATE VECTOR OF FACTORIALS
C
       FACT(1)=1.
       DO 6 I=2,57
     6 FACT(I)=FACT(I-1)*DFLOAT(I-1)
C
C CARRY OUT ITERATIVE PROCEDURE FOR ALL N SUB-DIVISIONS
C
       DO 1 I=1,N
       TIMLIM=FLOAT(I)*DT1
       DT=1./L(I)
       IF(I.EQ.1)GOTO5
C
C IF R & L DO NOT CHANGE, TR(I,J) WILL NOT CHANGE, THEREFORE NO CALL IS
C      MADE TO TRNMAT
C
       IF(L(I).EQ.L(I-1).AND.R(I).EQ.R(I-1))GOTO4
     5 CALL TRNMAT(R(I),DT)
C
C CALCULATE CROSS-PRODUCT OF P & TR TO DETERMINE NEW STATE PROBABILITIES
C
     4 DO 2 K=1,M
       X(K)=0.
       DO 3 II=1,M
     3 X(K)=X(K)+P(II)*TR(II,K)
     2 CONTINUE
       DO 8 K=1,M
     8 P(K)=X(K)
       TIME=TIME+DT
       A = -R(I)*DT
       NN = 0
    20 NN = NN+1
       SUM = 0.
       IF(NN.EQ.1)GO TO 40
       KMAX = NN-1
       DO 30 K = 1,KMAX
    30 SUM = SUM+(-A)**(K-1)/FACT(K)
    40 TEMP = -A-NN*(1.-((1.+A/NN)*SUM+(-A)**(NN-1)/FACT(NN))*EXP(A))
       DELTA = P(M-NN+1)*TEMP
       RJECTS = RJECTS+DELTA
       IF(DELTA.GT.0.001) GO TO 20
```

```
      IF(TIME.LT.TIMLIM)GOTO4
C
C AT THE END OF A SUB-DIVISION, CALCULATE QUEUE STATISTICS
C
      PMAX = 0.
      MX = 0
      SUM=0.
      DO 7 J=1,M
      IF(P(J).LT.0.) P(J) = 0.
      IF(P(J).LT.PMAX) GO TO 7
      PMAX = P(J)
      MX = J
    7 SUM=SUM+P(J)
      IF(SUM.EQ.1.) GO TO 17
      DELT = SUM-1.
      P(MX) = P(MX)-DELT
   17 CONTINUE
      SDQ=0.
      EQ=0.
      DO 9 K=2,M
      EQ=EQ+FLOAT(K-1)*P(K)
    9 SDQ=SDQ+((FLOAT(K-1))**2)*P(K)
      SDQ=SQRT(SDQ-EQ**2)
      DQ = EQ
      EHOLD = 30.*DT1*(DQ+DQL)
      DQL = DQ
      THOLD = THOLD+EHOLD
      PRINT102,TIME,EQ,SDQ,EHOLD,THOLD,RJECTS
      PRINT112,R(I),L(I),SUM,MX
      PRINT104,(K,K=1,14),(P(K),K=1,15)
      PRINT103,(K,K=15,29),(P(K),K=16,30)
      PRINT103,(K,K=30,44),(P(K),K=31,45)
    1 PRINT103,(K,K=45,59),(P(K),K=46,60)
  100 FORMAT(24F3.0)
  101 FORMAT('-',5X,'RESULTS FOR THE TRUNCATED BATCH SERVICE QUEUEING
     1SYSTEM',/,10X,'TIME VARYING POISSON ARRIVALS AND DETERMINISTIC
     2SERVICE',/,15X,'MAXIMUM QUEUE SIZE = ',I2,5X,'BATCH SIZE = ',I2)
  102 FORMAT('-',5X,'TIME = ',F6.3,5X,'EQ = ',F5.2,5X,'STD DEV = ',F5.2,
     15X,'EHOLD = ',F7.2,'MIN.',3X,'THOLD = ',F9.2,3X,'CUM REJECTS = ',
     2F6.1,/)
  112 FORMAT('-',15X,'ARR = ',F3.0,5X,'SER = ',F3.0,5X,'SUM = ',F6.3,5X,
     1'MX = ',I2/)
  103 FORMAT(' ',1X,'STATE # :',3X,15I7,/,1X,'P(N) :',7X,15F7.2)
  104 FORMAT(' ',1X,'STATE # :',9X,'0',14I7,/,1X,'P(N) : ',6X,15F7.2)
      STOP
      END
      SUBROUTINE TRNMAT(Y,DT)
C
C THIS SUBROUTINE CALCULATES THE TRANSITION MATRIX TR(I,J) GIVEN THE
C      LENGTH OF THE EPOCH, T, AND THE MEAN ARRIVAL RATE DURING THE
C      EPOCH, Y.
C
C VARIABLES USED ARE:
C      PR(I)= PROBABILITY OF I-1 ARRIVALS IN AN EPOCH;   I=1,2,...MM
C      CUM(I)= PROBABILITY OF I OR MORE ARRIVALS DURING AN EPOCH
C
C
      COMMON TR(60,60),FACT(60),M,KI
      DIMENSION PR(59),CUM(59)
      DOUBLE PRECISION FACT
      MM=M-1
      X=-(Y*DT)
```

```
C
C FILL OUT PR VECTOR BY CALCULATING POISSON PROBABILITY OF I-1 ARRIVALS
C
      PR(1)=EXP(X)
      DO 1 I=2,MM
    1 PR(I)=EXP(X)*((-X)**(I-1))/FACT(I)
      TOT=0.
      DO 2 I=1,MM
      TOT=TOT+PR(I)
    2 CUM(I)=1.-TOT
C
C FIRST KI+1 ROWS OF TR CONTAIN TRANSITION FROM THE ZERO STATE. ANY
C CUSTOMERS PRESENT UP TO MAX. BATCH SIZE KI WILL BE SERVED BY THE
C UP-COMING SERVICE EVENT.
C
      KMAX = KI+1
      DO 8 I=1,KI
      DO 3 J=1,MM
    3 TR(I,J) = PR(J)
    8 TR(I,M) = CUM(MM)
      DO 4 I=KMAX,M
      DO 4 J=1,M
    4 TR(I,J) = 0.
C
C OTHER ROWS OF TR CONTAIN TRANSITIONS FROM NON-ZERO STATES.  ALL
C     ENTRIES BELOW THE KI-SUB-DIAGONAL ARE ZERO, SINCE AT MOST ONE
C     SERVICE CAN TAKE PLACE DURING AN EPOCH.
C
      DO 5 I=KMAX,M
      JJ=I-KI
      DO 6 J=JJ,MM
    6 TR(I,J)=PR(J-I+KMAX)
    5 TR(I,M)=CUM(M-I+KI)
      RETURN
      END
/*


 5. 5. 5. 5. 5. 5. 5. 3. 2. 1. 1. 1. 2. 2. 2. 2. 2. 2. 2. 3. 3. 3. 3. 4.
 4. 4. 9.13.19.25.31.37.43.50.50.50.50.50.50.50.51.53.51.49.47.45.45.46.
47.47.48.48.49.50.51.52.53.54.55.56.58.60.62.61.61.60.61.62.64.66.65.63.
61.59.57.55.54.52.50.48.46.45.43.41.39.37.35.34.32.30.26.22.17.13. 9. 5.
45.45.45.45.45.45.45.45.45.45.45.45.45.45.45.45.45.45.45.45.45.45.45.45.
45.45.45.45.45.45.45.45.45.45.45.45.45.45.45.45.45.45.45.45.45.45.45.45.
45.45.45.45.45.45.45.45.45.45.45.45.45.45.45.45.45.45.45.45.45.45.45.45.
45.45.45.45.45.45.45.45.45.45.45.45.45.45.45.45.45.45.45.45.45.45.45.45.
 12524.
/*
//
```

SAMPLE OUTPUT

```
RESULTS FOR THE TRUNCATED BATCH SERVICE QUEUEING  SYSTEM
    TIME VARYING POISSON ARRIVALS AND DETERMINISTIC  SERVICE
         MAXIMUM QUEUE SIZE = 25       BATCH SIZE =  2
```

TIME = 0.267   EQ = 0.11   STD DEV = 0.33   EHOLD =   0.84MIN.   THOLD =   0.84   CUM REJECTS =   0.0

ARR = 5.   SER = 45.   SUM = 1.000   MX = 1

STATE #	0	1	2	3	4	5	6	7	8	9	10	11	12	13	14
P(N)	0.89	0.10	0.01	0.00	0.00	0.00	0.00	0.00	0.00	0.00	0.00	0.00	0.00	0.00	0.00
STATE #	15	16	17	18	19	20	21	22	23	24	25	26	27	28	29
P(N)	0.00	0.00	0.00	0.00	0.00	0.00	0.00	0.00	0.00	0.00	0.00	0.0	0.0	0.0	0.0
STATE #	30	31	32	33	34	35	36	37	38	39	40	41	42	43	44
P(N)	0.0	0.0	0.0	0.0	0.0	0.0	0.0	0.0	0.0	0.0	0.0	0.0	0.0	0.0	0.0
STATE #	45	46	47	48	49	50	51	52	53	54	55	56	57	58	59
P(N)	0.0	0.0	0.0	0.0	0.0	0.0	0.0	0.0	0.0	0.0	0.0	0.0	0.0	0.0	0.0

TIME = 0.511   EQ = 0.11   STD DEV = 0.33   EHOLD =   1.67MIN.   THOLD =   2.51   CUM REJECTS =   0.0

ARR = 5.   SER = 45.   SUM = 1.000   MX = 1

STATE #	0	1	2	3	4	5	6	7	8	9	10	11	12	13	14
P(N)	0.89	0.10	0.01	0.00	0.00	0.00	0.00	0.00	0.00	0.00	0.00	0.00	0.00	0.00	0.00
STATE #	15	16	17	18	19	20	21	22	23	24	25	26	27	28	29
P(N)	0.00	0.00	0.00	0.00	0.00	0.00	0.00	0.00	0.00	0.00	0.00	0.0	0.0	0.0	0.0
STATE #	30	31	32	33	34	35	36	37	38	39	40	41	42	43	44
P(N)	0.0	0.0	0.0	0.0	0.0	0.0	0.0	0.0	0.0	0.0	0.0	0.0	0.0	0.0	0.0
STATE #	45	46	47	48	49	50	51	52	53	54	55	56	57	58	59
P(N)	0.0	0.0	0.0	0.0	0.0	0.0	0.0	0.0	0.0	0.0	0.0	0.0	0.0	0.0	0.0

TIME = 0.756   EQ = 0.11   STD DEV = 0.33   EHOLD =   1.67MIN.   THOLD =   4.18   CUM REJECTS =   0.0

ARR = 5.   SER = 45.   SUM = 1.000   MX = 1

STATE #	0	1	2	3	4	5	6	7	8	9	10	11	12	13	14
P(N)	0.89	0.10	0.01	0.00	0.00	0.00	0.00	0.00	0.00	0.00	0.00	0.00	0.00	0.00	0.00
STATE #	15	16	17	18	19	20	21	22	23	24	25	26	27	28	29
P(N)	0.00	0.00	0.00	0.00	0.00	0.00	0.00	0.00	0.00	0.00	0.00	0.0	0.0	0.0	0.0
STATE #	30	31	32	33	34	35	36	37	38	39	40	41	42	43	44
P(N)	0.0	0.0	0.0	0.0	0.0	0.0	0.0	0.0	0.0	0.0	0.0	0.0	0.0	0.0	0.0
STATE #	45	46	47	48	49	50	51	52	53	54	55	56	57	58	59
P(N)	0.0	0.0	0.0	0.0	0.0	0.0	0.0	0.0	0.0	0.0	0.0	0.0	0.0	0.0	0.0

TIME = 1.022   EQ = 0.11   STD DEV = 0.33   EHOLD =   1.67MIN.   THOLD =   5.85   CUM REJECTS =   0.0

ARR = 5.   SER = 45.   SUM = 1.000   MX = 1

STATE #	0	1	2	3	4	5	6	7	8	9	10	11	12	13	14
P(N)	0.89	0.10	0.01	0.00	0.00	0.00	0.00	0.00	0.00	0.00	0.00	0.00	0.00	0.00	0.00

REFERENCES

1. Abramowitz, M., and Stegun, I. A. *Handbook of Mathematical Functions.* Washington D.C.: National Bureau of Standards, Applied Mathematics Series 55, 1964.

2. *Airport Capacity—A Handbook for Analyzing Airport Designs to Determine Practical Movement Rates and Aircraft Operating Costs.* Long Island, New York: Airborne Instruments Laboratory, June 1963.

3. Ashcroft, H. "The Productivity of Several Machines Under the Care of One Operator." *Journal of the Royal Statistical Society, Ser. B,* 12, (1950) 145-151.

4. Bailey, N. T. J. "A Continuous Time Treatment of a Simple Queue Using Generating Functions." *Journal of the Royal Statistical Society, Ser. B.,* 16, (1956) 288-291.

5. ————. "On Queueing Processes with Bulk Service." *Journal of the Royal Statistical Society, Ser. B,* 16, (1954) 80-87.

6. ————. *The Mathematical Theory of Epidemics.* New York: Hafner Publishing Company, 1957.

7. Beckman, P. *Introduction to Elementary Queueing Theory and Telephone Traffic.* Boulder, Colorado: Golem, 1968.

8. Bhat, U. N. "Sixty Years of Queueing Theory," *Management Science.* 15, (1969) B280-B292.

9. ————. "Transient Behavior of Multi-Server Queues with Recurrent Input and Exponential Service Times." *Journal of Applied Probability,* 5, (1968) 158-168.

10. Billingsley, P. "Statistical Methods in Markov Chains." *Annals of Mathematical Statistics,* 32, (1961) 12-40.

11. ————. *Statistical Inference for Markov Processes.* Chicago: Univ. of Chicago Press, 1961.

12. Bowker, A. H., and Lieberman, G. J. *Engineering Statistics,* 2nd Ed. Englewood Cliffs, New Jersey: Prentice Hall, 1972.

13. Box, G. E. P., and Jenkins, G. M. *Time Series Analysis: Forecasting and Control,* Rev. Ed. San Francisco: Holden-Day, 1976.

14. Brockmeyer, E., Halstrom, H. L., and Jensen, A. *The Life and Works of A. K. Erlang.* Copenhagen: Copenhagen Telephone Co., 1948.

15. Bundy, W. W. *Queues with Delayed Probabilistic Feedback as a Model of Air Traffic Control Communications.* Unpublished M.S. Thesis, The Ohio State University, 1976.

16. Burke, P. J. "The Output of a Queueing System." *Operations Research,* 4, (1956) 699-704.

17. Churchill, R. V. *Complex Variables and Applications,* 2nd Ed. New York: McGraw-Hill, 1960.

18. ———. *Operational Mathematics,* 3rd Ed. New York: McGraw-Hill, 1972.

19. Clarke, A. B. "Maximum Likelihood Estimates in a Simple Queue." *Annals of Mathematical Statistics,* 28, (1957) 1036–1040.

20. Conte, S. D. *Elementary Numerical Analysis.* New York: McGraw-Hill, 1965.

21. Conway, R. W., and Maxwell, W. L. "A Queueing Model with State Dependent Service Rates." *Journal of Industrial Engineering,* 12, (1962) 132–136.

22. Conway, R. W., Maxwell, W. L., and Miller, L. W. *Theory of Scheduling.* Reading, Mass.: Addison-Wesley, 1967.

23. Cooper, R. B. *Introduction to Queueing Theory.* New York: Macmillan, 1972.

24. Cox, D. R. "The Analysis of Non-Markovian Stochastic Processes by the Inclusion of Supplementary Variables." Proceedings of Cambridge Philosophical Society, 51, (1955) 443–441.

25. ———. "Some Problems of Statistical Analysis Connected with Congestion," Appearing in Smith, W. L., and Wilkinson, W. E., *Proceedings of the Symposium on Congestion Theory.* Chapel Hill: The University of North Carolina Press, 1965.

26. ———. *Renewal Theory.* London: Methuen, 1967.

27. Cox, D. R., and Lewis, P. A. W. *The Statistical Analysis of Series of Events.* London: Methuen, 1966.

28. Cox, D. R., and Miller, H. D. *The Theory of Stochastic Processes.* New York: Wiley, 1968.

29. Cox, D. R., and Smith, W. L. *Queues.* London: Methuen, 1961.

30. Crommelin, C. D. "Delay Probability Formulae." *P. O. Electrical Engineers Journal,* 26, (1934) 266–274.

31. Downton, F. "Waiting Time in Bulk Service Queues." *Journal of the Royal Statistical Society,* Ser. B, 17, (1955) 256–261.

32. Duncan, A. J. *Quality Control and Industrial Statistics,* 3rd Ed. Homewood, Ill.: Irwin, 1965.

33. Dunlay, W. J. "Stochastic Properties of Total Air Traffic Control Voice Communication Time." *Transportation Research,* 9, (1975) 275–278.

34. Edie, L. C. "Traffic Delays at Toll Booths." *Operations Research,* 4, (1956) 107–138.

35. Epstein, B. "Tests for the Validity of the Assumption that the Underlying Distribution of Life is Exponential." *Technometrics,* 2, (1960) 83–101 and 167–183.

36. Erlang, A. K. "The Theory of Probabilities and Telephone Conversations." *Nyt Tidsskrift Matematik, B,* 20, (1909) 33–39.

37. Feller, W. *An Introduction to Probability Theory and Its Applications,* 2nd Ed. New York: Wiley, 1957.

38. Finch, P. D. "The Output Process of the Queueing System M/G/1." *Journal of the Royal Statistical Society, Ser. B,* 21, (1959) 375–380.

39. Fry, T. C. *Probability and Its Engineering Uses.* Princeton, N.Y.: D. Van Nostrand Company, 1928.

40. Giffin, W. C. *Transform Techniques for Probability Modeling.* New York: Academic Press, 1975.

41. Gordon, W. J., and Newell, G. F. "Closed Queueing Systems with Exponential Servers." *Operations Research,* 15, (1967) 254–265.

42. Gross, D., and Harris, D. M. *Fundamentals of Queueing Theory.* New York: Wiley, 1974.

43. Hartman, M. G. *The Truncated Time Varying Queue as a Model of an Air Traffic Control System*. Unpublished M.S. Thesis, The Ohio State University, 1971.

44. Hillier, F. S., Conway, R. W., and Maxwell, W. L. "A Multiple Server Queueing Model with State Dependent Service Rates." *Journal of Industrial Engineering*, 15, (1964) 153–157.

45. Hillier, F. S., and Lieberman, G. J. *Introduction to Operations Research*, 2nd Ed. San Francisco: Holden-Day, 1974.

46. Howard, R. A. *Dynamic Probabilistic Systems Vol I: Markov Models*. New York: Wiley, 1971.

47. Hunter, J. S., and Hsu, D-A. *Simulation Model for New York Air Traffic Control Communications*. Report No. FAA-RD-74-203, February, 1975.

48. IBM Corporation. *System/360 Scientific Subroutine Package (360A-CM-03X) Version III*. White Plains, New York: IBM Corporation, 1968.

49. Jackson, J. R. "Jobshop-Like Queueing Systems." *Management Science*, 10, (1963) 131–142.

50. ———. "Networks of Waiting Lines," *Operations Research*, 5, (1957) 518–521.

51. Jackson, R. R. P. "Queueing Systems with Phase Type Service." *Operations Research Quarterly*, 5, (1954) 109–120.

52. Jaiswal, N. K. *Priority Queues*. New York: Academic Press, 1968.

53. Jewell, W. S. "A Simple Proof of $L = \lambda W$." *Operations Research*, 15, (1967) 1109–1116.

54. Johannsen, F. W. "Waiting Times and Number of Calls." *P. O. Electrical Engineers Journal*, (1907).

55. Kaplan, W. *Ordinary Differential Equations*. Reading, Mass.: Addison-Wesley, 1958.

56. Kendall, D. G. "Some Problems in the Theory of Queues." *Journal of the Royal Statistical Society, Ser. B*, 13, (1951) 151–185.

57. ———. "Stochastic Processes Occurring in the Theory of Queues and Their Analysis by the Method of Markov Chains." *Annals of Mathematical Statistics*, 24, (1953) 338–354.

58. Khintchine, A. Y. "Mathematisches über die Erwortung vor einemöffentlichen Schalter." *Mat. Sbornik*, 39, (1932) 73–84.

59. Kielson, J. and Kooharian, A. "On Time Dependent Queueing Processes." *Annals of Mathematical Statistics*, 31, (1960) 104–112.

60. Kingman, J. F. C. "The Heavy Traffic Approximation in the Theory of Queues." *Proceedings of the Symposium on Congestion Theory*, University of North Carolina, Chapel Hill, 1964, 137–169.

61. Kleinrock, L. *Queueing Systems, Vol. 2: Computer Applications*. New York: Wiley, 1976.

62. Koenigsberg, E. "Cyclic Queues." *Operations Research Quarterly*, 9, (1958) 22–35.

63. Kolomogorov, A. N. "Sur la problème d'attente." *Mat. Sbornik*, 8, (1931) 101–106.

64. Koopman, B. O. "Air-Terminal Queues under Time Dependent Conditions." *Operations Research*, 20, (1972) 1089–1114.

65. Lilliefors, H. W. "Some Confidence Intervals for Queues." *Operations Research*, 14, (1966) 723–77.

66. Little, J. D. C. "A Proof for the Queueing Formula $L = \lambda W$." *Operations Research*, 9, (1961) 383–387.

67. Molina, E. C. "Application of the Theory of Probability to Telephone Trunking Problems." *Bell System Technical Journal*, 6, (1927) 461–494.

68. Mood, A. M. *Introduction to the Theory of Statistics*. New York: McGraw-Hill, 1950.

69. Morse, P M. *Queues, Inventories and Maintenance.* New York: Wiley, 1958.

70. Nelson, R. T. "Waiting-Time Distributions for Application to a Series of Service Centers." *Operations Research,* 6, (1958) 856–862.

71. Newell, G. F. *Applications of Queueing Theory.* London: Chapman and Hall Ltd., 1971.

72. Palm, C. "Analysis of the Erlang Traffic Formulae for Busy-Signal Arrangements." *Ericsson Tech.,* 6, (1938) 39–58.

73. Palm, D. C. "The Assignment of Workers in Servicing Automatic Machines." *Journal of Industrial Engineering,* 9, (1958) 28–42.

74. Peck, L. G., and Hazelwood, R. N. *Finite Queueing Tables.* New York: Wiley, 1958.

75. Pollaczek, F. "Concerning an Analytic Method for the Treatment of Queueing Problems." *Proceedings of the Symposium on Congestion Theory,* University of North Carolina, Chapel Hill, (1964) 1–42.

76. Raisbeck, G., Koopman, B. O., Lister, S. F., and Kapadia, A. S. *A Study of Air Traffic Control System Capacity.* Arthur D. Little, Inc., Interim Report, Oct. 1970, Report No. FAA-RD-70-70.

77. Ralston, A., and Wilf, H. S. *Mathematical Methods for Digital Computers.* New York: Wiley, 1960.

78. Reich, E. "Waiting Times When Queues Are in Tandem." *Annals of Mathematical Statistics,* 28, (1957) 768–773.

79. Reza, F. M. *An Introduction to Information Theory.* New York: McGraw-Hill, 1961.

80. Riordan, J. *Stochastic Service Systems.* New York: Wiley, 1962.

81. Ross, S. M. *Applied Probability Models with Optimization Applications.* San Francisco: Holden-Day, 1970.

82. Saaty, T. L. *Elements of Queueing Theory.* New York: McGraw-Hill, 1961.

83. Schmidt, J. W., and Taylor, R. E. *Simulation and Analysis of Industrial Systems.* Homewood, Ill.: Irwin, 1970.

84. Schmotzer, R. E., et al. *Future Air Traffic: A Study of the Terminal Area.* NASA Langley Research Center, West Virginia University, Morgantown, West Virginia, 1970.

85. Spiegel, M. R. *Theory and Problems of Complex Variables.* (Schaum's Outline Series). New York: McGraw-Hill, 1964.

86. Taha, H. A. *Operations Research.* New York: Macmillan, 1971.

87. Takacs, L. *Introduction to the Theory of Queues.* New York: Oxford University Press, 1962.

88. Walters, A. H. *The Application of an Analytical Model to the Analysis of Air Traffic in the Terminal Area.* Unpublished M.S. Thesis, The Ohio State University, 1976.

89. Wolff, R. W., "Problems of Statistical Inference for Birth and Death Queuing Models." *Operations Research,* 13, (1965) 343–357.

INDEX

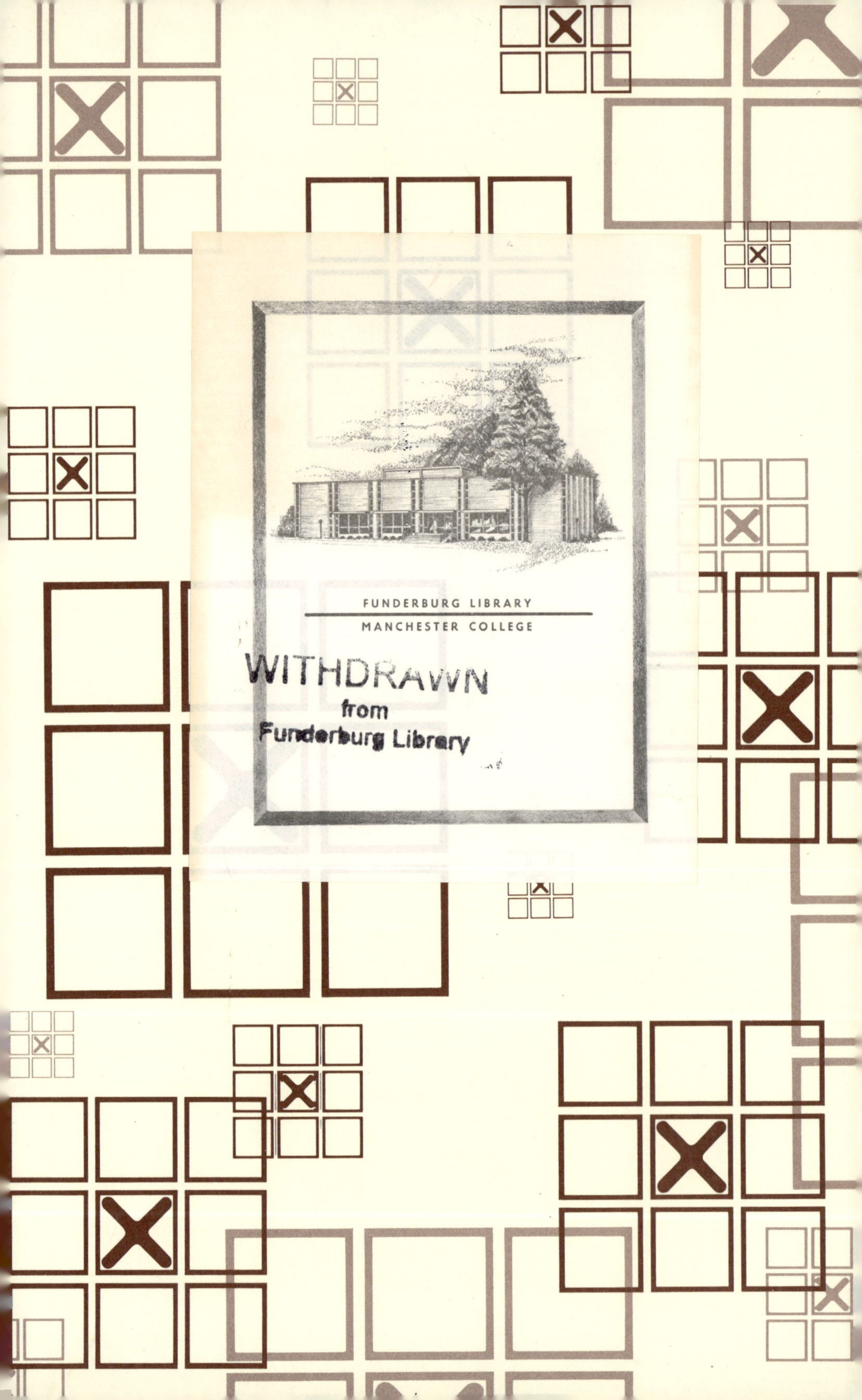

FUNDERBURG LIBRARY
MANCHESTER COLLEGE